Physiology

Fainting Felines and Faulty Physiology

Mark Milanick

Disclaimer: Do NOT use this book to solve health issues. The book uses cases as a way to motivate understanding physiology; it is NOT appropriate to use these approaches or information to diagnosis or treat diseases. If you have any health issues, please see your health care provider.

All material in this book is copyrighted except for the figures taken or modified from Wikipedia. August 2012.
Please contact the author for permission to reproduce any material. megandmark@mail.com

The use of figures from Wikipedia does NOT imply that Wikipedia in any way endorses this work or my use of their work. The figures directly taken, or modified, from Wikipedia, retain their Wikipedia copyright license, including the following: You are free:
- **to share** – to copy, distribute and transmit the work
- **to remix** – to adapt the work
    - Under the following conditions:
- **attribution** – You must attribute the work in the manner specified by the author or licensor (but not in any way that suggests that they endorse you or your use of the work).
- **share alike** – If you alter, transform, or build upon this work, you may distribute the resulting work only under the same or similar license to this one.

I have done my best to present the science as accurately as possible and to indicate where my ideas are different than the current mainstream. Please contact me if you detect any errors or instances of plagiarism so that I can correct the material in future editions.
megandmark@mail.com

Physiology

Physiology

Table of Contents

Chapter Topics	8
Acknowledgments	15
Orange you glad?	17
Blood, Sweat and Tears	33
Impressive Horses and Fainting Goats	43
Do you hear what I hear?	65
The Good, the Bad and the Spicy	79
Ouch That Hurts	95
Sleep tight, don't let the bed bugs bite	109
Can Viagra™ kill you?	119
In a Heart Beat	141
Simon Says, "Move"	169
Blowing in the Wind	187
Soy Sauce and Licorice as Medicine	203
From Anemia to Vampires	226
Got Cows?	249
How Sweet it is	271
Bubbling Beverages Bad for Bones?	299
Steroids Make the World go Round	327
Cycles in Synch?	355
Skin Games	365
Black and White (and Banning Tanning?)	383
Is there Proof?	399
Appendix. References for case	421
Chapter Summaries	425

Chapter Topics

Orange you glad?
- the discovery of the role of vitamin A in humans,
- the role of vitamin A deficiency in night blindness,
- how vitamin A and carotene are absorbed
- the role of retinol-binding-protein in transporting vitamin A,
- the problem with excess vitamin A
- why some groups advocate against putting vitamin A into over-the-counter-vitamin pills,
- the story about polar bears and vitamin A

Blood Sweat and Tears
- What is acetylcholinesterase?
- Why does a nerve need a chemical to talk to its neighbor?
- How does the listener nerve cell realize it is being talked to?
- Are all acetylcholine receptors the same?
- An aside about how nicotine and muscarine got their names.
- How are the nicotinic and muscarinic acetylcholine receptors different?
- Can't you get that nerve to shut up?
- How do acetylcholinesterase inhibitors work as insecticides or nerve gases?
- Why is location important for compounds that bind to receptors?

Impressive Horses and Fainting Goats
- Acetylcholine
- Acetylcholine to calcium
- How do we get from sodium to releasing calcium from the storage containers?
- The problem with Impressive
- Impressive's gene
- Testing Impressive's offspring as well as humans
- How do we know that the mutation in the sodium channel causes the symptoms we see in the horse?
- Patch Clamp
- Related Human Diseases
- Fainting goats and cats
- Related Human Diseases

Physiology

Do you hear what I hear?
- Hearing structures
- How can loud sounds lead to loss of cells?
- Drugs and hearing loss
- How loud is too loud?
    - Population studies
- The three cases
- Bats and dogs and ultrasonic
- Earaches
- The general plan of the sensory system

The Good, the Bad, and the Spicy
- Sensing Hot Food
- Why peppers make us sweat
- What about birds and peppers?
- Why does brandy feel warm when swallowed?
- Double ring burners
- Relief from spicy foods
- Does sensing cold temperatures work the same was as hot temperatures?
- Why does saccharine taste sweet?
- Why does saccharine have an aftertaste?
- How many basic tastes are there?
- Cats, sweetness, and antifreeze

Ouch That Hurts
- What is pain?
- What is the purpose of pain?
- How do the neurons that sense pain work?
- What is the relationship between itching and pain?
- Are physical and social pain related?
- How do pain killers work?
- What is the difference between abuse, dependence, and addiction?

Sleep tight, don't let the bed bugs bite
- Why do we sleep?
- What are the stages of sleep?
- How do sleeping pills work?

- How much sleep do we need? What happens if we don't get it?
- Sleepwalking
- Sleep tight, don't let the bed bugs bite
- Unusual sleepers

Can Viagra™ Kill You?
- How does dilation of arteries help the heart?
- What is the mechanism by which dilation occurs?
- What is it about nitroglycerin that allows it to alter the smooth muscle contraction process?
- What are the roles for ATP in smooth muscle contraction?
- What is the series of steps involved in relaxation? Who cares?
- What happens if Harry had both nitroglycerin and Viagra™?
- How does Viagra™ work?
- Why isn't Viagra™ marketed to women?
- Another case that raises an additional way for drug-drug interaction.
- Does marijuana lead to arterial dilation?
- Nitric Oxide and Nobel Prize

In A Heart Beat
- What influences cardiac output?
- What regulates heart rate?
- What else besides heart rate influences cardiac output?
- How do pressure and volume change in the cardiac cycle?
- How do Pressure Volume Curves work?
- Is the elite athletes' and trout's ability to increase stroke volume due to their hearts having better contraction strength?
- What is the mechanism for elite endurance athletes' and trout's ability to increase stroke volume?

Physiology

Simon Says "Move"
- Limitations and strengths of different types of studies
- Basic science for exercise
- Supplement 1: Creatine
- Supplement 2: Carb loading
- Supplement 3: HMB
- Supplement 4: Whey protein hydrolysates and carbohydrates, for example, maltodextrin
- Response to exercise, oxygen, and Antarctic fish
- Types of exercise and physical activity

Blowing in the Wind
- Wind knocked out
- Acute mountain sickness
- Oxygen saturation
- Special cases: alligators, birds, and fetuses
- More red blood cells: Athletes, Andeans, Tibetans
- Red Roger
- Cough, sneeze (and taste?)
- Hiccup

Soy Sauce and Licorice as Medicine
- Fluid losses
- Water regulation
- Salt regulation
- What happens if one becomes dehydrated?
- Can you get too much water?
- Fish and aldosterone
- Aquaporins and the Nobel Prize
- More on Solomon

From Anemia to Vampires
- Blood components and hematocrit
- Menstrual blood loss
- Regulation of body iron
- Review of major steps regulating iron absorption
- Iris after 2 months
- Vitamin B12
- Main points so far
- Vampires are in the title, why is that?
- Males vs. females and red blood cells

Got Cows
- What is it in milk that causes the farting and loose stools?
- How do those that can drink lots of milk avoid the problem?
- What is this genetic basis for the difference in milk digestion across populations?
- Do you think lactose persistence is a dominant or recessive trait?
- Why not just use glucose and galactose, why bother to make lactose?
- Do other foods have related problems?
- Do animals have related problems?
- What about the enzyme to make lactose?
- GI Water balance
- How does cholera cause diarrhea?
- What caused Cynthia Francis's diarrhea?

How Sweet it is
- renal regulation of urine formation
- blood sugar regulation
- glucose, glycogen, and fat
- theories about the causes of type II diabetes
- is there a type 1.5 diabetes
- drugs for treating type 2 diabetes

Bubbling Beverages Bad for Bones?
- Population studies
- Calcium threshold and how much calcium needed
- Parathyroid hormone
- Vitamin D
- Calcium regulation
- What is wrong with high blood calcium
- Soda, phosphate and calcium
- Acid and calcium
- Caffeine and calcium

Steroids Make the World go Round
- Steroid name confusion
- General principles of steroids and their receptors
- Cathy's cat Andy

Physiology

- Tessa, androgen effects and "normal" range
- Testosterone, Acne and Hair Growth
- Steroid binding proteins, carriers, and "free"
- Tori and testosterone creams
- Paul as well as phytoestrogens
- Caveat on case studies
- Testosterone and muscle strength and aggression
- Do high concentrations of androgens cause cardiac side effects?
- What about androgens and psychological and behavior problems?
- Bumps on the Penis

Cycles in Synch
- What are some differences between ovulation in animals and humans?
- What occurs during the ovulation cycle in humans?
- How do oral contraceptives work?
- Synchronization?

Skin Games
- Skin microbes
- Why does sex spread HIV?
- What about circumcision and the spread of HIV?
    - What is circumcision and foreskin?
    - Epidemiology and correlation between circumcision and STD transmission.
    - Possible mechanisms by which having foreskin might make one more susceptible to STD infection.
- What are the odds of getting HIV?
- Summary of HIV
- Herpes Simplex I and II
- Human papillomavirus
- Mononucleosis
- Analogy for virus entry and selectivity
- Bacteria: Chlamydia and Gonorrhea.
- Crabs
- What can be caught from a toilet seat?
- Intimate behavior

Black and White (and Banning Tanning ?)

- Chemicals for color
- Genes for Color?
- Melanocytes
- Ultraviolet light and DNA damage
- Why do UVA wavelengths penetrate more deeply than UVB wavelengths?
- Sunburn
- Skin Cancers
- Sunless Tanning
- Addiction

Is there Proof?
- Absorption
- Liver, #1
- Lungs
- Brain
- Alcohol breakdown/liver #2
- Hangovers
- Alcohol and mice and flies
- Fetal Alcohol Syndrome and data scrutiny

Acknowledgements

My wife, Meg, and my son, Bill, have provided key input and support. They succeeded in convincing me to reduce the number of lame jokes and improved or eliminated some of the most convoluted analogies, but I still managed to keep a few of the bad puns because I enjoy them and I am solely responsible for all the parts you'll find annoying.

Melissa Bushman, Philip Leak, and Emily Pintel provided key insights in reading my 5 initial chapters (the order has since changed dramatically).

Ya-Wen Cheng, Chris Murakami, and Parker Stuart are Science Education graduate students who provided key feedback on early drafts of the chapters and suggested several interesting topics to include.

I am also very grateful for the undergraduate students in the honors course, How and Why Does My Body Do That? who provided critical feedback in all areas. Some of the students in the course were: Gideon Berdahl, Alyssa Bujnak, Christy Brethorst, Christine Coyle, Drake Duckworth, Laura Ebone, Courtney Everts, Joel Fitts, Matt Flanigan, Tyler Jennings, Elizabeth Johnson, Madeline Komes, Charlie Landis, Dan Morris, Caitlin Morrison, John Spackler, Martin Sutovsky, and Alyson "Evan" Walker.

Physiology

Orange you glad?

Literature cases

1) A thirty two year old man, Vincent[1], was referred to an eye specialist because his vision was getting worse. He complained of irritated, gritty, dry eyes. He was diagnosed with xerophthalmia, dry eye syndrome. When questioned, Vincent said he ate white bread rolls, French fries, and potato chips. He denied eating vegetables, fruit, or meat. Because the physician suspected a nutritional deficiency, the patient's blood was tested. Because of the lack of iron in his diet, you might suspect he could have anemia (see the chapter, From Anemia to Vampires). He was anemic. The anemia, however, does not account for the eye problems. It turns out that Vincent's blood was low in vitamin A.

2) Night blindness resulted in a 24-year-old New York City woman, Valerie, going to her physician. She admitted that she had a poor diet; it consisted of white onions, white potatoes and red meat. Would you expect her to be anemic? The red meat in her diet should supply her with adequate iron and B12, so there is no reason to expect anemia and she didn't. Valerie was found to be vitamin A deficient.

3) A family physician in Tennessee saw a patient whose skin looked a very reddish orange. Neither the man nor his wife mentioned his skin color; he had come for another complaint. When asked, they both initially said he looked a normal color. The physician determined that the patient, Frank, an avid vegetable farmer, was eating probably a pound of tomatoes and carrots daily from his garden. The physician assured the patient that if it was just the pigments in tomato and carrots that caused the color, there was no need to worry, the condition would be harmless. [2] <u>Why would overeating to the point of a skin color change be harmless?</u>

---

[1] I made up all the patient names in this book for both the literature cases and the fictional cases. The references for the literature cases are in Appendix A.
[2] You might be able to find some pictures on the web, searching the term "carotenemia" in images. There is also an image in this article, Mazzone A, Dal Canton A. Hypercarotenemia. N Engl J Med 2002;346:821-821. PMID: 11893793, which should be freely available at: http://www.nejm.org/doi/pdf/10.1056/NEJMicm950425

4) Gerrit de Veer was with a group of explorers who had stopped at Nova Zembla Island on their attempt to find a northern route from Holland to Indonesia. While they were stopped, among the foods they ate was polar bear liver. De Veer's 1597 diary described symptoms that we think are the result of ingestion of too much vitamin A. It probably included severe headaches due to intracranial hypertension, bleeding gums, vomiting and blurred vision.

In this chapter we will discuss
- the discovery of the role of vitamin A in humans,
- the role of vitamin A deficiency in night blindness,
- how vitamin A and carotene are absorbed
- the role of retinol-binding-protein in transporting vitamin A,
- the problem with excess vitamin A
- why some groups advocate against putting vitamin A into over-the-counter-vitamin pills,
- the story about polar bears and vitamin A

Vitamin A discovery
The original deduction that night blindness and xerophthalmia were caused by a lack of a key dietary substance was done during World War I. A physician, C.E. Bloch, overseeing a children's home in Copenhagen noted that 8 children had night blindness and xerophthalmia. He looked at their charts and all 8 were in one ward (B1). Of the 32 children in Department B, 16 were randomly assigned to ward B1 and 16 to ward B2. None of the B2 children got xerophthalmia. Initially, it seemed both wards were treated the same, but upon close analysis, he found that they were getting different breakfasts. Ward B2 got Danish beer-and-bread soup made with whole milk. Because a few children in Ward B1 at one time had loose stools, the matron in charge had changed their breakfast to oatmeal gruel with rusks and fat free milk. Having discovered this difference, Bloch gave the 8 children with eye problems cod liver oil as he was probably aware that W. Wilde had noticed night blindness during the Irish famine of 1851 and had recommended cod liver oil as a treatment. By that evening, the children's night sight had improved; by 8 days, the xerophthalmia was gone.

This has the hallmarks of a well-designed experiment since the patients were placed in random groups that were treated equally except for 1 variable (breakfast). After the symptoms developed, the only change was cod liver oil and there was a quick recovery. At that time, several investigators suggested that night blindness was due to lack of fat in the diet. Bloch did not feel this explanation worked in his case as the children in both wards ate margarine with bread in the afternoon as well as with supper. Therefore, Bloch felt both groups had adequate fat. Bloch reasoned that it was not the fat itself, but something dissolved in the fat of whole milk or cod liver oil that accounted for the difference between the two groups of children. After the war, he became aware of animal experiments in the US where it had been found that something called fat-soluble accessory factor A prevented xerophthalmia in animals. We now call this factor, vitamin A.

Eating some vitamin A can quickly restore the vitamin A to the eyes and fix night blindness. Vitamin A is also a hormone and regulates the expression of some proteins from their respective genes. This accounts for the xerophthalmia; in xerophthalmia, epithelial cells make more keratin and this causes the problem. Once vitamin A is restored, it takes much longer for the cells to change which genes are transcribed and to break down the keratin, so it takes about a week for the xerophthalmia to go away.

How does vitamin A deficiency cause night blindness? The molecule that actually absorbs the light in your eye is rhodopsin. Rhodopsin is composed of a protein, opsin, and a pigment vitamin A. Vitamin A has several related chemical forms. The form important in the eye is called retinal. Retinal has 20 carbon atoms. Six of the carbon atoms are linked in a ring, which is often depicted as a hexagon. The other 14 form a tail. The tail has 2 primary different shapes and an analogy might help this a bit. Imagine that connecting each carbon atom is a pipe and that the carbon atom is a 45° connector. As shown in the top of Figure 1, the tail can have all the connectors set so that the pipes zig and zag and that structure is called "all trans-retinal". In the other important form, one of the connectors is turned the other way. To know which connector is turned the other way, we need to number the carbon atoms; the convention is that the ring carbons are numbered first and then the first carbon off the ring is 7 and so on. With this convention, it is the 11th carbon that zags instead of zigs and this is call a "cis" shape, so the other form is call "11-cis-retinal".

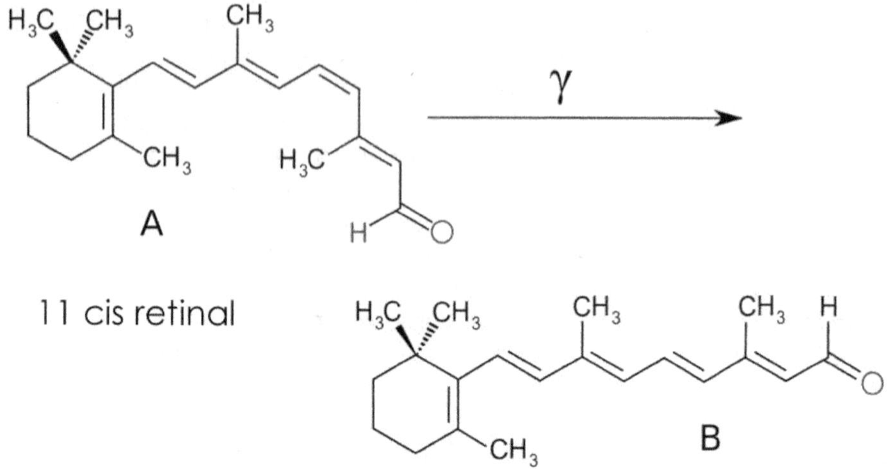

A

11 cis retinal

B

all trans retinal

Figure 1 Eleven cis retinal has zag at carbon 11; when light hits it, it changes so that there is a zig at carbon 11 becoming all trans retinal. Each vertex is a carbon atom. Figure is from Wikipedia, http://en.wikipedia.org/wiki/File:RetinalCisandTrans.svg

When light hits the 11-cis-retinal in rhodopsin, it changes shape to all-trans-retinal. This makes the protein part, opsin, also change shape and then through a series of steps, the rod cells tell the nerve cells that light has been detected and this is passed onto the brain. All-trans-retinal does not fit well in the protein, opsin, unlike 11-cis-retinal. Thus all-trans-retinal comes off, leaving opsin. The all-trans-retinal has to move around the cell, change to 11-cis-retinal, and be put together with opsin to form new rhodopsin before the cycle can go again, see Figure 2. This process takes time. Therefore, if you are in bright light, many of your rhodopsin molecules have lost their retinal. The opsin cannot respond again until it is rejoined with retinal (vitamin A). This is what happens as your eyes adapt. It takes almost 30 minutes of dark for most of rhodopsin molecules to be reformed. Thus as you stay in the dark, your eyesight gets better.

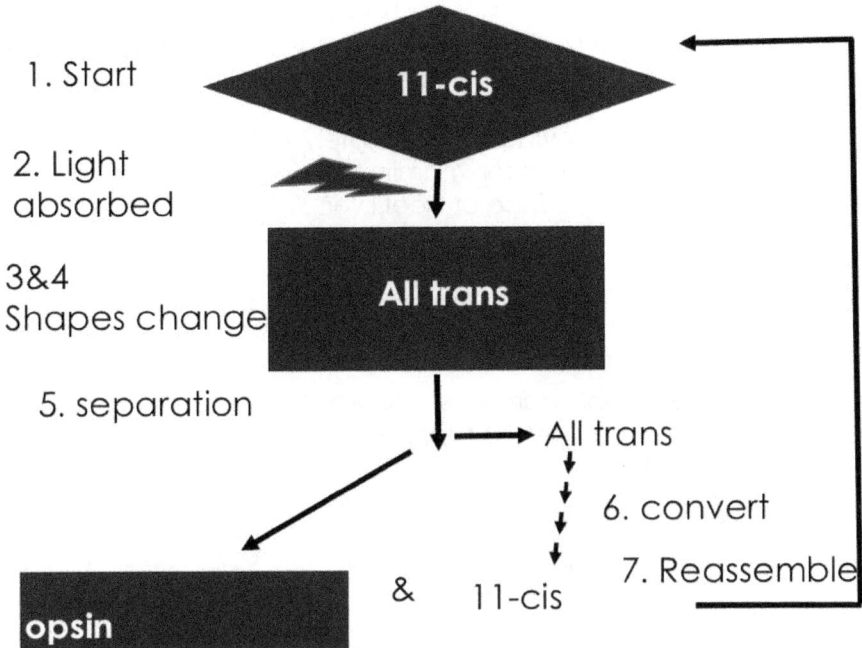

Figure 2. Light cycle. Diamond shaped opsin contains 11 cis retinal; the combination is rhodopsin. When light is absorbed by 11 cis retinal, it changes shape to all trans retinal and this causes a shape (conformational) change in opsin (black square). The retinal and opsin separate. In the cell all trans retinal is converted back to 11 cis retinal and rhodopsin reassembled from opsin and 11 cis retinal. The retinals are forms of vitamin A.

Retinal is one form of vitamin A. If you are vitamin A deficient, you have a lower amount of vitamin A to make retinal to form rhodopsin. This can be a real problem in modern life when driving at night, as the bright headlights of an oncoming car will cause most of your rhodopsin to lose retinal, and if you are vitamin A deficient, it will take you a much longer to get back your night vision.

Poor diet is a common cause of night blindness. In wealthy countries it is rare that someone has such a poor diet that they become vitamin A deficient although the 2 real cases of Vincent and Valerie show that it does happen. In wealthy countries, vitamin A deficiency is usually caused by a digestive tract problem, in particular a liver, pancreas or small intestine problem. In contrast, in poor countries, low dietary vitamin A is a major cause of night blindness, and eventually total blindness in children, affecting millions of children worldwide.

The following would provide approximately 100% of the daily needs for vitamin A.
¾ ounce of beef liver
5 cups of ricotta cheese
10 cups of milk
½ stick (4 tablespoons) of butter
1 ¼ sticks (10 tablespoons) of margarine

The following supply carotene and when converted to vitamin A in the body, would supply approximately 100% of the daily needs for vitamin A.
1/5 of a sweet potato
¼ cup spinach
¼ cup carrots
1 cup cantaloupe
1 cup sweet red peppers
2 mangoes
20 apricots
2 cups broccoli

The WHO recommends giving all children in poorer countries a vitamin A shot; in contrast some other organizations (e.g., Centers for Disease Control, the Teratology Society, and the Council for Responsible Nutrition) have recommended that no vitamin pills contain vitamin A.

Why is there a difference in opinion about including vitamin A in over-the-counter vitamin pills?

There are 2 main reasons. One is that vitamin A has a narrow therapeutic index[1]; the amount that is toxic is only 2 or 3 fold greater than the amount needed for optimal health, so there is not much room for error. The second reason relates to how vitamin A is absorbed. Carotene is absorbed in a manner similar to Vitamin A.

---

[1] Therapeutic index is the ratio of the toxic dose to the therapeutic dose. The higher the therapeutic index the safer, in general, is the drug.

Carotene is essentially 2 vitamin A's put together. In contrast to vitamin A, carotene has no known toxicity.

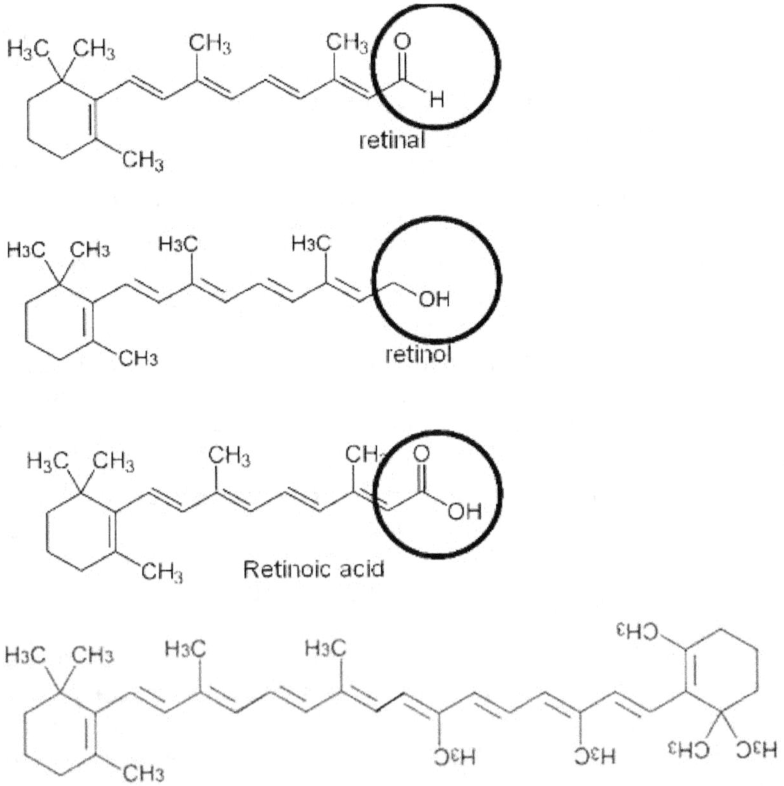

Figure 3. When I first see chemical diagrams, I pretend I'm 5 years old and have a kid's menu where I have to look at 2 pictures and identify what's missing or different. Hopefully you can see that the right hand side of retinal, retinol, and retinoic acid that are different. They are all forms of vitamin A. I've also included the carotene structure. Can you see how to combine 2 vitamin A's and get carotene (with a little subtraction)?

How are vitamin A and carotene absorbed?

Both vitamin A and carotene do not dissolve well in water but do dissolve in oil. The membrane around a cell is oil-like as it is made up of lipids. Therefore both vitamin A and carotene have no problem getting across this oil, or lipid, layer. A problem is getting vitamin A and carotene to this membrane and then, once they have crossed the membrane, getting them into the water phase inside of the cell.

In most of our foods, vitamin A and carotene are in the fat portion (fat is a type of oil). In the intestine one essentially has very little tiny droplets of fat and these droplets contain all the fat-soluble vitamins, vitamins A, D, E, and K; carotene and other fat-soluble compounds are also dissolved in the fat. To break up these fat droplets, the body uses soap-like compounds called bile acids. The liver secretes bile acids; these are stored in the gallbladder.[1] The bile acids also allow lipid breakdown enzymes secreted by the pancreas to break down the small fat droplets. As a lipid breakdown enzyme is a mouthful, biochemists use the word lipase[2] for lipid breakdown enzyme. In a healthy person, all the dietary fat is broken down in the intestine. Since fat likes to be dissolved in fat, dietary fats (lipids) have no problem going across the lipid membrane and getting into the cell. However, when all the fat is gone, where will the vitamin A go? It now has a choice of being in water (which it hates) or in the cell membrane (which it loves).[3] Therefore, after all the fat has been absorbed, vitamin A (and the other fat soluble vitamins E, D, and K) and carotene will be in the intestinal cell membrane.

You now also understand why the pancreas, liver, and small intestine are key organs for vitamin A absorption. It is because these 3 organs are critical for fat digestion and absorption. If the fat is not digested, the fat stays in the digestive system and is lost in the feces along with most fat-soluble vitamins, including vitamin A.

The liver is important for secreting bile salts, the soap to disperse the fats, the pancreas is important for secreting lipases, the enzymes to breakdown the fat, and the small intestine needs to have lots of cell membrane surface area to allow all the fat to be absorbed (Figure 4).

---

[1] Bile acids were first found in the gall bladder, hence their name.

[2] "Ase" is a common suffix for enzymes (particularly breakdown enzymes). Have you heard about the enzyme lactase that breaks down the milk sugar lactose? What do you think sucrase does?

[3] You might think the words hate and love should not be applied to chemicals, but the jargon words for vitamin A are lipophilic and hydrophobic, which are just Latin or Greek for lipid loving and water fearing or hating.

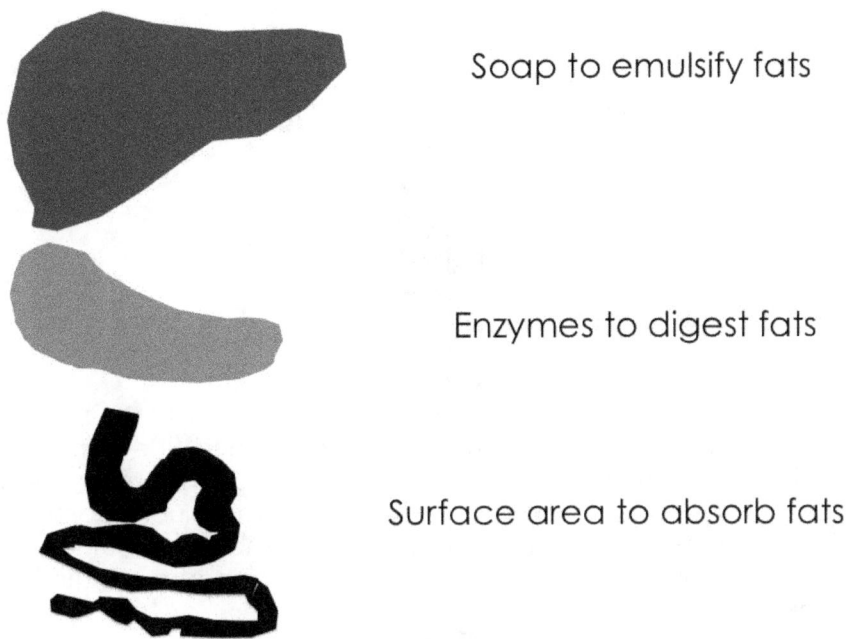

Figure 4. The liver, pancreas and small intestine are key organs for fat digestion.

How does vitamin A move from the cell membrane into the cell? There are proteins that will bind the vitamin A. Consider an analogy: a fat-soluble compound is like a cat; they both do not like water, but like to live on land or in fat.

A cat (fat) will have no problem getting across the piece of land (the cell membrane), but a cat does have a problem being in the water on either side of the land. There are proteins that are like rafts and help the fat-soluble compounds when they are in the water phases. (Figure 5)

Figure 5. Cats have no problem getting across land (the dark stripe in the middle) but they need a raft in order to be on the water (the light gray on either side). This is an analogy for fat-soluble compounds which don't need assistance in crossing membranes, but do need carrier proteins when they are in water compartments (inside or outside the cell).

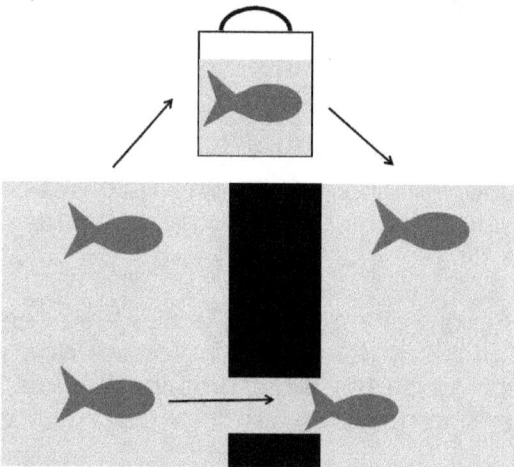

Figure 6. Transport of water-soluble compounds across a lipid (oil) membrane.
A bucket or a tunnel can serve as a carrier or transporter that helps water-soluble compounds (fish) cross the fat, or lipid, cell membrane (land). These carrier or transport proteins are found in membranes, such as the cell surface membrane and in cell organelles.

Here, we have a point of confusion as both the raft proteins and the bucket proteins are called transport proteins. You have to be careful when reading the literature to determine whether transport protein[1] refers

- to a protein in the membrane that carries water-soluble compounds across the membrane, a bucket
OR
- to a protein that is in the water, that helps to carries the fat-soluble compounds, a raft.

Inside the cell, vitamin A is collected with some fat and put into particles called chylomicrons. The sugars, amino acids and iron that you absorb go out of the cell and into capillaries, but the chylomicrons are too big to do this easily and they move into the lymph. The lymph system has a whole set of vessels, somewhat in parallel to the blood vessels. A big difference is that the whole lymph system is at low pressure, in contrast to the high pressure of the arterial system. Moreover, the lymph and blood have a different chemical make-up. Lymph vessels bring lymph to, and take it away, from lymph nodes. However, eventually the lymph does drain into the blood stream.[2]

Once in the blood stream, the chylomicrons are taken up by the liver. The liver stores excess vitamin A in special cells in lipid droplets. However, some of the vitamin A is allowed to circulate. Of course, it cannot be in the water phase (because it hates it). Most of the vitamin A in humans is bound to a protein called retinol-binding-protein. Retinol is the chemical name for a specific form of vitamin A and is slightly different from retinal, see Figure 3.

Most of the rest of the vitamin A in blood is in tiny little fat particles, called VLDLs, very low density lipoproteins. Lipoproteins are characterized by their fat content, which correlates with their density and size. You probably remember that fat is less dense than water.

---

[1] Just to make matters worse, in both cases, a synonym for transport protein is carrier.
[2] Maybe you have been sick and had a lymph node swollen. The node swelled because the drainage from the lymph nodes was blocked.

Can you guess which particles have more fat: low density lipoproteins (LDLs) or high density lipoproteins (HDLs)? Of course, very low density lipoproteins (VLDLs) would have even more fat. You might recognize some of these abbreviations, as HDL is considered to be the "good" cholesterol. This term is a bit of a misnomer, as cholesterol is cholesterol is cholesterol. Nevertheless, it is healthier if most of your cholesterol is carried in HDLs and not VLDLs and LDLs, hence the short hand term, "good cholesterol". Carotene is not bound to retinol binding protein and most stays bound to the lipoproteins.[1]

What happens if we eat too much vitamin A?

De Veer and his men presumably suffered from an acute overdose of vitamin A and one of the acute affects involves extra fluid inside the skull. Acute effects of vitamin A include severe headaches due to intracranial hypertension, bleeding gums, vomiting and blurred vision.

Chronic overconsumption of vitamin A also leads to other problems, including problems with the bones. This is because Vitamin A transcription factors are also important for bone growth. When calves are fed too much vitamin A, they get Hyena Disease [2]. Some anthropologists have studied the skeletal remains of a Homo Erectus and have noted that he had the bone changes expected from too much vitamin A. They speculated that this occurred as the species was beginning to eat carnivores and that carnivore liver is very high in vitamin A. From my limited reading, only a few carnivores have high liver iron stores, so I find that explanation not very likely.

Note that high doses of vitamin A during pregnancy can lead to malformations and other birth defects.

Most of the chronic effects of high vitamin A are due to vitamin A's role as a hormone. Vitamin A, like vitamin D, binds to nuclear

---

[1] Chylomicrons are even larger than VLDL particles and contain the least amount of protein. The sequence, HDL, LDL, VLDL, and chylomicrons would be easier to remember if chylomicrons were renamed VVLDL (very, very low density lipoproteins).

[2] It is called Hyena Disease because with the bone malformations, the calves walk like hyenas.

receptors (see the chapter Bubbling Beverages Bad for Bones?, for more about Vitamin D). The complex of the vitamin and its nuclear receptor regulates gene (DNA) transcription. Remember that to go from DNA to protein, there are 2 steps. First, the cell must transcribe from DNA to mRNA and then it translates from mRNA to protein. We have ~30,000 genes. Which genes are transcribed and translated into proteins in a particular cell is a function of time and place. The cell has many different transcription factors that regulate transcription. The vitamin A transcription factors are important in regulating a number of processes; if one has too much vitamin A, then one is turning on these processes in the wrong cells at the wrong time. The development of an embryo and fetus obviously requires a particular sequence for expression of genes and too much vitamin A interferes with this process.

Can the body solve the problem of too much vitamin A as it does for iron? In the case of iron, increases in the hormone hepcidin lead to a decrease in intestinal absorption of iron. For iron, the transport in the intestinal cells involves proteins and you can regulate how much protein you make and where they are located by changing hepcidin levels. However, vitamin A just moves through the cell membrane as it is fat loving; you cannot control that, just as you cannot control how much air comes through a screen door. So the problem for vitamin A is that you absorb all you eat.

If you are eating a reasonable amount of vitamin A, the blood concentration remains normal and some vitamin A is stored in the liver. If you now start to eat too much vitamin A, it is all absorbed. For a while, the extra is stored in the liver and the blood concentration remains normal. If the blood concentration is normal, then the concentration in the cell is also normal. Eventually, the liver store is full and a further increase in dietary vitamin A results in an increase in blood concentrations. This activates some transcription factors that should not be activated and one gets sick. Because vitamin A is fat soluble, it is not normally excreted in the urine. It is also broken down slowly. So it is hard to get rid of extra vitamin A.

Carotene is also fat soluble, so you also absorb all you eat. However, a key difference is that the conversion from carotene to vitamin A requires an enzyme, which is a protein. You can regulate the amount of this enzyme. Thus if the body has plenty of vitamin A, it can turn off the expression or activity of the enzyme that converts carotene to vitamin A. The details are not yet known. Does vitamin A bind to transcription factor and turn on expression of a protein that inhibits the enzyme? Alternatively, does vitamin A bind to a repression transcription factor and turn of making the conversion enzyme?

If we examine Figure 3, we can make some educated guesses about why carotene is not toxic, but vitamin A at high concentrations is. There are 3 forms of vitamin A-like material that have important functions in the body. Each structure plays a different role. Retinal, the visual pigment, is made from the retinol form of vitamin A. Retinal is an aldehyde, which can be chemically abbreviated as CHO. Retinol is the alcohol, CHOH, so it has gained a hydrogen compared to an aldehyde. Retinoic acid is a carboxylic acid, COOH, which has gained an oxygen compared to an aldehyde. Retinol and Retinoic acid bind to different transcription factors; this means the transcription factor must have a binding site that distinguishes between retinol and retinoic acid. Can you see what part is different? You'll note that carotene has neither an alcohol nor an acid group, so it is not surprising that it cannot bind to either of the vitamin A transcription factors; in addition, it may be too big to fit into the binding site.

Interestingly, cats lack the conversion enzyme and so they cannot convert carotene to vitamin A. However, carotene is a plant compound and is not normally in a cat's diet. The cat gets its vitamin A from meat (presumably liver). There has been a report of a human who had a mutation and her enzyme also could not covert carotene to vitamin A. This means she needs to rely on vitamin A to get all her needs and because vitamin A has a narrow therapeutic range, she has to be careful about how much vitamin A she eats. She can eat all the carotene she wants as high carotene is not known to cause any problems except for skin color. Carotene does not bind to the vitamin A transcription factor nor interfere with the visual process.

Why carotene may be better to consume than vitamin A
For the majority of the population who have an active carotene-to-vitamin-A-conversion-enzyme, it seems much more effective to consume carotene than vitamin A. Our bodies then can regulate the amount of vitamin A we make if we have excess carotene. That is why some organizations want to replace all vitamin A in vitamins with carotene. It would reduce the possibilities of vitamin A overdoses. A

few children have eaten too many vitamin pills because the vitamins seem like candy. The children often present with symptoms consistent with either vitamin A or iron overdose.

Polar Bear Liver
As mentioned previously, the liver is a storage area for vitamin A. Polar bears have high amounts of vitamin A stored in their liver. But why would bears store so much in their liver? Think about a polar bears' diet: do you think they can get fresh carrots year around? (Fish and other animals also have vitamin A, of course, but except for liver, lean animal sources are probably not as rich as yellow and orange plants.) So my guess is that this extra high vitamin A helps the polar bears when they go a long time between vitamin A sources.

What if the amount of vitamin A stored in the liver is under genetic control? If so, did the mutation for an increase in vitamin A only happen in polar bears or did it happen before polar bears became a species? It turns out that several artic animals have livers with high vitamin A, including arctic foxes and seals. Dogs, which are more closely related to foxes than polar bears, have "normal" vitamin A levels in their liver. I have been unable to find a reference for vitamin A levels in black bear liver. I think there are two primary theories that remain to be sorted out. One possibility is that an early member of Carnivora family had the mutation for increased liver vitamin A stores and all artic members have been selected to maintain that mutation, whereas other members that have normal levels have lost that mutation or have another one which counteracts it. Another possibility is that the mutation arose independently several times in Carnivora and was selected for in those animals living in the artic. As far as I can tell, no one has examined these possibilities by looking at vitamin A stores in most of the Carnivora species. In the chapter The Good, the Bad and the Spicy, we will briefly discuss studies done which show that the mutation to taste sweet has apparently been independently lost in many, but not all, Carnivora species.

Chapter Summary
- Vitamin A is like many nutrients. We need enough, but not too much.
- Vitamin A deficiency is a cause of night blindness, because a vitamin A derivative is part of the photoreceptor.

- Too much vitamin A is a problem, because vitamin A is also a transcription factor and high levels at the wrong time and place alter cell growth and death; this can account for birth defects as well as bone defects in adults.
- Vitamin A has a narrow therapeutic index. We are unable to regulate its absorption because no proteins are directly involved.
- While we also cannot control carotene absorption, carotene is a much better way to supply the body with vitamin A. We can control the rate of conversion from carotene to vitamin A because an enzyme is required. Too much carotene only results in orange skin, so carotene's therapeutic index is large
- Proper fat digestion is important for proper vitamin A absorption and carotene absorption (and the absorption of other fat soluble vitamins).

Blood Sweat and Tears

In one episode of CSI:SVU, the response team thinks, at one point, they discover a barrel of a chemical that makes them think they may be dealing with a terrorist cell. Before this point, Officer Benson is interviewing a boy in his apartment as the mother observes. As Benson talks to the boy, he eyes keep closing, as if he is getting sleepy. Benson starts to feel sweaty and sends the mother for a glass of water. She hears the mother drop the glass and notices that the boy has passed out. As she walks to the kitchen to find the mother, she stumbles and has vision problems. She eventually drags both the boy and the mother to the hall, breaks the windows to let in fresh air and just before Benson passes out, tells a neighbor to call 9-1-1. At the hospital Benson is told she has low levels of acetylcholinesterase in her blood.[1]

In a small town in Nebraska, several people got sick within a short period. The major symptoms included diarrhea, vomiting, sweating, blurred vision, and nausea. It appeared that these people got sick shortly after eating some locally grown hydroponic cucumbers. One middle aged woman ate part of a cucumber. Within 15 minutes she experienced many of the above symptoms and she was taken to the hospital and recovered after several hours. She apparently told a male friend of hers and he scoffed at the idea that her symptoms were caused by a cucumber. Presumably a discussion followed and both sides remained adamant. He eventually ate the rest of the exact same cucumber. Guess what? Within 20 minutes, he had the same symptoms and had to be rushed to the hospital.[2]

---

[1] CSI:SVU. Season 8. Episode 13. Loophole.

[2] From Goes EA, Savage EP, Gibbons G, Aaronson M, Ford SA, Wheeler HW. Suspected foodborne carbamate pesticide intoxications associated with ingestion of hydroponic cucumbers. Am J Epidemiol. 1980 Feb;111(2):254-60. PubMed PMID:7355886.

The sections in this chapter are:
- What is acetylcholinesterase?
- Why does a nerve need a chemical to talk to its neighbor?
- How does the listener nerve cell realize it is being talked to?
- Are all acetylcholine receptors the same?
- An aside about how nicotine and muscarine got their names.
- How are the nicotinic and muscarinic acetylcholine receptors different?
- Can't you get that nerve to shut up?
- How do acetylcholinesterase inhibitors work as insecticides or nerve gases?
- Why is location important for compounds that bind to receptors?

What is acetylcholinesterase?
"Ase" is a common suffix that biochemists use to indicate an enzyme. Acetylcholinesterase is an enzyme that breaks down acetylcholine. So naturally, you want to know what acetylcholine is. Acetylcholine is a chemical that allows a nerve to talk to its neighboring nerve or a muscle and is one example of a neurotransmitter.

Why does a nerve need a chemical to talk to its neighbor?
There are really two ways that cells can communicate: one is by chemicals and the other is by electrical signals. Electrical signals only occur when two cells actually touch each other, a bit like communicating between two hotel rooms that have connecting doors. (The connecting doors between cells are called gap junctions.)
Nerve cells in general have a space between one nerve cell and the next or between a nerve cell and a muscle cell. This space is large enough that an electrical signal cannot pass from one cell to another cell, rather a chemical leaves one cell and goes to the next. I have a hard time making a clear analogy like I did with the hotel doors between adjacent rooms, so here is another analogy. Electrical signals between cells are like communicating by touch between two people; they obviously have to "touch". Chemical signals between two people can occur if they are close to each other and one person whispers (releases a chemical) and the other person can hear and understand that sound. For a chemical, a receptor is a protein that binds the chemical and changes shape.

Physiology

If you step on a tack, there is not one continuous nerve cell that runs from your foot up to your spinal cord and back down to your leg muscles, there are two or three nerve cells; one talking to the next. "Talking" for a cell means the "speaker" releases a chemical (words or sounds) and the "listener" receives the sound (receptor or ear).
Acetylcholine is the first chemical discovered that allows the signal to be transmitted from one nerve cell to the next, that is, it was the first neurotransmitter discovered.

How does the listener nerve cell realize it is being talked to?
In this analogy, the sounds the speaking cell makes are analogous to the speaking cell releasing acetylcholine. Acetylcholine spreads out from the speaking cell, just like your voice spreads out across a room. But your voice has no effect on the sofa because the sofa lacks ears. Hopefully there is a person in the room with ears (receptors) that can hear the sound. The adjacent nerve cell has proteins that receive the message, that is, bind acetylcholine. These proteins are called receptors. Some receptors have helpful names; the protein that binds acetylcholine is called, drum roll,... the acetylcholine receptor. (Asking the name of the protein that binds acetylcholine on an exam is about as likely as asking on an exam, when was the War of 1812 or who fought in the Spanish-American War.)

Are all acetylcholine receptors the same?
There turn out to be different types of acetylcholine receptors; indeed, most proteins have families, that is, several proteins are very similar and we consider them related. (Some protein families have a common ancestor which makes using the term "family" more appropriate.) Acetylcholine receptor proteins can be divided into two categories, nicotinic acetylcholine receptors and muscarinic acetylcholine receptors. Within each category there are multiple subtypes.

You probably know that nicotine is a chemical found in tobacco. Muscarine is a chemical originally found in some mushrooms. As you might guess, nicotinic receptors bind nicotine better than muscarine and muscarinic receptors bind muscarine better than nicotine. It turns out that all the acetylcholine receptors on skeletal muscles are nicotinic which I find confusing because I think of muscles when I see muscarinic, but as I mentioned, muscarinic has nothing to do with muscles.

An aside about how nicotine and muscarine got their names.
Both nicotine and muscarine have been used as insecticides; indeed, there are theories that the insecticide properties of these compounds gave the plants that make these compounds (tobacco, some mushrooms) a selective advantage.[1]

The mushroom that makes muscarine has a common name of fly agaric and was used long ago as an insecticide; the mushroom was sprinkled on milk. This was mentioned in writing as early as 1256. Agaric is a term that refers to fungus or mushrooms. The common name influenced the decision on the plant's scientific name, *A. muscaria*, because muscaria comes from the Latin for fly. When chemists in the mid-1800s isolated a compound from these mushrooms that was biologically active, they named the compound after the plant, hence, muscarine.

Tobacco's scientific name, *Nicotiana tabacum*, derived from the name of one of the early Europeans, Jean Nicot, who brought tobacco from the New World to Europe and promoted its medicinal effects. (Nicotine type compounds are used today for some medical applications!) When chemists in the early 1800s isolated a compound from tobacco that they considered poisonous, they named the compound after the plant, hence, nicotine.

You will note that muscarine and nicotine rhyme because they have the same suffix; chemists tend to give the same suffix to compounds with the same chemical property or structure. In this case, I **think** "ine" refers to the fact that the compound contains nitrogen.

---

[1] A dominant theory in biology is evolution and it is often misunderstood. Part of the misunderstanding comes from a bit of sloppiness in how information is presented. My first draft of the previous sentence was: indeed, there are theories that the reason some plants (tobacco, mushrooms) make these compounds is because of their insecticide properties. Do you see how the two phrases are different?

Physiology

How are the nicotinic and muscarinic acetylcholine receptors different?

The nicotinic receptor is a protein in the category called a ligand gated ion channel and the muscarinic receptor is a protein in the category called a G protein.

Ligand gated ion channel is a mouthful. Ligand just means a chemical that binds to the protein, in this case, nicotine is the ligand. Channel refers to a protein that essentially makes a water filled tunnel through the lipid (oil) membrane bilayer. Ion refers to the fact that an ion channel lets ions, that is charged atoms or molecules, flow through the tunnel. There are some channels that only let uncharged ions flow through. It is a lipid membrane bilayer that separates the inside of the cell from the outside of the cell; if one wants to keep the inside and outside separate, one obviously does not want to have all the tunnels, or channels, open, all the time, just like you don't have the door to your house open all the time; in this analogy I consider walls of your house to be analogous to the lipid bilayer surface (plasma) membrane of the cell. Rather than call the piece that opens and closes the tunnel/channel entrance a door, scientists call it a gate. So a ligand gated ion channel is a protein that, when a ligand binds (nicotine, acetylcholine), the protein changes shape, causing the "gate" to open, and allowing ions to flow into or out of the cell. For nicotinic acetylcholine receptors, the ions that flow are sodium into the cell (and potassium out of the cell).

The nicotinic receptor is like the handle on a garden hose faucet- when the first messenger (acetylcholine) binds to the handle, the handle changes shape (the AChR changes shape, or conformation) and water (sodium) can run into the cell.

The muscarinic receptor is not an ion channel but rather a G protein coupled receptor. These proteins have parts on the inside that bind guanosine triphosphate (GTP) and guanosine diphosphate (GDP). These are very similar to ATP and ADP, except one pair includes adenosine and the other guanosine. Adenosine and guanosine are also the A and G of the coding portions of DNA. The muscarinic receptor is more like the first messenger (ACh) turning a door handle, (the AChR changes shape, or conformation) and the foreman inside the greenhouse who has his hand on the other side of the handle lets go, and tells his assistant (a second messenger, e.g., cAMP, Ca, IP3, cGMP) to tell the gardener to turn on the hose. Often there are many assistants that have to be told in sequence and sometimes the gardener grabs a rake or a shovel and does not turn on a hose.

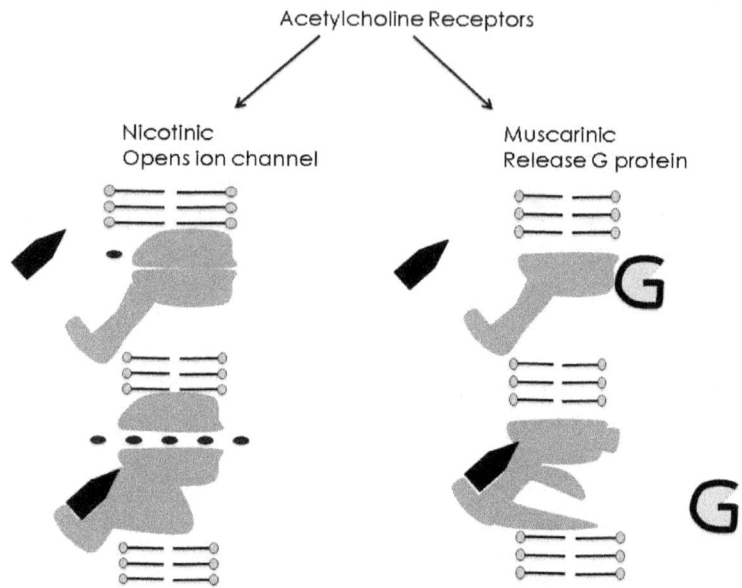

Figure 1. There are two general types of acetylcholine receptors. One binds nicotine better than muscarine and the other type binds muscarine better than nicotine. The receptors also differ in their action. When the nicotinic receptors are activated, they change shape and that opens an ion channel. The open ion channel allows ions to move across the cell membrane (left panel). In contrast, when muscarinic receptors are activated, they change shape and release a G protein (right panel).

Can't you get that nerve to shut up?
My speaking/listening analogy breaks down here. If someone stops speaking, there are no more sounds waves and the listener gets peace and quiet. But if the speaking nerve cell stops releasing acetylcholine, the acetylcholine is still in the space between the two nerves and can still activate the acetylcholine-receptor on the listening cell.

There are two ways to remove the signal. One is to destroy it, that is to break it apart. The second is to sweep it away. Breaking it apart might ring a bell from p. 1 of the chapter - remember acetylcholinesterase? It breaks down acetylcholine. So that is how the signal is stopped at cholinergic junctions. The breakdown products are acetate (vinegar) and choline. In order for this process to be useful, the acetylcholine-receptor better not bind acetate or choline, only the combo and this is true.

Figure 2. The top gray blob is the terminal of a nerve cell releasing acetylcholine (the black objects). The bottom black line is the outline of the muscle which as gray acetylcholine receptors and pie shaped acetylcholinesterase. The acetylcholine can diffuse across the space between the nerve and the muscle (the neuromuscular junction). The acetylcholine can also diffuse away. Acetylcholine that makes it to the muscle membrane can bind to, activate, and change the shape of the acetylcholine receptor. Acetylcholine could also be broken down by acetylcholinesterase. In the left panel, the acetylcholinesterase works so only 2 of the acetylcholine receptors are activated. On the right panel, the acetylcholinesterase inhibitor has blocked (black X's) the acetylcholinesterase. Thus there is less breakdown and more acetylcholine can bind to more receptors and activate them.

Sweeping away a neurotransmitter also happens; the second neurotransmitter discovered was norepinephrine. It is removed by transporters in the speaking cell membrane and can actually be reused-since it is recycled, I guess you might say that adrenergic (noradrenaline is a synonym for norepinephrine) is "greener" than cholinergic receptors. ☺ Cocaine has many of its effects by blocking this reuptake of norepinephrine. We'll come back to cocaine later because there is a connection with acetylcholinesterase and cocaine.

How do acetylcholinesterase inhibitors work as insecticides or nerve gases?

An acetylcholinesterase inhibitor blocks acetylcholinesterase from working. If acetylcholinesterase does not work, then acetylcholine is not broken down. If acetylcholine remains in the space between the junction (synapse) of two nerves or a nerve and a muscle, then acetylcholine continues to bind to the acetylcholine receptor and activate it. This makes nerves and muscles fire when they shouldn't and eventually this can kill an insect, an animal or a human.

Why is location important for compounds that bind to receptors?

Acetylcholine is released by lots of nerves; the specificity is that only the receptors very close to the releasing nerve are exposed to the acetylcholine. The acetylcholine is broken down before it activates any receptors that aren't very close.[1]

This is a critical point about physiology and pharmacology that often gets lost. In our body, one nerve can precisely activate the next nerve for two reasons. The first is that the speaking nerve releases a chemical and the listening nerve has a receptor for that same chemical, the two nerves, if you will, understand the same language. The second is

[1] Also as acetylcholine diffuses away from the release point, its concentration gets lower and lower. The volume of the synapse might be 1 femtoliter. (There are a quadrillion (10^{15}) femtoliters in a liter. A gigabyte is 10^9, a terabyte is 10^{12} and a petabyte is 10^{15} bytes) If there are 1 million acetylcholine molecules in 1 femtoliter, the concentration is roughly 1 millimolar, enough to activate the receptors on the adjacent nerve or muscle which is roughly 1 micron away from the release point. If the next closest cell with a receptor is 10 times away, that means that the acetylcholine has spread out and is now dissolved in a larger volume. Remember that volume goes as the cube, so increasing the space from 1 micron x 1 micron x 1 micron (which is 1 femtoliter) to 10 micron x 10 micron x 10 micron (which is 1 picoliter) decreases the concentration by 1,000, to 1 micromolar and this is probably too low to bind to many receptors.

that the speaking nerve is whispering and only the listening nerve can hear. When a drug is given to a patient, the drug goes throughout the body and could activate all the receptors which is why drugs have side effects. It is like being at the UN and nerve communication are people whispering to their neighbor; many speak the same language, but the communication is specific because each person whispers. A drug is like someone shouting at the UN-not everyone responds because not everyone understands that language, but it is much less specific or private than whispering.

Cocaine and Cholinesterase.
Cocaine blocks the reuptake of neurotransmitters, including dopamine, serotonin, and norepinephrine. One of the toxic effects of high cocaine is an increase in heart rate. This should make sense to you because norepinephrine increases heart rate.

It turns out that an enzyme related to acetylcholinesterase breaks down cocaine. The related enzyme is called butyrylcholinesterase; note the common cholinesterase part of both names. Butyrylcholinesterase is found in blood and we don't know its exact function. There are differences in the activity of this enzyme in different people when measured with different drugs. Some people have a form of butyrylcholinesterase that does not break down cocaine. They are completely normal in the absence of cocaine, but if they happen to take cocaine, the blood concentration becomes higher and stays high for longer than most other people, because they cannot break it down. One theory is that some of the people who die when taking cocaine for the first time have this mutation so that butyrylcholinesterase does not work. The study of how people differ in response to drugs because of their different proteins is called pharmacogenetics. Pharmacogenetics is usually limited to the study of different protein mutations that have no effect on people in the absence of that particular drug.

Summary of key points.
1. Nerves communicate to adjacent nerves or muscles by releasing a chemical, a neurotransmitter.
2. Acetylcholine and norepinephrine (noradrenaline is a synonym) are two neurotransmitters.

3. The signaling is halted by either breaking down the neurotransmitter (e.g., acetylcholine) or by transporting it out of the synaptic space (e.g., norepinephrine)
4. Acetylcholinesterase breaks down acetylcholine. Inhibitors of acetylcholinesterase will lead to an increase of acetylcholine at active nerves.
5. Specific proteins bind specific chemicals; these are often called receptors.
6. There are two classes of acetylcholine receptors: one opens ion channels, one activates G proteins.
7. In the body, neurotransmitters are specific because their concentration is high only in a very local region. Drugs that mimic neurotransmitters have side effects because the drug concentration is high throughout the body and can activate all their receptors, not just the local ones that need "to be fixed".

Physiology

Impressive Horses and Fainting Goats

Many people spend a lot of time making their body look "good". There are even competitions where one is judged based solely on one's appearance. We do this not only with humans, but also with some animals. Horses are magnificent animals and there are some shows where part of the judging is based upon the horse's physique. So naturally, horse breeders work to develop horses that are most likely to win more shows. Impressive was a quarterhorse that was aptly named as his muscles rippled. However, in spite of his appearance, he could not race very well. How did Impressive get such good muscles?

At a national Biophysics Society meeting, I heard a scientist describe his work determining the gene that caused an interesting behavior in goats. He spoke about going to a farm where they bred these types of goats. The scientist was just as amazed by the farmer's reaction as the goats. The farmer would run up toward some goats and wave his arms. The goats would look up, be startled, and fall over, hence the name, "fainting goats". Within a few minutes, they would stand back up and continue to be just fine. What surprised the scientist was the smile on the farmer's face when the goats fell over, as the farmer had witnessed this many times. I bet the scientists repeated pleasure with an experiment that works yet again would be equally surprising to the goat farmer. Why do the goats fall over?

In 1998, two cats were brought to a clinic because they kept falling over. They were siblings from two different litters and one was male and one was female. Clinical examination and close testing of their muscles suggested that had a similar problem as the goats. How could we be sure it is a similar problem?

Two factors are important for getting muscle to get bigger. One is that they need building blocks and signals to make more muscle. The other is what makes a skeletal muscle contract? "What" has at least 2 parts. What are the actual proteins involved in contraction? And what is the signal that allows the muscle to contract? By analogy, consider the question, what makes a car run? The engine makes the car run just as the contractile proteins actin and myosin make the muscle contract[1]. However, to have a car run, you also need a key, the signal that allows the engine to run. For skeletal muscle, as for smooth muscle, calcium is the signal or trigger. In smooth muscle, there are other ways to control contraction, but in skeletal muscle, calcium is the primary signal.

In this chapter we will cover,
- Acetylcholine
- Acetylcholine to calcium
- How do we get from sodium to releasing calcium from the storage containers?
- The problem with Impressive
- Impressive's gene
- Testing Impressive's offspring as well as humans
- How do we know that the mutation in the sodium channel causes the symptoms we see in the horse?
- Patch Clamp
- Related Human Diseases
- Fainting goats and cats
- Related Human Diseases

Acetylcholine
The problem for Impressive occurs in the steps that lead to an increase in intracellular calcium. There are a series of steps that go from a nerve signal to a rise in muscle intracellular Ca to muscle contraction. The end of the nerve is close to, but not touching, the muscle cell. To communicate across this space, the nerve releases a chemical; in the case of nerve to muscle transmission, the chemical is acetylcholine. Because this chemical transmits the signal from nerve to muscle it is called a neurotransmitter. Acetylcholine is also a neurotransmitter for some nerve-to-nerve communication, too[1]. The acetylcholine diffuses

---

[1] Contraction occurs with 2 proteins, actin and myosin. Many actins line up together to form a polymer; much like many beads can line up together to form a necklace. One end of the polymer is attached to a wall. Myosin is a bit like a person with a paddle. The myosin paddle attaches to one actin molecule. For contraction, the myosin paddle moves causing the actin molecule to move.
[1] Acetylcholine is the only neurotransmitter for nerve to muscle communication. Nerve-to-nerve communication is achieved by different neurotransmitters, depending upon which nerve is sending and which receiving. Some of the common neurotransmitters include noradrenaline, serotonin and

away from where it is released. Because the space between the nerve and muscle is small and narrow, this results in some acetylcholine arriving at the muscle membrane.

The muscle membrane has special protein receptors that bind acetylcholine. The body has two basic types of acetylcholine receptors, muscarinic and nicotinic. Unfortunately, the "mus" in muscarinic does not refer to muscle, but to muscarine, a compound isolated from *Amanita muscaria* mushrooms. On muscle cells, the receptor type is nicotinic. And this does refer to nicotine, the compound in tobacco. While both receptors are designed in the body to respond to acetylcholine, they each prefer to bind only one of these two different plant compounds[1]. Muscarinic receptors are in the G protein coupled class of receptors[2]. Nicotinic receptors are in a very different class of receptors, called ligand gated channels. Ligand just means that something binds to the channel to make it open; in this case, it is acetylcholine. An ion channel is like a short hose through the cell membrane. For many channels, when not in use, the hose is closed. Neuroscientists use the phrase gate to refer to how the opening and closing of the channel is controlled, just like a gate in a fence is used to control who (or which animals) are allowed to go from one side of the fence to the other.

Most cells have slightly more negative charge inside and slightly more positive charge outside. The charge difference between inside and outside is called the membrane potential. In cells like nerves and muscles, changes in membrane potential happen rapidly and are important for signaling. Hence these cells are called excitable. When

---

dopamine.

[1] Why do we have protein receptors that interact with plant compounds? Well, that question actually the wrong way round; the better question is why do plants make compounds that bind to our proteins? It is thought these compounds offer the respective plants some protection from being eaten. There is a bit more information about plant compounds and human receptors in the chapter "Ouch That Hurts".

[2] G protein coupled receptors are transmembrane proteins; on the inside of the cell they are coupled to G proteins. G proteins bind GTP and GDP, compounds similar to ATP and ADP, but containing guanosine instead of adenosine.

these cells are not "excited" they are resting, so the basal or resting state has the resting membrane potential. You have probably learned in many places that solutions are electroneutral, that is, that the number of positive charges equals the number of negative charges. That is true on a chemical scale. Consider a cell with 150 millimolar potassium chloride inside. For the typical resting membrane potential, this cell would have about 6 micromolar less K+ inside than chloride inside. There are 1000 micromoles in 1 millimole, so this means that when there are 150,000 potassium ions inside there are about 150,006 chloride ions inside. To a chemist 150,000 and 150,006 look like essentially the same number (and I agree with them). But the charge difference can have profound effects. Let me give you an analogy for where the pairs are almost equal, but the small difference can cause a big effect. 6 out of 150,000 is the same ratio as 3 out of 75,000. Consider 75,000 people watching a football game in a stadium. There were about 75,000 pairs in the stadium. Pairs of what, you might ask. I think all 75,000 people were paired with 75,000 sets of clothing. Do you think anyone would notice if there were 75,003 clothed people instead of 75,000. Not likely. Do you think if 3 people were not paired with clothing, there would be a big effect? That's a bit like having 3 chloride ions not paired with potassium ions out of the ~75,000 potassium chloride pairs in the cell.

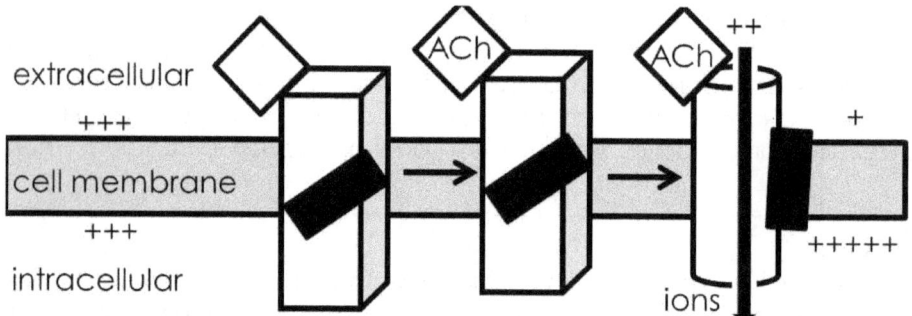

Figure 1 Nicotinic acetylcholine receptors, when activated by acetylcholine, change shape and open a gate which allows positive ions to enter the cell and change the ratio of charge inside and outside the cell. Other proteins can sense this change in the ratio.

Physiology

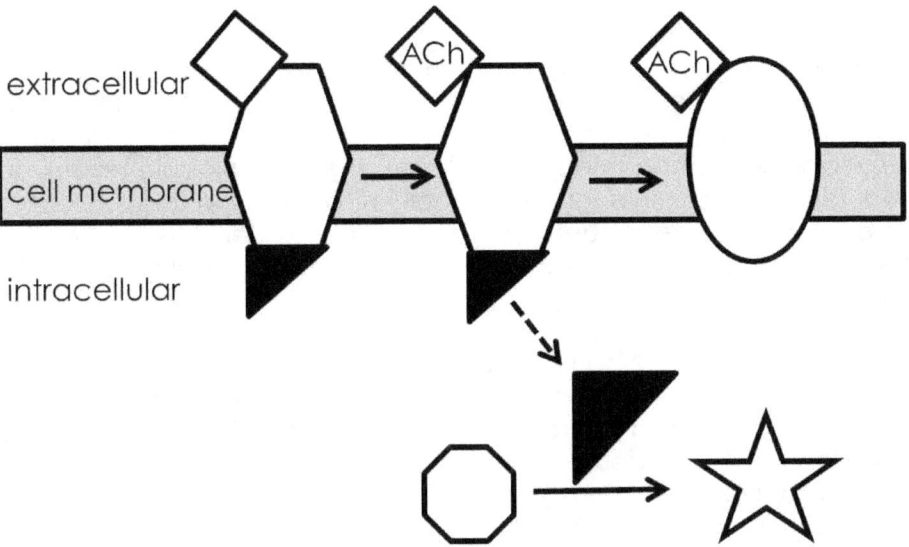

Figure 2. Muscarinic acetylcholine receptor, when activated by acetylcholine, change shape which releases the G protein (black triangle) which can then bind and activate an amplifier enzyme (stop size to star).

Acetylcholine to calcium
Acetylcholine binds to the nicotinic type of acetylcholine receptor and this allows the protein to change shape, opening the gate, and ions rush through. These ions rushing through lead to a change in the cell. This in turn leads to an opening of the gates in sodium channels and this eventually leads to the increase in calcium. Even without all the details, the process is pretty complicated, isn't it? But the complication, I think, is mostly that there are many steps, the concepts you have dealt with in real life. Here are analogies for each step:
Acetylcholine binding to its receptor is like putting a key in a door and opening the door and allowing people to enter, just as acetylcholine binding to its receptor opens a gate and allows ions in.
When enough ions go in, that triggers the increase in sodium entry, causing an action potential. This is a bit like flushing a toilet. You can push on the toilet handle a little and get no response, but when you hit the threshold point, not only does some water move, all the water

in the tank moves. So with sodium, one you've hit threshold all the sodium channels open.

The calcium being released from the intracellular stores might be like turning on a faucet that has a soaker hose attached, now the whole lawn has water, just as the whole cell has calcium. Muscle cells have a specialized compartment for calcium, the sarcoplasmic reticulum, and it also threads throughout the cell just as you might lay a soaker hose all around the garden. No part of the garden (cell) is very far from the hose and its water (the sarcoplasmic reticulum and its calcium)

How does sodium rushing in lead to an increase in cell calcium? First, let's talk about where the calcium is. In skeletal muscle, almost all the calcium that is used to raise cell calcium is in the sarcoplasmic reticulum closet. Because there are many closets in the house and they are distributed everywhere, this means that calcium can quickly get to all parts of the cell/house.

The cell is a bit like a battery; the membrane allows it to separate charge inside and outside. When the gate opens and sodium goes in, less charge is separated. Proteins in the sarcoplasmic reticulum membrane are able to sense this change of charge distribution and they in turn lead to the opening of calcium channels that allow calcium to move from the sarcoplasmic reticulum to the intracellular space. The change in charge distribution changes the electric field, the sensing protein itself has a charge, and so it feels the change in electric field. The idea of an electric field troubles many students because it is explained so poorly in many classes and books and because it is something you can't directly see. But you can easily understand how a field works by using a magnet. A magnet sets up a magnetic field. If you put a magnet close to a compass, the compass needle shifts; the compass needle is responding to a shift in the magnetic field because the needle contains molecules that have a magnetic sensitivity. If you try to shift a piece of plastic with a magnet, it won't work because there is nothing in the plastic that can sense the magnetic field.

The key feature here is that at rest, the muscle has a particular difference in charge between the inside and outside. This is the resting membrane potential and it is probably helpful at this point to give it a number. Unfortunately, the convention is that this number is negative, -70 millivolts. Think about it like the temperature inside a very cold freezer. When acetylcholine binds to its receptor, sodium rushes in and the membrane potential changes. This would be like opening the door to the freezer and warm air rushes in. In this case, the freezer would warm to -50 millivolts. There is another class of doors on the freezer, (B doors). These have a sensor and when the freezer is -50 millivolts or warmer, the B doors suddenly pop open and let in even more warm air. This is like the sodium rushing in the specialized sodium channels (B doors).

How do we get from sodium to releasing calcium from the storage containers? (Note that I've now shifted the analogy; let's make the freezer the cell and storage containers the sarcoplasmic reticulum.) The storage containers have a temperature sensor (membrane potential sensor) and when the freezer gets warm, say, zero, then the storage containers let out the calcium.

Changes in membrane potential happen in a number of cells; membrane potential is used to regulate not only calcium (and therefore muscle contraction, including heart and skeletal muscle) but also nerve activity and gland cell activity and even transport in the intestine and kidney. Because membrane potential is such an important process and because many students turn part of their brains off when they hear the word electricity (or electric field, or membrane potential), I have an analogy to show you how simple the concepts really are; it's just the words that get in the way for some people.

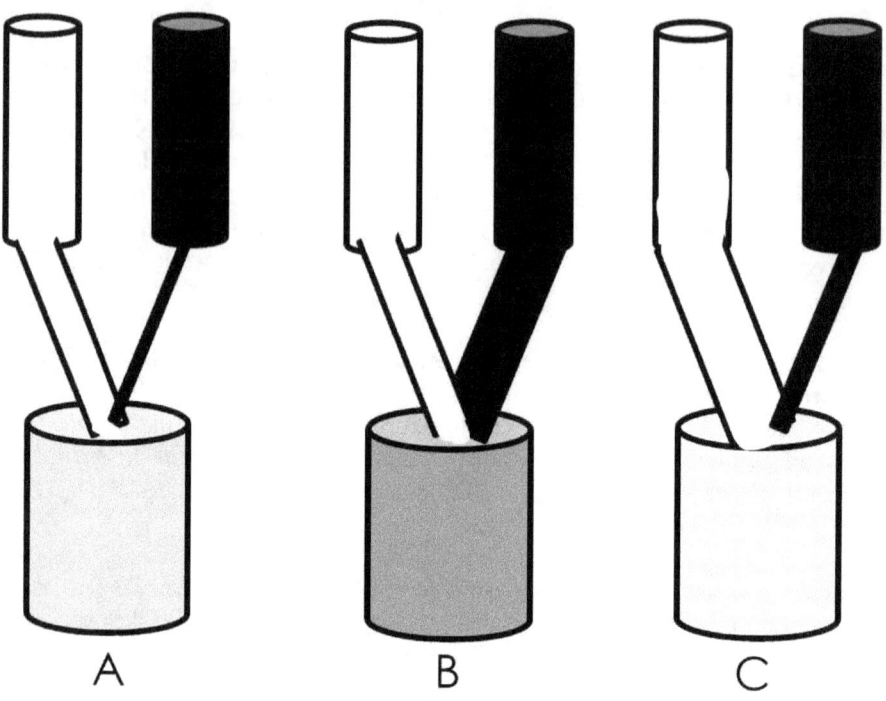

Figure 3. Changes of the ratio of inside and outside charges (membrane potential) by analogy with changes of color in a bucket. The bottom bucket color is the membrane potential, all white would be the potential if just potassium ions moved (about -90) and all black would be the potential if just sodium ions moved (about + 60). The color in the bucket reflects the size of the white hose (potassium) and the black hose (sodium). A is the resting condition where the bucket is light gray, because the white hose (potassium) is much larger than the black hose (sodium). B is what happens when the sodium channels open; the black hose is much larger and the bucket becomes dark gray. To return toward rest, the white hose opens more and the black hose closes, giving a very pale gray in C and then the white hose returns to the A condition.

In this analogy, the membrane potential is the color of the water in a bucket. The color of the water is influenced by two hoses, the potassium, or white water hose, and the sodium or black water hose.

Physiology

At rest, the potassium hose is open, so the water in the bucket is mostly light gray. For the muscle cell, when acetylcholine binds to the receptor, the sodium/black hose opens, and now the water in the bucket is gray; the membrane potential has changed. Now of course, we need to get the potential back to light gray in order to get calcium back to normal and relax the muscle. We can do this by either turning off the black, sodium hose, or by opening up the white potassium hose and both of these happen to restore the skeletal muscle's membrane potential.

The problem with Impressive

It turns out that the problem with Impressive is that his black, sodium hose, has a defect. Instead of closing completely after the channel has opened for a short time, the sodium channel remains slightly open. This means that the water in the bucket (the membrane potential) is not as white (at resting membrane potential) as it should be. This means some calcium can still be released and the muscle contracts again. And again. So even when Impressive is resting and his brain is not telling is muscle nerves to activate his muscles, his muscles are still contracting and getting exercise.

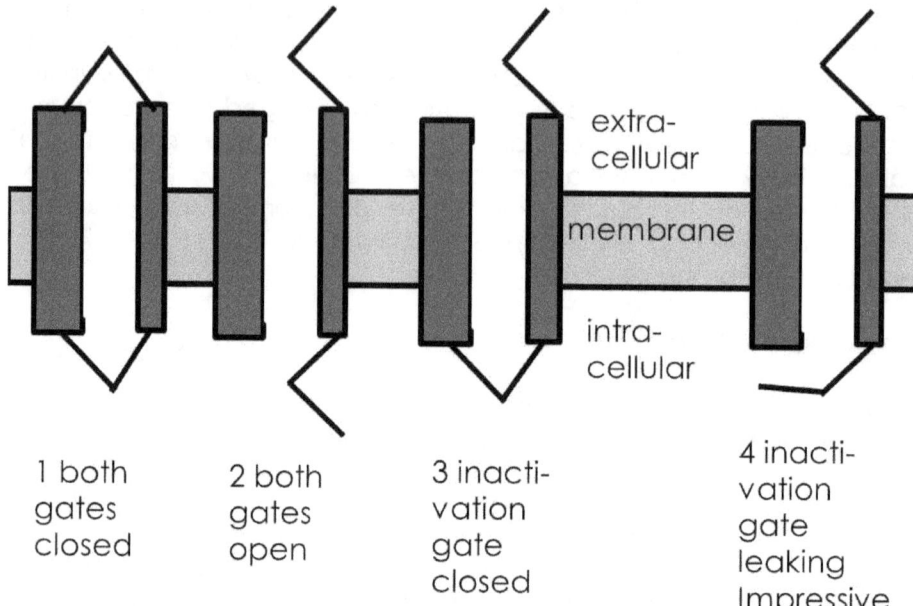

Figure 4. Sodium channel gates and the problem with impressive. The black bars and lines are the sodium channels. In 1, both the intracellular and extracellular gates are closed. In 2, both the gates are open and sodium ions can flow. In 3, the intracellular inactivation gate has closed as normally happens after a muscle fires. 4 shows what happens in Impressive, the inside inactivation gate does not completely close and some sodium can still enter the cell and trigger contraction.

Sometimes Impressive's muscles are contracting even without a nerve signal and sometimes they stop contracting and so he falls over. Why the different responses? The differences are thought to be due to slight changes in how much the membrane potential changes (or using our freezer analogy, how warm the freezer gets). If there is only a small change from -80 then the muscles become hypersensitive and they can contract even when the nerve accidentally releases only a little bit of acetylcholine. In a normal person, this accidental release of acetylcholine also goes on. But in a normal person, it doesn't cause a problem, because it doesn't make the membrane potential "warm enough" to open normal sodium channels. Therefore, there is no calcium released and the muscle doesn't contract. But in Impressive, that's enough warming to open some of the sodium channels and the muscle does contract after the calcium goes in. However, if for any reason, the muscle potential changes a lot from -80 (the freezer gets quite warm), then the Na channels auxiliary gate also closes and they are unable to open. Thus, there is no signal for calcium to be released and the muscle can't contract. Without muscle contraction in his posture muscles, the horse can fall over.

Impressive's gene
The change in the sodium channels in Impressive is due to a change in a gene, a mutation in the DNA. Thus, it can be inherited. And since Impressive won so many awards, he was used to father many offspring. Is the mutation recessive or dominant? Remember that you have two genes for every protein you make, one from your father and one from your mother[1]. Typically, you make about the same amount of the protein using your father's genes as your mother's. A dominant inheritance pattern is observed when it only requires one of the two genes to be expressed in order for you to show the trait. A recessive inheritance pattern is observed when both genes have to have the problem in order for you to show the trait.

If we think about what happens in this case, we can predict the type of inheritance. The basic question is, if Impressive has only 1 copy of the mutated sodium channel, would he show symptoms? This mutation could be classified as a gain-of-function mutant, because the sodium channel has gained the function of working under conditions where the normal channel does not work. So even one copy of the gene being mutated means that about half of his sodium channels will be open when they shouldn't; this will make it more likely that the muscle will misbehave. In terms of our analogies, we could consider there to be 2 hoses carrying the blue sodium water reflecting the 2 copies of the gene for the skeletal muscle sodium channel. Even one hose open when it shouldn't be, would lead to greener than normal water. Or for the freezer analogy, there are two sodium doors. If only one is open when it shouldn't be, it will still let warm air into the freezer and the freezer won't be quite as cold. This analysis predicts a dominant inheritance pattern.

---

[1] Except for most genes on the X chromosome.

To test the theory that the gene is dominant, we could look at Impressive's offspring and see how many offspring have the trait. As we predict that the gene is dominant, then Impressive would only need to have 1 copy of the mutated sodium channel. Thus about half of his offspring will have that copy. Since we predict that the mutation is dominate, the one mutated copy should be enough to see the effect, so about half of his offspring should also show the symptoms. Indeed, that is what is observed, about half of Impressive's offspring have a similar symptoms.

When I do cases like this, I also like to consider the other case. So what would a recessive sodium channel disease look like? Well, suppose the mutation just made the channel not work. The mutated hose doesn't let blue sodium water flow or the mutated sodium door on the freezer won't open. If you have one normal copy and one mutated copy, that might slightly slow down how fast the water turns green, or the freezer gets warm, but it is unlikely to cause much of a difference in contraction. But if you had two copies of the mutation, then neither blue sodium hoses would let blue water flow or both sodium freezer doors won't open. In this case, the water won't turn green or the freezer won't warm, and so there will be no muscle contraction. (Unless there is a back up system of other pipes or freezer windows that can do the job.)

Testing Impressive's offspring as well as humans
Suppose you just got a foal that was sired by Impressive. How could you easily test for whether he has the mutation? You could take some of his DNA and sequence the sodium channel to look for the mutation. But a quicker and less expensive way is to give the horse a precise potassium injection. One needs to be very careful, because even in normal horses, too much potassium can cause death. A horse with a mutated sodium channel will become temporarily paralyzed after the injection and often fall over. During extreme exercise, blood potassium can also increase and this can also trigger a paralysis attack.

Humans also get a similar disease with a mutation in their muscle sodium channel. A more convenient test in humans is that when the muscle is cooled, it also gets paralyzed; we don't know why, yet. But if you give a person with this disease cold ice cream or milk shake, their tongue becomes stiff and they have slurred speech.

Physiology

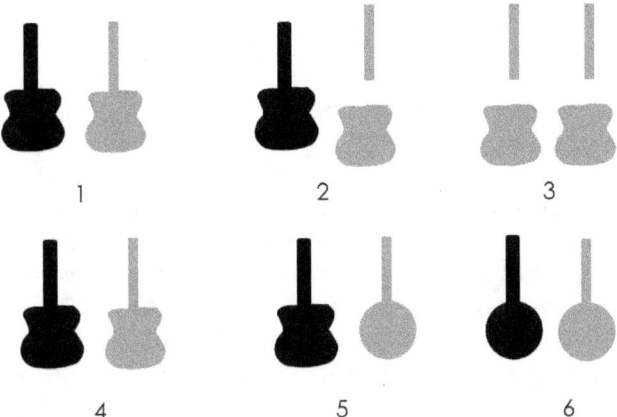

Figure 5. Dominant and recessive genes and loss-of-function vs. gain-of-function. Consider the guitar gene, you have two copies, black and gray, one from Mom and one from Dad. In #1, you have two functional genes and "normal" guitar playing. In #2, you have one functional guitar and one broken guitar. You can still make music. In #3, you inherited two broken guitars and you can not make guitar music. A broken guitar is loss of function and it usually takes two broken guitars to cause problems.[1]

#4-6 illustrate the gain of function case. #4 is just the normal situation, just like #1. In #5, you have inherited one normal guitar gene and one mutated gene, which has now gained the function of a banjo. So #5 will now be "abnormal" with only 1 mutated gene. #6 is the case where one inherits two banjo genes.

How do we know that the mutation in the sodium channel causes the symptoms we see in the horse?

One piece of evidence is that the sodium channels behave as we describe is that they are open under conditions when normal sodium channels are closed. The measurement of sodium channels opening and closing relies on a very clever technique developed in the late 1970s and which resulted in a Nobel Prize for the developers. This

---

[1] There are situations where you need to have loud guitar music and having only one functional guitar would be inadequate for "normal" function and therefore #2 would also show symptoms.

technique is called the patch clamp technique, because only a small patch of the membrane is studied.

Patch Clamp

The patch clamp technique is so sensitive, that you can observe the opening and closing of a single channel, which is a protein, about 100 Angstrom's (or 10 nanometers) in size. The technique works because every time a sodium channels gate opens, about 1000 sodium ions go through in 1 millisecond. One thousand seems like a lot of sodium, but if you think in terms of moles, it is only 10^{-20} moles, which is teeny-tiny in chemical terms. But if we have a good insulator, then 1000 sodium ions represent a significant amount of current and we can measure the current. To be a reasonable size to measure, the resistance needs to be about 1 gigaohm.

Let's consider a field and fence and gate analogy. We have a big field (cell) with a fence around it (membrane) containing many gates (sodium channels). We want to know how often one of the gates opens and closes, but the gate is too small to see. When the gates open, dogs get through, but too few for us to measure, because there are lots more dogs in the field than come through a single open gate. But every dog in the field has his own food dish. The new dogs do not. (Every Na in the cell has a chloride ion, which balances the charge, but only sodium ions go through the channel.) So when the new dogs come in, there aren't enough dishes to go around and some of the dogs start barking. There is extra sodium over chloride and the charge difference is as easy to measure as hearing dogs bark. In an active cell, there are lots of gates opening and closing and so dogs coming into the field from everywhere, so always dogs barking. If you want to know about one specific gate, you need to isolate that part of the field. So we'll build a wall that goes to the fence and encloses only 1 gate. We can't let dogs sneak out between the fence and the wall; otherwise they can go into the field some other way. This is why we need a tight resistance.

The scientists found that with the right type of glass tubing, and polishing the edges well, they could get a seal between the glass and the cell membrane that was about 1 gigaohm in resistance. You are used to the prefix Giga for computer memory, but you've heard it so much, you may not realize how big it is. In the early days of computers, a kilobyte was considered a lot of memory. It is 1,000 bits

Physiology

(like kilometer is 1,000 meters). Megabyte memories were considered large just a few years ago; megabyte is 1 million bytes. A megameter would be 1 million meters or 1,000 kilometers which is a long ways! A gigabyte is 1 billion bytes, or a gigameter is 1 billion meters or 1 million kilometers. So a Gigaohm is 1 billion ohms, a very high resistance indeed. So now it is routinely possible to watch a single channel open and close. The mutated sodium channels have a little extra current going through them than the normal channels under some conditions and this explains the disease.

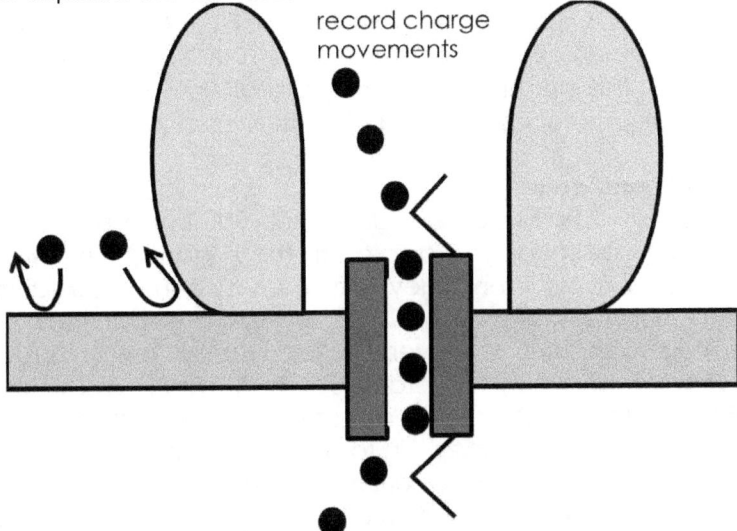

Figure 6. Cross section of a patch clamp electrode on a membrane. The membrane is the horizontal gray rectangle, containing one ion channel, the black bars and lines. Ions (black spheres) can flow through the open channel. If the patch clamp electrode (the top gray blobs) seal tightly against the membrane, then only the ions flowing through the channel are recorded as charge movements and no ions from outside the patch of membrane contribute to the signal.

Related Human Diseases

In humans, the diseases that defective skeletal muscle sodium channels cause are called hyperkalemic periodic paralysis, paramyotonia congenita, and potassium-aggravated myotonias. The exact disease depends upon the exact mutation of the sodium channel. These names, if you translate the Latin and Greek, make some sense. Hyperkalemic means excessive potassium; hyper meaning high or excess and kalium is the Latin for potassium and hence the chemical symbol K for potassium. Periodic basically means it comes and goes. And paralysis makes sense, right? What about Paramyotonia? Well myotonia refers to the fact that it is hard to relax the muscle. "Para" means it's related to other types of myotonia. Congenita has the same root as congenital as it is present from birth. Now that you know what myotonia means, potassium-aggravated myotonias makes perfect sense, right? Unfortunately the sodium channel's name is SCN4A-really helpful, huh? About as useful as saying Highway B1D. But you can use the number to look up information on the channel. In horses the disease is Equine Periodic Paralysis which just means Horse Paralysis that comes and goes.

Fainting goats and cats
If you go to visit a few selected farms, you will find a special kind of goat. Be very careful as you approach these goats; they won't hurt you. But if you surprise them, they'll fall over as if they are fainting. If you watch closely, they are not actually fainting, but their muscles become rigid and they fall over. If you wait a few minutes, their muscles recover and they can stand up. (You can find a video clip if you search fainting goats.) The cats have similar symptoms, but have not been as extensively studied. In fact, the opening case was the first description in cats, in 1998, whereas the fainting goats have been known for over a century. In 2010 a number of YouTube videos were popular that featured fainting cats born in England. So far as I have been able to determine, no one has published whether the mutation is actually in the same gene as the goats, as we would expect. However, Miniature Schnauzers are at high risk for the disease and a genetic test is now available and it is for the same gene as in goats.

When the goats' muscles go stiff, they are unable to balance and they fall over. You don't realize it, but when you are standing, your leg muscles are constantly adjusting the tension to keep you upright.

Why do these muscles become rigid? As you might guess, rigid muscles are muscles that are contracting; but why are they contracting?

It turns out that these muscles become rigid because they are too sensitive; when you surprise the goats, they aren't really frightened, but they are surprised and this slightly stimulates the flight or fight response. In this response, your adrenals release catecholamines. Perhaps the suffix, amines, sounds a bit familiar. It is used in amino acids and basically means nitrogen containing. Catechol refers to a particular organic chemical structure with a couple of chemical rings. Examples of catecholamines are adrenaline, noradrenaline and dopamine. Look at the word, "adrenaline" comes from adrenal, so this compound was first isolated from the adrenal glands. Where are the adrenals located? "Ad-renal" that is, above the renal, or kidney. Epinephrine is the same compound as "epi" means "above" and "nephrin" is another word for kidney. "Nor" is just chemical short hand prefix for having 1 less methyl group than the name without nor.

So how does the release of epinephrine and norepinephrine lead to contraction in these goats? As we learned with Impressive, the nerve going to the muscle releases acetylcholine when the nerve is activated. The acetylcholine binds to the acetylcholine receptor on the muscle membrane, which allows its channel to open. As ions go in this triggers sodium channels to open, more sodium ions go in and calcium is released, leading to muscle contraction.

As I hinted when talking about Impressive, even when the nerve is not activated, it sometimes releases a bit of acetylcholine. In normal muscle, this is not enough to start the cascade in the muscle cell that leads to contraction. In some ways, these goat's muscle membranes seem extra sensitive. It is sort of like the horses, but not quite the same. It turns out that the problem with the goats is not their skeletal muscle sodium channels. It is their chloride channels. Now we haven't mentioned chloride channels yet; we've talked a lot about sodium channels and a bit about potassium and calcium. Calcium, potassium and sodium are all are cations[1]; chloride is an anion. The role of the chloride channel is actually to keep the skeletal muscle membrane from being too sensitive, a bit like wearing ear muffs so sounds aren't too loud.

Let's go back to our color water analogy for membrane potential. Let me review it first before adding chloride to the mix. We had white water potassium hoses and black water sodium hoses going into a

---

[1] Cations have a positive charge; anions a negative charge.

tank. When the tank because gray enough, that caused calcium to be released and the muscle contracted.

To understand the fainting goats, we need to enlarge the analogy a bit. It is a bit like adding a light gray water hose that is also emptying into the tank. This light gray water hose is the skeletal muscle chloride channel. Whether or not this hose is open under resting conditions does not change the light gray color of the tank, since it is exactly the same light gray color you'd get for the resting about of white water (lots) and black water (very little). However, when the light gray water hose is open and contributing a lot of light gray water to the tank, you have to open the black water sodium hose a lot to get the water in the tank to become more gray. This is the situation in a normal muscle and this way, you don't contract your muscles when you get a little excited. But if the light gray water chloride hose is not working, then small changes in the black water sodium hose can give you a big change in the gray-ness of the tank. In a muscle, this would mean you'd trigger calcium release and the muscle would contract, which is exactly what happens with the goats.

Some humans also have defects in their skeletal muscles chloride channels and they have myotonia congenital.

Physiology

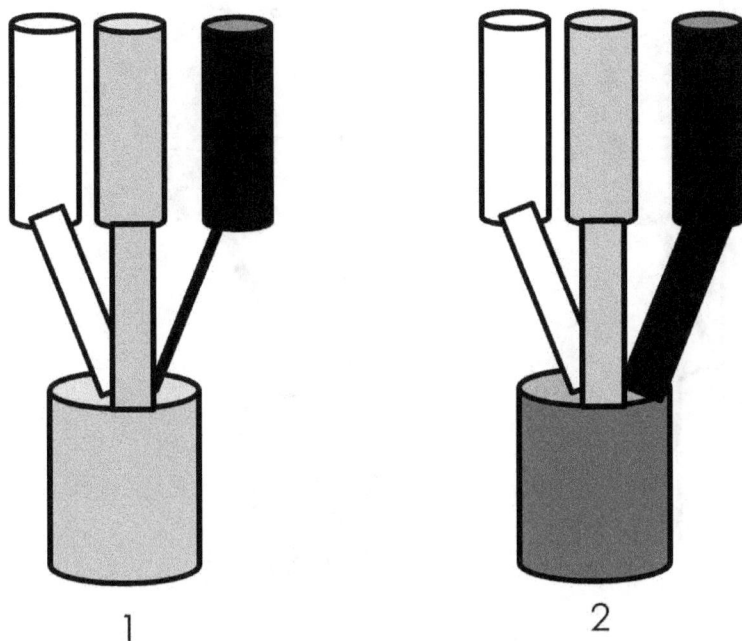

Figure 7. Adding chloride channels to regulation of the muscle resting membrane potential. In #1 we have the white potassium solution and the black sodium solution as in Figure 3. We have added a light gray chloride channel. When this channel is open at rest, there is no obvious effect as it is the same color as the bottom bucket. For a muscle to contract, the bucket needs to become a dark gray and this occurs by opening the black sodium channels a lot. Just a little extra sodium flow won't make a difference and so you need a "real" signal to contract.

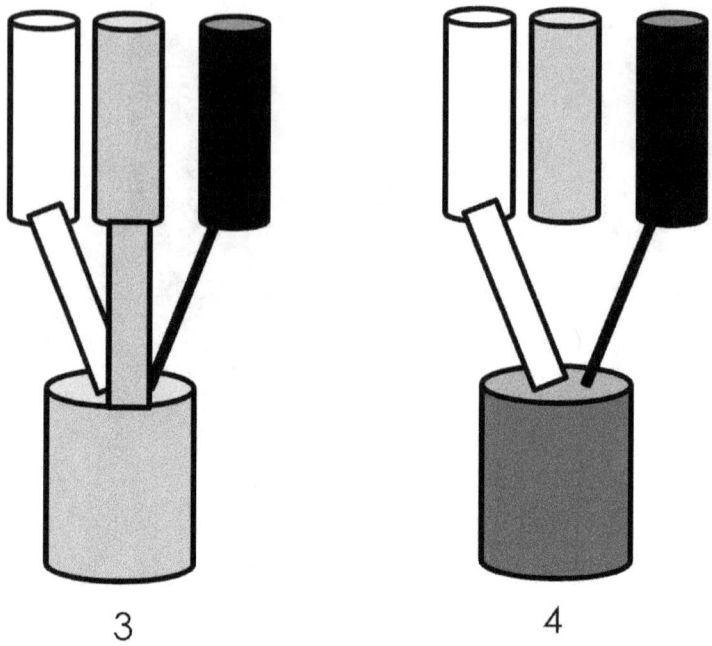

Figure 8. What is going wrong in the goats and presumably the cats. #3 is the normal condition where the gray chloride hose prevents small changes in the black sodium hose from changing the color of the bucket very much. In the goats, the gray chloride hose is broken so it does not contribute to keeping the bucket color stable. Thus any small changes in black sodium flow can lead to large changes in how gray the bucket is; if it gets dark gray, the muscle contracts.

Physiology

To summarize this chapter,
- we have learned that mutations in sodium and chloride channel can alter the signaling from the nerve to skeletal muscle calcium release.
- These mutations can occur in humans and domestic animals.
- skeletal muscle contraction involves the sequence:
- acetylcholine activating the nicotinic acetylcholine receptor which opens its gate and allows positive ions to enter
- the positive ions shift the balance of charge between the inside and outside of the cell
- when a threshold in charge has changed, sodium channels open and an even larger change in charge occurs
- the change in the balance of charge between the inside and outside of the cell causes intracellular stores (sarcoplasmic reticulum) to release their calcium
- in Impressive, the sodium channels do not close tightly after opening and this slow leak causes more calcium entry and muscle contraction
- in the fainting goats, the chloride channels do not work well as rest, making the muscle cells more excitable

Physiology

Do you hear what I hear?

Fictional Cases:
Bob likes to go to the shooting range with his friends. They have to wear ear protection. Bob is not happy with the idea. Is he at greater risk for hearing loss without the ear protection if he only goes to the shooting range a few times a year?

Sue goes to at least one loud concert or sporting event each week. Her roommates have noticed that when she returns, she can't hear very well for a few hours. Recently, she when returned from a particularly exciting men's basketball game, they have to yell at her before she can hear what they are saying. Finally, by the next morning, she seemed to have nearly normal hearing. Does this frequent exposure to loud music or cheering increase her risk of hearing loss?

Pat is a high school student and his parents have forbidden him to wear ear buds at home. He has talked a friend into letting him listen to the friend's iPod, with ear buds, at school. If Pat turns up the volume too high, will he risk hearing loss? If so, what is "too high"?

Hearing is one way our body gets information about the external environment. Taste is another. In this chapter, we will look at hearing and determine some of the similarities and differences between hearing and tasting.

We will discuss these 3 fictional hearing cases from 2 points of view. The first point of view is what could be the biological basis for hearing damage. The second point of view examines the epidemiological information about sound exposure and hearing loss.

The sections in this chapter are:
- Hearing structures
- How can loud sounds lead to loss of cells?
- Drugs and hearing loss
- How loud is too loud?
    - Population studies
- The three cases
- Bats and dogs and ultrasonic

- Earaches
- The general plan of the sensory system
- Summary

Hearing structures

I find it convenient to divide the hearing processes into 3 categories when thinking about what can cause hearing loss. The middle category is the cochlea. The other two categories are all the processes from air to the cochlea and all the processes after the cochlea. The former would include eardrum perforation and the latter includes damage to the nerves in the brain. In general, the damage from loud sounds occurs in the cochlea, so we will focus on that step. In permanent hearing loss due to loud noises, cells in the cochlea are destroyed; without these cells, the nerves don't get information about the sounds coming to the ear.

How can loud sound lead to loss of cells?

This requires understanding how these cells work to detect normal sounds. At the top of the ear's sound sensing cells are little cilia (hair-like strands). As sound is transmitted to the inner ear or cochlea, it causes the fluid in the inner ear to move and that causes these cilia to move. If you start out in perfect quiet, the sound sensing cells are at rest. When a sound arrives, the cilia move and this causes proteins in the membrane to change shape. When their shape has changed, they open little holes that allow potassium to move. This leads to a signal that causes the calcium in the cell to increase. The increase in calcium leads to an increase in the release of glutamate from the sound sensing cell. The glutamate diffuses across the short distance to the adjacent nerve cell and causes that nerve cell to fire. The signal goes from there to the brain for processing. Different hair cells respond to different pitches of sound and the brain processes the signals from both ears and the different responding cells to determine the pitch of the sound and the location.

Loudness is detected by how much the hair cells bend; the more they bend, the more potassium goes in, the more calcium goes up, the more glutamate is released and the more the nerve fires. The cilia on the sound sensing cells are very sensitive. One can record changes in one isolated sound sensing cell that reflect the cilia moving 5 Angstroms, about the size of a modest sized atom! In fact, the detection limits are set by the diffusion of ions that hit the hair cells;

they represent the background signal that occurs even in the complete absence of sound.

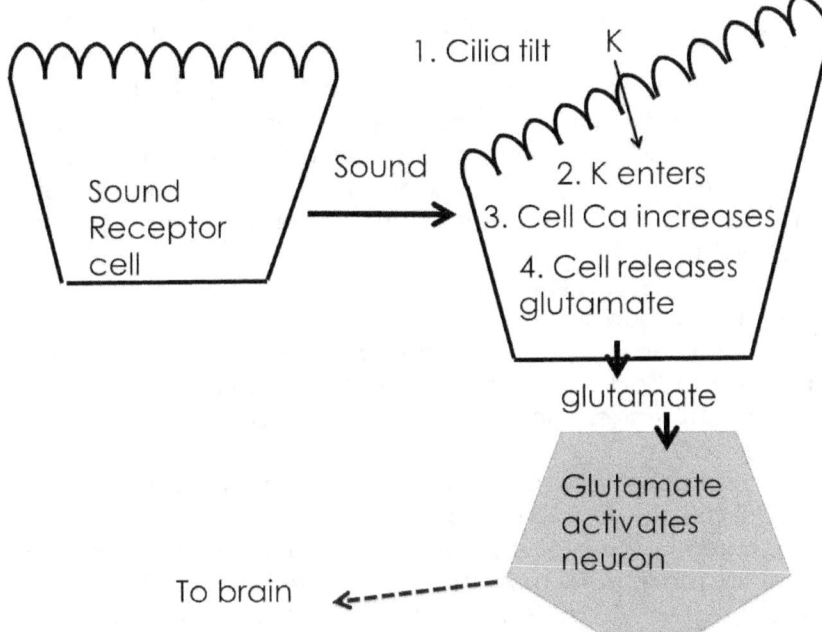

Figure 11. Top cell is the sound receptor cell, called a hair cell because its top has lots of little hairs, or cilia, or microvilli.
Sound will cause the cilia to bend, this opens a channel and potassium (K) enters. This causes the cell calcium (Ca) to increase. This leads to release of glutamate. Glutamate diffuses to the adjacent nerve cell (or neuron) and the signal is carried to the brain.

Of course, when the sound stops, the cells need to go back to resting conditions, so the cell has to move the potassium back and take the calcium inside the cell back to basal levels. Potassium and calcium are moved uphill, so the process requires energy.

Very loud sounds might move the cilia so far that the break off or remain bent; this is a bit like what happens when you stomp on grass too much and it doesn't spring back. The cell cannot repair this and eventually one can lose all the cilia and the cell won't be able to respond to sound. Mammalian sound sensing cells cannot regenerate, so once you loss a cell, it's gone for good. Interestingly, avian sound sensing cells are able to regenerate. Several scientists are actively studying the difference between avian and mammalian sound sensing cells in the hope that they can find a way to change human sound sensing cells to they can be made to regenerate.

Fortunately, most people aren't exposed to sounds this loud; the most frequent places for this sort of immediate, permanent hearing loss are battlefields. People that go to loud concerts or sporting events often report short term hearing loss. It is thought that if this happens enough, it can lead to permanent hearing loss. What is going on the cochlea in this case?

For moderately loud sounds, the cilia bounce back and the cell does not die. But when a sound is too loud, the cell will have moved a lot of potassium and had a very large increase in calcium. The very high calcium might trigger some events that damage the cell. The more popular theory at the moment is that the cell has to use a lot of energy to pump out the calcium and move the potassium back. This means the cell has to make a lot of ATP, adenosine triphosphate, and it uses oxygen in the process. As a byproduct of using oxygen (and glucose) to make ATP, oxygen radicals are formed. These are very reactive species of oxygen. Oxygen radicals can react with most molecules including proteins and DNA. Normally the cell is able to protect itself from the oxygen radicals by using molecules called anti-oxidants. But if the sound is too loud, the amount of energy required to restore the potassium and calcium levels to normal is large. Then the oxygen radical production exceeds the anti-oxidant capacity and the cell is injured.

Repeated oxidant damage eventually leads to cell death. The details have not been conclusively understood. But here are some possibilities and it might be a combination that causes the problem. 1. Oxidative damage to a particular protein or molecule might be a signal that tells the cell to die. 2. Oxidative damage to a set of proteins might happen in such a way that the cell can't get rid of these proteins, they accumulate, and eventually interfere with the normal cell processes and it dies. 3. The DNA is damaged by the free radicals. While these cells don't divide (and don't regenerate), the DNA is used as the blueprint to make proteins; the cell is continually removing old proteins and making new ones. If the DNA is damaged, it can change the genetic code and therefore lead to the production of a faulty protein. If too many proteins are faulty, the cell can't survive.

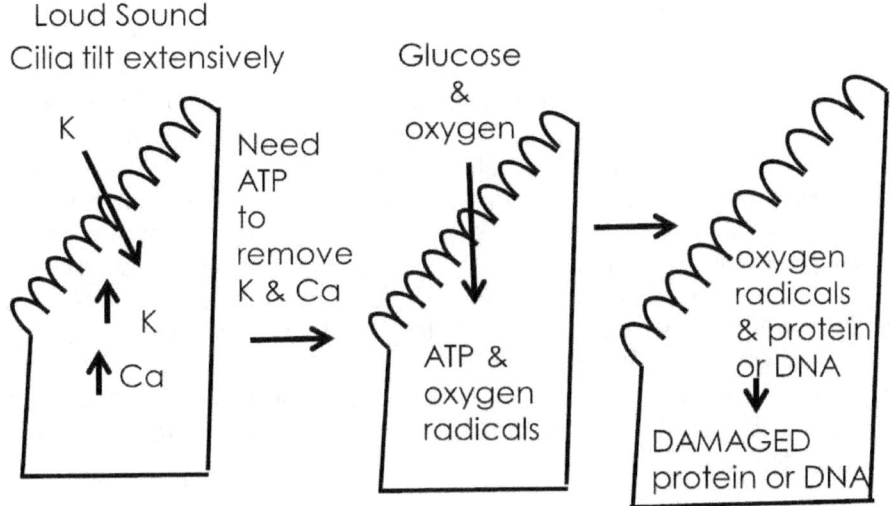

Figure 2. When sounds are very loud, the cilia tilt far and this leads to excessive potassium (K) entry and extra high cell calcium (Ca). The cell must pump the calcium and potassium back to normal. This requires energy which is from ATP. In order to make more ATP, the cell uses glucose and oxygen. A byproduct of this reaction is the formation of oxygen radicals. (shown, top, right). Shown in the bottom frame is that oxygen radicals can damage protein. If this protein accumulates, this can kill the cell. The oxygen radicals can also damage the DNA and accumulated DNA may also lead to cell death as the cell uses the DNA as a template to make proteins; if the template is wrong, the proteins may not work.

Some preliminary experiments in animals suggest that providing the cochlea with high doses of antioxidants before the loud noises can prevent the oxidative damage to the cell.

In summary, very loud sounds probably cause immediate cilia damage and perhaps cell death. Sounds that are loud enough to cause temporary hearing loss probably also cause excessive production of oxygen radicals as a by-product of producing the energy to restore the potassium and calcium to their resting levels. The excessive oxygen radicals overwhelm the antioxidant system. This damages the cell. If the damage is repeated often enough, the cell dies and is not replaced. If enough sound sensing cells die, there is permanent cell loss.

Drugs and hearing loss
Excess amounts of aspirin and some antibiotics can also damage the ear. For the antibiotics in the aminoglycoside class, the mechanism appears to involve overproduction of oxygen radicals. Aspirin can cause both tinnitus and hearing loss. The hearing loss is probably due to aspirin effects on one of the proteins on the cilia of the sound sensing cells. In contrast, the tinnitus is probably because aspirin leads to an activation of receptors on the auditory nerves. Normally, the sound sensing cells release glutamate, which diffuses to the adjacent nerve, binds to a receptor, and activates the nerve. A few experiments suggest that aspirin also leads to activation of that same receptor; for example, inhibitors of that receptor prevent aspirin induced tinnitus in mice.

One of the causes of ringing in the ears, tinnitus, is that the cilia of many of the hearing cells are bent excessively or broken. This leads to continuous activation of the adjacent nerve and so a signal gets to the brain that we interpret as sound. Some drugs probably also lead to nerve activation. The fact that no matter how you stimulate the nerve, your brain interprets it as sound is an example of an important principle.

How loud is too loud?
Sound is usually measured in decibels[1], which is a log system, and I think it is misleading. Sound is caused by a pressure change in air. The minimum pressure that we can hear is about 20 **micro**pascals.

---

[1] Deci comes from 10 and Bel from Alexander Graham Bell. A pascal is a measure of pressure. Micro- is a prefix which means one millionth. Other common prefixes are:

tetra-a trillion	pico-one trillionth
giga-a billion	nano-one billionth
mega-a million	micro-one millionth
kilo- a thousand	milli-one thousandth

Physiology

Our normal conversations are usually about 1000 times more pressure, 20 **milli**pascals. Sound gets painful at about 20,000 times greater than normal conversation, that is, 20 pascals. Decibel is a unit that is 20 times the log of the ratio of the pressure of a sound to the minimum pressure we can hear. For normal conversation, the ratio is 20,000/20 = 1000. The log of 1000 is 3. So 20,000 micropascals is 20*3 = 60 dB. A telephone dial tone is 80 dB. How much more pressure is this? It is NOT 80/60 = 1.33. Rather the pressure is 10 times greater; see the bolded print in the table below. Hearing loss can start at 90 dB which is about 30 times more pressure than normal conversation. A power mower or power saw at 3 feet is about 110 dB which 300 times more pressure than normal conversation.[1]

Table of Decibals and Pressure

decibels	ratio of pressure	micropascals
1	1	22
10	3	63
20	10	200
30	32	632
40	100	2,000
50	316	6,325
60	**1,000**	**20,000**
70	3,162	63,246
80	**10,000**	**200,000**
90	31,623	632,456
100	100,000	2,000,000
110	316,228	6,324,555
120	1,000,000	20,000,000

Population studies

---

[1] Here's an analogy for the decibel scale. Imagine a stadium with rows of seats. In row 50, there are 316 fans. In row 60 there are 1000 fans. In row 70 are 3162 fans. In row 80 are 10,000 fans. And so on. The number of fans in the row is the linear scale for the pressure ratio. The row number is the decibels.

OSHA has set limits on workplace noise. These are partly based on studies examining hearing loss at different levels of sound. Hearing also declines with age. Figure 3 shows the combination of hearing loss and age and decibel exposure. Studies examining whether MP3 players cause hearing loss have been equivocal. One problem is that the damage may take a long time to become evident. An even bigger problem is that doing a random, prospective study is probably unethical; you can't ask people to be exposed to loud sounds that you think will cause permanent hearing damage.

Let's return to our 3 hearing cases.
Bob is probably being exposed to at least 1 million times more pressure than the minimum to hear, that is at least 120 dB, Sue is probably in the range from 100,000 to 1,000,000 times more at 100-120 dB and Pat is at 3,000 to 100,000 times more pressure than minimum at 70-100 dB. Perhaps it would be better to compare this to a normal conversation, at 60 dB or 1000 times the minimum. In that case Bob is still 1,000 times more pressure than a normal conversation or 60 dB more, Sue is at 100 to 1,000 times more than a normal conversation at 40 to 60 dB more, and even Sue is at 3 to 100 times more at 10 to 40 dB more.

Without ear protection, if Bob is too close to a rifle being shot, he risks immediate hearing loss. As far as I know, all target ranges require ear protection for this reason.

At loud concerts or basketball games, Sue may well have temporary hearing loss or tinnitus. These are warning signs of too much exposure. She will probably be fine in the morning, but repeated exposure could lead to permanent damage[1].

I would think its ok to go to the occasional concert or basketball game but the problem is how we determine what occasional means. Does it mean once a week, once a month or once a year? This is hard to determine from studying people. If you survey 25 or 50 year olds about their exposure to loud noises the last 10 or 35 years, how accurate are the answers likely to be? Even a 50 year old that had

---

[1] An analogy for Sue might be driving a car that overheats. You pull over, let it cool off, and drive again maybe thinking everything is ok. But every time you overheat the car, you are doing a bit of damage.

Physiology

avoided exposure to any loud noises has a risk for some hearing loss just due to the changes of aging, so this is a background signal. That makes it harder to determine whether there is additional damage in others due to the loud noise[1].

Even Pat may experience hearing loss with her iPod, especially if she listens to loud music for many hours each day. But putting numbers onto "loud" and "many" is tough. In fact, some studies claim that they don't measure any significant change in hearing with folks using their iPods. In the sound field, an accepted rule of thumb is that, when you are listening to your iPod, you should still be able to hear someone talking to you in a normal voice if you can touch them with your arm outstretched. If you can't, then you are probably at risk for long-term hearing loss.

Bats and dogs and ultrasonic
Why would bats use ultrasonic frequencies to locate insects? What are ultrasonic sounds? Ultrasonic means that the frequency of the sound waves is higher than humans can detect. It is not because bats don't want us to hear them. Rather it has to do with the size of the insects. Insects are small, say about 1 centimeter. If you want to find something as small as 1 centimeter, would you use a 1-foot grid? No, you'd want a grid of about 1 cm. Wavelength and frequency are inversely relate; the higher the frequency, the small the wavelength. So for a bat to find a 1 cm insect the frequency has to be 35,000 per second; our hearing maximum is about 20,000 per second. In contrast bats can hear frequencies up to about 100,000 per second.

---

[1] The ratio of the signal to the background variations is an important constraint on being able to detect changes in experiments. Let's say I'm trying to determine if you are texting on your cell phone in your pocket during class. That's the signal. The background variation is every other sound. If we are in a perfectly quiet classroom and no one is talking, I might be able to hear you texting. That is a situation where the signal is slightly above the background. But if one student is talking, I will probably have a 50% chance of hearing the cell phone texting; the signal and the background variation are the same size. If several students are talking, it is highly unlikely that I'll hear the texting as the signal is smaller than the background variation. Note that, no matter whether the signal is sound or not, background variation is usually referred to as noise.

As an analogy for wavelength and frequency, consider watching cars on the highway. Suppose the cars are spaced uniformly. The frequency is how many pass your point in a given amount of time, for example, 1 per second. The wavelength is how far apart each car is, for example, 3 car lengths. In this analogy the cars need move at the same speed and be single file. What happens if we double the frequency, so now there are 2 cars per second? That implies that the distance between the cars has halved, to 1.5 car lengths between cars.

The conversion from wavelength to frequency for sound and vice versa can be found at http://www.sengpielaudio.com/calculator-wavelength.htm.
And a comparison between species and the difficulties in determining the maximum frequency can be found at http://www.lsu.edu/deafness/HearingRange.html.

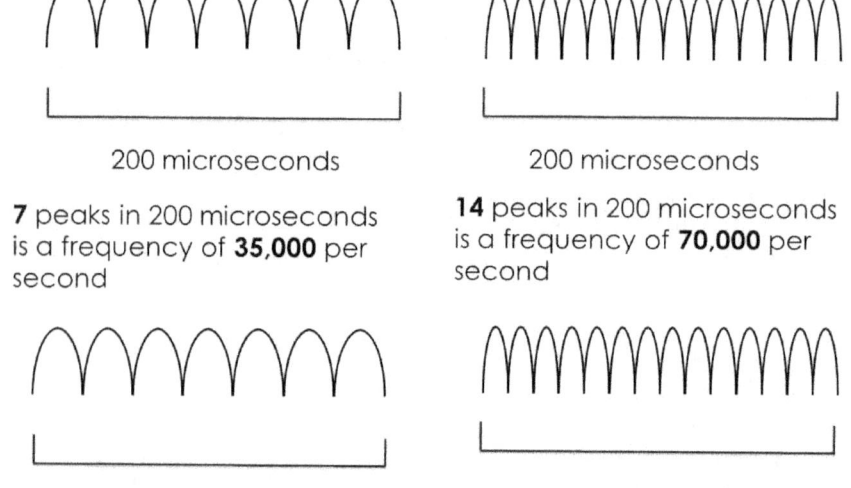

Figure 3. This figure illustrates frequency and wavelength. On the top, left the x-axis is time and there are 7 peaks within 200 microseconds, so the frequency is 35,000 per second. For the same wave, if you

Physiology

stopped it in time, you'd get a plot like the bottom, where the x-axis is distance, in this case 7 centimeters. So the wavelength is 1 cm. On the right side is a wave with twice the frequency and half the wavelength.

How can dogs hear ultrasonic sounds? Their cochlea presumably has a bit different architecture; since high frequencies are detected at the basal end, one would think this region would be extended in the dog.

Have you noticed that young children get earaches more often than teens or adults?
The most common earache is a middle ear infection/inflammation, which in Latin or medicalese is Otis media. Figure 4 shows the structure of the ear. From the eardrum out is the external ear. The middle ear is between the eardrum and the oval window of the cochlea. It is air filled, in contrast to the inner ear. Normally, any pressure or fluid build up in the inner ear drains out the Eustachian tubes.
Two types of microbes can cause middle ear infections, bacteria and viruses. In both cases, more fluid is produced and also sometimes the Eustachian tubes get narrow. This means pressure and fluid build up in the middle ear. When the infection is caused by bacteria, then antibiotics[1] will help. In young children, the Eustachian tubes are more horizontal, so they don't drain as well. In addition, they are narrower. Children's immune system isn't as mature. Also adults, just by living longer, have been exposed to more microbes than children and so have had a chance to develop immunity.

---

[1] I think antibiotics should really be called antibacterial because they act against bacteria. Since viruses are biological, some think that antibiotics should attack them, but that is not they way medical professionals use the word antibiotic.

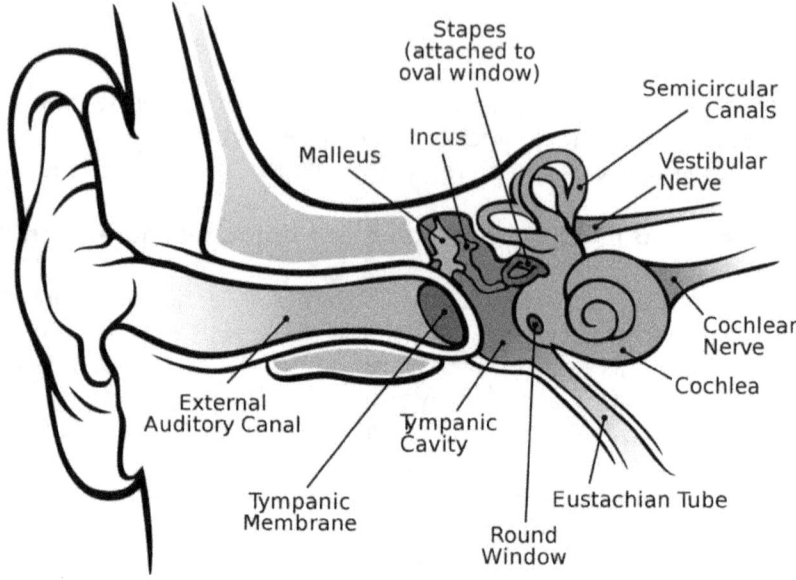

Figure 4. Anatomy of the ear and shows the Eustachian tube draining the middle ear. The figure is from http://en.wikipedia.org/wiki/File:Anatomy_of_the_Human_Ear.svg.

The general plan of the sensory system is that you have a cell that is particularly sensitive to a specific input:
- sound pressure at a distinct frequency for hearing cells,
- light at certain frequencies for cones,
- different chemicals for taste and smell, and
- pressure or temperature for some touch systems.

In each case, the stimulus to these specialized cells causes them to communicate with their adjacent neuron. The neuron then communicates with the brain. As far as the brain is concerned, if a nerve from the eye retina fires, then our perception is that there is light. If you apply high pressure to your closed eyes (don't try it!), you can "see" sparks or flashes of light. But of course, there has been no light detected, but your brain interpreted the firing of your retinal nerves as light. Similarly, if you use a cold hammer and hit an area of your thumb that has heat sensitive cells, your brain will interpret that as

Physiology

warmth, even thought that area didn't get warm. For your ears, if the cilia are bent, then the nerve fires and your brain "hears" sounds. In the chapter The Good, the Bad, and the Spicy, we talked about how this applies to the perception that eating jalapeno peppers feels hot and why sipping brandy also gives you a warm feeling.

Consider this analogy. The brain seeing light when you apply high pressure to your eye is like your parent saying clean up the muddy footprints. The sensory nerve is the muddy footprints. The appropriate stimulus is you making the muddy footprints. But your parent still says, "Clean up the muddy footprints" whether the muddy footprints are from you, your dog, or your cat.

This principle also explains how cochlear implants work. The system has a microphone and amplifier outside the ear that detects the sound and converts this into an electrical signal. The signal is sent, essentially via a radio transmitter, from outside the ear to the implant inside the ear. This implant is able to activate the neurons for the appropriate pitch (frequency) of sound hitting the microphone. When the neuron is activated, it sends the signal to the brain and the brain assumes that it "heard" the sound.

Summary
- Sequence for hearing:
    - sound waves bend cilia
    - bent cilia open K channels
    - balance of charge inside and outside changes
    - cell calcium increases
    - cell releases glutamate
    - glutamate activates adjacent neuron
    - signal goes to brain via a series of neurons
- When sounds are too loud, extra energy is required to regenerate the K and Ca gradients; the extra energy leads to high levels of oxygen radicals which damage proteins and DNA
- Hearing cells do not regenerate so if damage is severe and too many cells die, hearing is lost
- Decibels are a measure of how loud a sound it; decibels are based on a log scale so that going from 60 to 80 decibels means the pressure has increased ten fold
- wavelength and frequency are reciprocals, so high frequency sounds have small wavelengths which explains why bats use ultrasonic frequencies to find small insects
- **The general plan of the sensory system** is that you have a cell that is particularly sensitive to a specific input:
    - sound pressure at a distinct frequency for hearing cells,
    - light at certain frequencies for cones,
    - different chemicals for taste and smell, and
    - pressure or temperature for some touch systems.
- Sequence to brain:
    - specialized cells change charge balance
    - often cell calcium changes
    - cell releases a neurotransmitter
    - adjacent neuron is activated
    - a series of neurons take signal to appropriate part of brain
    - brain interprets signal as from the stimulus

Physiology

The Good, the Bad and the Spicy

Fictional Cases
Bob eats some spicy food and starts to sweat. Is his body really overheating? Barbara soaks her bird seed in cayenne pepper before putting it into the bird feeder; she thinks the pepper will deter the squirrels but not the birds. Joyce drinks some diet drink and spits it out. Her friend asks, "Isn't it sweet enough?" Joyce says it is sweet, but has a metallic after taste. John's dog and cat had to go to the vet because they drank some antifreeze; why would they drink that stuff?

In this chapter we will explore the sense of taste and talk about receptor proteins and ion channels.
- Sensing Hot Food
- Why peppers make us sweat
- What about birds and peppers?
- Why does brandy feel warm when swallowed?
- Double ring burners
- Relief from spicy foods
- Does sensing cold temperatures work the same was as hot temperatures?
- Why does saccharine taste sweet?
- Why does saccharine have an aftertaste?
- How many basic tastes are there?
- Cats, sweetness, and antifreeze

Sensing Hot Food
We have all probably put some very hot temperature food in our mouth and burned our tongue. When our tongue burned, some of our cells died.[1] However, we do have specialized cells that warn us when the temperature is warm, before it gets too hot; so usually we sip a drink and know to stop before we get burned.

Our tongue has specialized nerve cells that are particularly sensitive to temperature; related specialized nerve cells are found elsewhere in

---

[1] In fact, all our cells are affected by changes of temperature; at least in the sense that if a cell gets too warm it dies. Most of the chemical and biochemical reactions in our cells are also temperature sensitive so that temperature alters what is going on inside the cell.

the body, particularly in our skin. The specialized heat sensitive nerve cells contain a particular protein, TRPV1,[1] which is very sensitive to changes in temperature. When studying the TRPV1 protein in isolated cells in the lab, scientists found that as the temperature increased, the probability of the TRPV1 changing shape (conformation) increased[2]. When this protein changes shape, it opens a hole (channel) and lets sodium go into the cell[3]. As the temperature changed from about 95 °F to about 125 °F, the probability of this temperature sensitive protein opening the sodium channel increased from about zero to about 1, that is, it was mostly closed at 95 °F and almost always open at 125 °F. The movement of sodium into the cell changes the relative amount of positive charge inside the cell relative to outside, since the sodium ion is positively charged. This leads to other changes in the nerve cell which cause it to communicate with a neighboring nerve cell. The neighbor nerve cell communicates with its neighbor and so on until various nerves in particular places in the brain are activated. Our brain interprets this signal as a sign that the original nerve got warm and sends out signals in response, including increasing how much we sweat.

---

[1] The initials TRPV1 stand for transient receptor potential, vanilloid type 1. Transient receptor potential just means that it only transiently changes the membrane potential. Vanilloid refers to the fact that this receptor binds vanilloid type compounds. Vanillyl refers to a particular chemical structure. If you've had organic chemistry you might be interested to know that these compounds contain an aldehyde, an ester, and a phenol. The main aromatic compound in vanilla beans is in the vanillyl class (hence the name). Capsaicin, a main hot molecule in cayenne peppers, is also in this chemical class.
[2] When proteins change shape it can change what molecules can touch the protein and this is important in signaling. Scientists use the fancy name conformational change to indicate that the protein has changed shape.
[3] Proteins that allow ions to flow across membranes are called ion channels. Typically, channel is used to refer to a protein that has a gate; when the gate is closed, no ions flow. When the protein changes shape, the gate opens and many ions flow. "Many" usually means hundreds or thousands. There are other proteins that transport ions across the membrane; these transporters typically have one or a few pockets to bind ions. When the transporter changes shape, the access to the pockets shifts from one side of the membrane to the other.

Physiology

The previous paragraph had 3 key concepts that we need to be sure are clear.

1. In general, our body has a variety of different cells to sense our environment, such as light sensing cells in the eye and sound sensing cells in the ear. The sensing cell then communicates to particular regions of the brain. The brain, for example, "interprets" the firing of eye cell x as light of ~ 500 nanometers hitting eye cell x and we call that light "green". As another example, the brain "interprets" the firing of ear cell y as sound of ~ 4 feet in wavelength (or 260,000 cycles per second) and we call that sound, middle C. Our brain can be fooled, for example, any way we activate ear cell y, makes our brain think we have heard middle C. This is the basis for cochlear implants, which have a microphone and electronic converters that stimulate the appropriate cell when middle C sound hits the microphone. (More details on hearing are in the chapter Do you hear what I hear?)

It turns out that a chemical in peppers[1] also activates the heat sensitive neurons and so our brain thinks we are hot. We'll explain that process in more detail shortly.

2. Changes in the relative amount of positive charge inside and out the cell are a signaling process in many cells. We will discuss this process in great detail in other chapters. For this chapter, we just need to understand:
- that moving sodium changes the relative amount of positive charges inside compared to outside the cell
- that other proteins in the cell sense this change in charge between inside and outside and change their shape,
- thus leading to a cascade of changes in the cell.

3. For nerve cell A to communicate with nerve cell B, part of one nerve cell A needs to be close to nerve cell B. For many nerves this place where they are close has a special structure and also a special name, synapse. Nerve cell A communicates with nerve cell B by a process that involves 3 key steps:

---

[1] The primary chemical in cayenne and jalapeno peppers that actives these cells is called **capsaicin**.

i. Nerve cell A is activated. This generally means that there is a change in the relative amount of positive charge inside nerve cell A compared to outside the cell. Generally, this is because positively charged sodium ions entered.
ii. The change in charge triggers a series of changes that result in nerve cell A releasing neurotransmitter[1] at the synapse. This neurotransmitter diffuses away from the release site.
iii. Some neurotransmitter binds to receptors on nerve cell B. This causes sodium ions to enter nerve cell B and it is now activated.

Why peppers make us sweat

When you eat jalapeno peppers, your mouth feels hot. If you eat enough you might even start to sweat. If you rub jalapeno peppers on your skin, your skin will feel like it is burning[2]. In these two cases, are you really warm or are you just perceiving the "taste" as warm?

The active compound in jalapeno peppers is capsaicin. Why does capsaicin make you feel warm? Capsaicin binds to TRPV1, the heat sensitive protein and causes the protein to change shape. This protein is located in specialized nerve cells in your tongue and your skin. The shape change opens a hole, or channel in the protein, allowing extra positive sodium ions into the cell, which activates the cell and causes it to communicate, by a series of nerves, to the brain. Our brain interprets this signal as "hot". An analogy might be if you got a text on your cell phone from your friend's number. You would assume they have called you, even if someone else actually used your friend's cell phone.

Part of the evidence for the involvement of TRPV1 in this response comes from mice experiments. If capsaicin is put into their drinking

---

[1] A neurotransmitter is merely a molecule that transmits the signal between two adjacent nerve cells. Common neurotransmitters include acetylcholine and noradrenaline.

[2] If you try this, only do a small part of your skin and be ready to use soap and water to wash the area. Some people that use peppers for home remedies have severely injured their skin; in fact a few have died either from the peppers for the home remedies or when being excessively sprayed by pepper spray during criminal activity.

water, normal mice will take one sip, make a facial expression, and then ignore that water. But mice in which the TRPV1 receptor has been knocked out[1] will continue to drink the water.

What about birds and peppers?
Birds can eat food treated with much more capsaicin than humans can stand. Rather than just provide you with the current answer, I want to go illustrate how a scientists might think about this issue. [2] Here are the two observations we want to explain: when mammals taste capsaicin, they respond as if their mouth are hot. When birds taste capsaicin, they don't seem to have any obvious response.

The response in mammals requires
- the TRPV1 to bind capsaicin and open holes for sodium,
- the cellular responses to opening the TRPV1 channel,
- connections to the brain that identify this neuron as a heat sensitive neuron,
- the brain to process this information,
- and the brain to make a response.

Which step might be different in a bird, that is, why doesn't a bird "taste" capsaicin?
Birds certainly have TRPV1, the heat sensitive protein channel.
Birds certainly have cells that the brain identifies as heat sensitive neurons.
One explanation is that bird TRPV1 does not respond to capsaicin. This indeed seems to be the case when examining capsaicin effects on

---

[1] Knock out mice are a type of transgenic mouse. In a knock out mice, a gene is changed so that it is not expressed. To knock out the TRPV1 gene scientists take some special mouse cells and selectively add a piece of DNA that inserts into the DNA for the gene for TRPV1. For example, they can put in a stop codon very early in the gene so only the first 10 amino acids are made and not the other 1000. Then this cell is returned to a very early stage of mouse development and the new mouse has some cells that make TRPV1 and some have the mutation so they can't make TRPV1. After appropriate breeding, it is possible to have mice that don't have any TRPV1.

[2] Much of science is like doing a jigsaw puzzle; you look for a possible pattern or clue and then see if the piece fits. Often the piece does not fit and you move on to the next piece.

bird TRPV1 in the lab; no effect is observed. However, there is one study that birds have nerve cells that are activated when capsaicin is added. We don't yet know if the cells that capsaicin activates are also the heat sensing cells, but it seems unlikely, since the birds don't behave like they are warm when they eat capsaicin. One possible explanation for this apparent paradox is that bird TRPV1 behaves differently in cultured cells compared to its native environment in cells in the mouth. Just like some people behave differently when dribbling a basketball round their backyard by themselves and when they are dribbling in a critical moment in basketball game in a filled arena. Another explanation is that, when tested in the mouth at high concentrations, capsaicin activated a different channel, not TRPV1 and this different channel was not expressed in the lab cell culture experiment.

Interestingly, birds don't like to eat food that contains a compound in grape flavoring, methyl anthranilate, whereas mammals are not bothered by it. Methyl anthranilate does activate nerves in the bird mouth at low concentrations; the bird mouth has more cells that response to methyl anthranilate and those cells tend to have a bigger response than it has cells that respond to capsaicin. Further work is required to work out the details of the response to methyl anthranilate and which step(s) are different in humans.

Why does brandy feel warm when swallowed?

TRPV1 also responds to alcohol. This partly accounts for the fact that alcohol, for example, in brandy, gives one a warm feeling.

Double ring burners
In one of my classes, students asked about why some spicy foods are double ring burners. Here the rings refer to the upper esophageal sphincter and the anal sphincter. **Sphincters** are special arrangements of smooth muscle cells that allow one to completely close a tube. So you close off your upper esophageal sphincter when you breathe, but open it to swallow. I don't think I need to describe when the anal sphincters are open and closed. From what we've discussed about capsaicin and taste, I would predict that there are heat sensitive neurons in both the upper esophageal area and the anal/rectal area that contain TRPV1 receptors. It is possible that the amount of these receptors, their exact locations, and perhaps even their sensitivity

varies between individuals. This might account, at least in part, for the different perceptions when eating, and getting rid of, capsaicin containing foods.

Relief from spicy foods
What if you eat too much spicy food and want immediate relief? Is there anything you can do? You have probably discovered that water doesn't do much. The hint here is that capsaicin is soluble in fat, not water. Drinking whole milk or eating sour cream or cream cheese helps reduce the sensation of heat. Any remaining capsaicin in the mouth is quickly dissolved in the fat thus is no longer available to activate the channel.

Does sensing cold temperatures work the same was as hot temperatures?
The processes are parallel. Other specialized nerve cells contain a cold sensitive channel. These specialized cold sensitive neurons communicate through a series of other nerves to the brain and the brain recognizes that the signal came from a cold sensitive neuron.

What does it mean to say that one channel is heat sensitive and one is cold sensitive? Aren't they both just temperature-sensitive? As temperature increases, the heat sensitive channel allows more sodium into the cell whereas the cold sensitive channel allows LESS sodium into the cell. So a heat sensitive channel will activate a cell as it gets warm but a cold sensitive channel will activate a cell as it gets cold. The cell is activated because when the channel changes shape and opens the sodium hole, sodium flows into the cell. For heat sensing cells, more sodium flows at warm temperatures; for cold sensing cells, more sodium flows in at cold temperatures.

Just as capsaicin can bind to heat sensitive channels and give the perception of heat, so other chemicals can bind to cold sensitive channels and give the perception of cold. Perhaps the most well known example is menthol[1], which was originally isolated from peppermint. In lab cell culture experiments, these cold activated

---

[1] Menthol is very different from methanol, in spite of their similar spelling. Methanol is a toxic alcohol; it has one less carbon than ethanol and two less than propanol, which is sold as rubbing alcohol.

channels begin to become active at about 65°F and reach maximal activity at about 45°F.

Why does saccharine taste sweet?
The response to heat, pepper, and alcohol is mediated by a protein found in nerve cells and these are not classified as "tastes". Most of the key cellular steps involved in taste are similar to sensing heat. Here are the steps that are similar:
- ion channel opens
- ions enter and change the cell properties
- proteins sensitive to the changed cell property cause other signals
- the sensing taste cell releases neurotransmitters that activate neighboring neurons which then, through a series of neurons, signal to the brain.

All tastes differ from heat and pepper sensing in that the sensing cell is a specialized epithelial derived taste cell for tastes whereas heat and pepper sensing cells are neurons. For salty taste one key step is different: what initially activates the cell. For sweet tastes, two steps are different: what initially activates the cell and how the taste cell signals the adjacent nerve cell.

Since salt taste only has 1 key different step, I thought I'd cover that first. As you recall, salt is sodium chloride. So it is probably not surprising that taste cells that respond to salt have sodium channels. These channels are usually open. If one eats salty food that leads to an increase in the sodium concentration near the taste cells. Therefore more sodium goes in through these particular channels and the cell is activated. After that, a similar cascade basically happens for the salt taste cells as for heat and pepper sensing neurons.

Sour taste is mediated, at least in part, by the proton (H^+) concentration. It appears to me that exactly how protons are sensed it still being worked out; in fact there remains some controversy about whether the sensing proteins sense protons outside the cell or inside the cell.

In all 3 of these cases, capsaicin, salt, and sour, it does appear that the sensing cell communicates with the neighboring nerve cell at a synapse-like structure and does involve the release of neurotransmitter

Physiology

from vesicles, just as in classical nerve-nerve or nerve-muscle communication. The nerve cell then communicates to the brain; much like the heat sensitive neuron and the brain "knows" that cell X responds to salt, cell Y to sour, etc.[1]

As I mentioned, sweet tastes differ in the way the cell is initially activated. Sugar does NOT bind to an ion channel and open its gate and allow sodium to enter; rather there are several steps between sugar binding to an extracellular protein and the release of neurotransmitter.

Extracellular sugar binds to a receptor on the taste cell. This receptor is in the Guanine Nucleotide Binding Protein Coupled Receptor family[2]. When receptors in this family are activated, proteins inside the cell that bind the guanine nucleotides, GTP and GDP change shape and are released into the cell. These changes can then lead to an increase in cAMP[3] as well as an increase in calcium. Because the name is interesting I'll mention one of the key first proteins involved in this cascade that is involved in the sweet taste, **gustducin**, which is essentially a contraction of gustation transducing.

Another difference between sweet tastes and salty tastes is that the cells that have the sweet receptors do not seem to have synapses. They release their neurotransmitter[4] through specialized proteins that open a channel for that neurotransmitter.

---

[1] Some cells may respond to more than one taste.
[2] Guanine Nucleotide Binding Protein Coupled Receptor is often abbreviated GPCR.
[3] cAMP is cyclic adenosine mono-phosphate.
GTP is guanosine tri-phosphate.
GDP is guanosine di-phosphate.
Adenosine and guanosine not only serve roles in these compounds but are also two of the 4 base pairs in DNA; thiamine and cytosine are the other two. You should be able to write out the full names for cGMP, ATP and ATP.
[4] ATP seems to be the neurotransmitter

Bitter taste involves a similar sequence, except that the compounds that taste bitter bind to a different receptor than the compounds that taste sweet. In fact, there are many different receptors for bitter tastes, each receptor having a different selectivity for the type of chemical it binds. The receptors for sweet and bitter are in the G protein coupled receptor family.

All sweet tasting compounds appear to bind to the sugar G protein receptor. Most small molecules that taste sweet bind roughly to the same place as sugar on this protein as they have enough chemical similarity to sugar. This includes artificial sweeteners as well as the primary ingredient in antifreeze, ethylene glycol. There are some sweet proteins and while these appear to bind to the same sugar G protein receptor, they apparently bind somewhere else on the receptor, but are still able to cause the shape change that activates the cell.

Why does saccharine have an aftertaste?

At low doses saccharine mostly gives a sweet taste, but at somewhat higher doses, many people report a metallic taste or after taste.

Suppose that, at one dose, saccharine gave both a sweet taste and a metallic taste. In theory, this could occur if saccharine bound to only 1 receptor and activation of that receptor and cell was perceived by the brain as both sweet and metallic. This seems unlikely because we know many compounds that taste either sweet or metallic but not both. Current molecule biology data are also consistent with the two tastes occurring because of two receptors.

The current theory is that saccharine binds to the sugar G protein receptor and as well as to second receptor. One interesting candidate for this purported receptor is the metallic receptor, which makes sense. Some mice show an inverted U response[1] to

---

[1] An inverted U response means that at low concentration there is little effect, at

Physiology

saccharine, preferring solutions with low saccharine to water solutions, but preferring water solutions to solutions with high saccharine. When mice had one of their genes for metallic taste knocked out, they didn't prefer the water solution quite as much compared to the high saccharine solution; that is, there was a small effect. In contrast, when their genes for the sugar G protein receptor were knocked out there was a drastic reduction in their preference for the low saccharin solution compared to water. The dramatic effect of the sugar G protein receptor is consistent with there being one major receptor for sweetness; the small effect of the metallic receptor could mean that there are other receptors that contribute to the aversion to high saccharine containing solutions.

A key point here is that at low concentrations, some chemicals may bind to only certain proteins, but at high concentrations, these same chemicals may bind to other proteins. Thus to say that a chemical or a receptor is specific is almost always implicitly qualified with a particular concentration range. You may have a strong preference for specific foods, but when you are hungry enough, you may not have such a great preference.

---

modest concentrations there is a great effect and at high concentrations, there is little effect. If you look at the figure, it looks like an upside down "U".

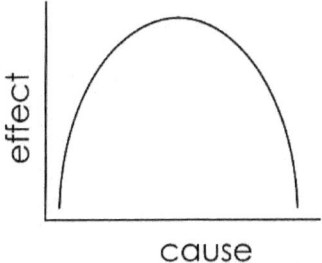

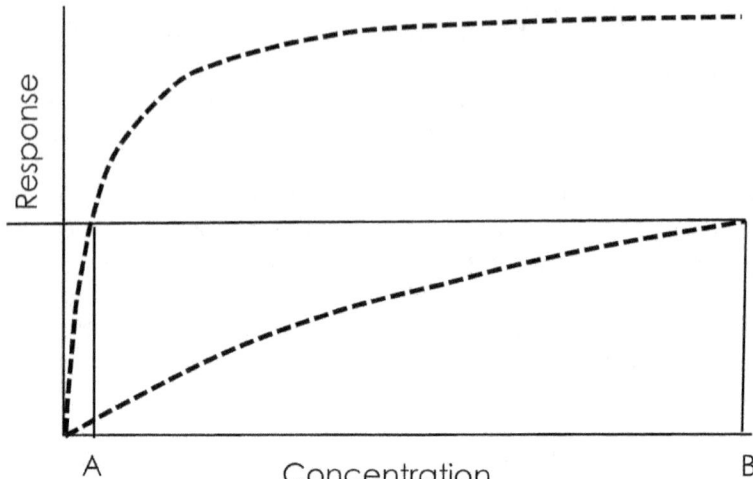

Figure 1. The same chemical can activate two different receptors. Saccharine at concentration A would mostly activate the sugar G protein coupled receptor (top curve).
Saccharine at concentration B would not only activate the sugar G protein coupled receptor (top curve) but also substantially activate the receptor responsible for the metallic case (lower curve).

How many tastes are there?

In older textbooks, you will find that there are 4 primary tastes: bitter, salty, sour, and sweet. Newer textbooks include umami; scientists in Japan about 100 years ago initially described this taste from their studies on taste perception. The receptor for umami was only biochemically and genetically identified much more recently and this made the concept of an umami taste more palatable for Western textbooks. Umami taste in a modern molecular sense refers to the enrichment of taste when glutamate[1] or related compounds are present.

There are a number of other "flavors" that may represent possible tastes including carbonation, fat, and calcium. For some scientists in order for something to be identified as a basic taste requires that there be a receptor in taste cells that responds to the chemical. Other

---

[1] For example, foods that contain MSG, monosodium glutamate.

scientists require that when the chemical is put into the mouth, nerves from the mouth are activated. Other scientists require that the chemical elicits a particular behavior. It may be that some chemicals will only met some of these possible definitions.

Do we decide there is a basic taste because:
- we have identified a protein that "specifically" binds the appropriate chemical?
- we have a particular response to an appropriate chemical?
- we observe a particular response to an appropriate chemical in animals?

Bitter, salty, sweet, sour, and umami seem to fulfill all 3 criteria. For fat, there is a search on to identify a protein. For calcium, a protein has been identified, but perception studies, that is, we observe a particular response to an appropriate chemical, aren't so clear. For carbonation, a protein is identified, but animal studies are lacking. In addition, for fat and carbonation, the relative contribution of non-taste factors remains to be determined. For example, how much of carbonation is the result of mechanical stimulation by the bubbles? How much of fat taste is due to the texture and the ability of fat to dissolve and disperse fat soluble, tasteable compounds?

This is a general dilemma in physiology: at what level do we define the words? For example, cortisol is a molecule that was identified in the blood when animals were stressed out. Suppose that there is NOT a one-to-one correspondence with the levels of cortisol and a person's perception of stress or an animal's stress-like behavior. Does the biochemical measure of cortisol define stress? Do the perceptions or behavior define stress? Both approaches have their strengths and weaknesses.

Why does antifreeze taste sweet?

Another illustration of the complexity of taste, consider the case of sweetness in cats. In1935, a scientist showed that cat tongue nerves could be activated in response to salt, bitter and acid solutions, but not to sugar solutions. Later work showed that almost all other animals responded to salt, bitter, acid AND sugar solutions. While most later experiments confirmed that cats didn't respond to sugar solutions, there were a few experiments that did find that cats responded to sweet solutions under appropriate circumstances; an example is in Figure 2.

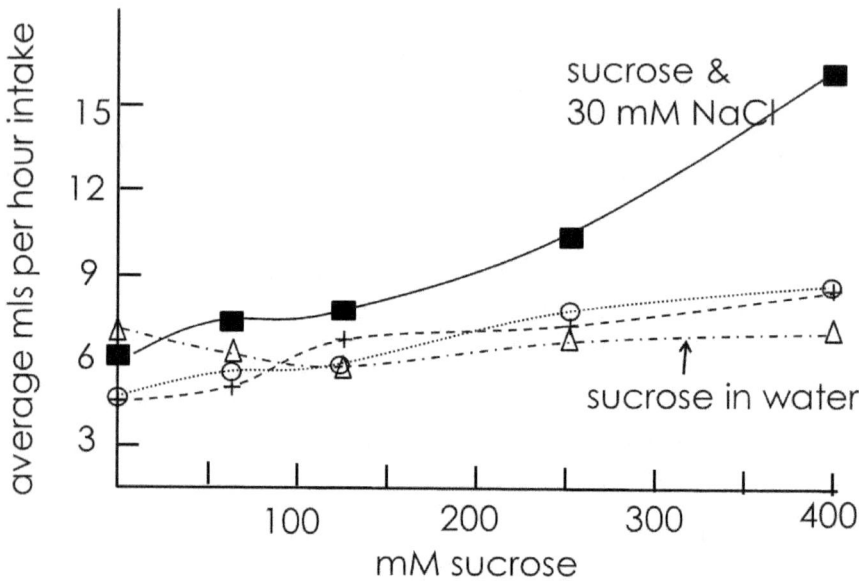

Figure 2. Cats respond to sucrose in salt solution but not in water. The top, solid line shows that cats drank more solution when it contained salt and sucrose than when it contained salt or sucrose or neither. It is important to note that this experiment gets the same result as previously found when sucrose is dissolved in water, that is, no increase in consumption; the novel result is the response observed when sucrose is dissolved in salt.

In the last decade, the proteins that we think are responsible for detecting sweetness, sugar G protein receptor, have been biochemically and genetically identified. Scientists then looked for these genes in cats; the gene has a mutation so that no protein is expressed. This fits with the predominant view that cats can't taste sweetness, but how do we reconcile the lack of a sweet receptor gene with the experiment in the figure? Was there some problem with the design? For example, did the sugar solution have some contaminant that tasted like pleasant to the cat? No one has published any objections to the data in the figure. Did the combination of sugar and salt allow the sugar to bind to another receptor that was in a "good-tasting" cell?

There is another possible concern about whether there only one possible sugar G protein receptor in cats is not expressed. Some people and animals drink antifreeze because it tastes sweet. Consuming antifreeze is life threatening because ethylene glycol is broken down in the body into toxic compounds (the details are in the chapter Is there Proof?). For this chapter, we have several pieces of information that don't quite seem to fit together: Ethylene glycol tastes sweet because it binds to the sugar G protein receptor. Cats are known to ingest antifreeze and get sick. Cats are not thought to have the sugar G protein receptor. Cats sometimes seem to respond to sweet solutions.

Summary
- The same protein can respond to different stimuli.
    - Some proteins can respond to both heat and the pepper chemical capsaicin.
    - Some proteins can respond to both cold and the mint chemical menthol.
- Once a protein is activated, the downstream cellular changes are the same independent of the stimulus, for example whether the stimulus was heat or capsaicin
- The brain "knows" whether a cell codes for heat or sweetness or other taste.
- A receptor changes shape when it binds a chemical; the change in shape leads to changes of the cell interior. A cell can communicate with a neighboring nerve cell by releasing a neurotransmitter.
- Often, at high concentrations, a chemical will bind to more than one type of receptor.
- There is not necessarily a one-to-one correspondence between molecular, perception, and behavioral responses to a chemical.
- Opening a sodium channel allows the positively charged sodium ions to enter and alters the relative balance of positive charges between the inside and outside of the cell.

- Some proteins change shape in response to a change in the relative balance of positive charges between the inside and outside of the cell.
- **A sweet compound** binds to the sugar G protein receptor, which undergoes a shape change, which leads to changes inside the cell, which causes the release of ATP, which signals the adjacent nerve and ultimately the signal is identified in the brain as arising from a "sweet-sensing -cell".
- **A bitter compound** binds to one of the bitter G protein receptor types, which undergoes a shape change, which leads to changes inside the cell, which causes the release of ATP, which signals the adjacent nerve and ultimately the signal is identified in the brain as arising from a "bitter-sensing -cell".
- **An umami compound** binds to one of the umami G protein receptor types, which undergoes a shape change, which leads to changes inside the cell, which causes the release of ATP, which signals the adjacent nerve and ultimately the signal is identified in the brain as arising from a "umami-sensing -cell".
- **A salty food** raises the local sodium concentration outside "salty-sensing cells", which means more sodium enters the cell, which changes the relative amount of positive charge inside vs. outside the cell, other proteins sense this change of relative charge, a cascade occurs and the "salty-sensing cell" releases neurotransmitters from stored vesicles, exciting the adjacent nerve and ultimately the signal is identified in the brain as arising from a "salty-sensing -cell".
- **A sour food** raises the proton concentration, which affects "sour-sensing cells" in ways that are not yet fully defined. This leads to a cascade and the "sour-sensing cell" releases neurotransmitters from stored vesicles, exciting the adjacent nerve and ultimately the signal is identified in the brain as arising from a "sour-sensing -cell".
- **A spicy hot food** has chemicals that bind to TRPV1, the heat sensitive channel, causing it to open and allow more sodium into the cell, which changes the relative amount of positive charge inside vs. outside the cell, other proteins sense this change of relative charge, a cascade occurs and the "heat sensitive neuron" releases neurotransmitters from stored vesicles, exciting the adjacent nerve and ultimately the signal is identified in the brain as arising from a "heat sensitive neuron".

Physiology

Ouch That Hurts

Fictional Cases
Emily goes to her physician because her feet get hot and red and feel very painful. No over the counter drugs relieve the pain, but she does get relief if she soaks her feet in cold water. She started having this problem when she was young.

Fred has intense pain around his eyes, his lower jaw and his rectal area. He also had this problem since he was young and one of his parents also had a similar problem.

Leslie's parents could not believe how lucky they were. She seldom cried as a baby. But when she was about 2 she severely burned herself on the stove. Her physician did some other tests and Leslie didn't seem to feel any pain, though the rest of her sensations were fine.

In this chapter we will discuss pain and medications that can relieve pain. While pain can be, well, a pain in the neck, it can also serve a useful purpose as those rare individuals who can't feel pain know. Sometimes when physicians prescribe pain medication that is too effective, the patients overdo some activities and actually cause themselves further harm.

- What is pain?
- What is the purpose of pain?
- How do the neurons that sense pain work?
- What is the relationship between itching and pain?
- Are physical and social pain related?
- How do pain killers work?
- What is the difference between abuse, dependence, and addiction?

What is pain?

"Normal" pain occurs because particular specialized neurons respond to their environment and send signals through the spinal cord to the brain; if particular regions are activated, the brain interprets that as pain. The pain sensing neurons are located in the skin and also throughout the body, but ironically, not in the brain. We will spend the bulk of this chapter examining the pain sensing neurons and how they work.

To understand what might lead to situations where there is inappropriate pain, one might consider an alarm system on a house. The motion detector would be the pain sensing neuron. The spinal cord might be the alarm box in the house. A signal is also sent from the house alarm box to the police station where the information is processed, just as the brain processes the information. The cases in the chapter deal with problems with actual motion detector-it is too sensitive. There are cases of patients with pain due to malfunctioning of the house alarm box and with the police station sending out a response team even when the alarm has not been tripped. While these are important situations, they are not as well understood as defects in the sensing system. However, most current treatments for pain are aimed at the police station response, that is, they act on the brain.

What is the purpose of pain?

Perhaps Leslie's experience is a reminder of one main purpose of pain- to signal is to stop doing something that is causing harm. If you accidently put your hand on a hot pan, the sensation of pain starts a reflex to have you move your hand. The unpleasantness of the sensation probably adds an emotional reinforcement so that you usually remember to grab a hot pad before you grab the pan.

But what about the pain that you have a few days after an injury? Does that pain have a purpose? That pain is too late to help reduce the initial injury. Maybe I need to clarify what I mean by "purpose". When I say "purpose", that is short hand for, from an evolutionary standpoint, what is the selective advantage of pain a few days after the injury? One idea is that your hand is still healing and there is an advantage to minimizing the use of the hand until the healing process is completed. During the initial stages of the healing process there is an inflammatory response. Some of the mediators of the inflammatory response activate the pain sensing neurons.

But there does seem to be pain that does not have an evolutionary advantage and that pain might be termed pathological pain. As an extreme example, some people who have had their left leg amputated still feel pain in their left foot ("phantom pain"). This does not seem to have a physiological or evolutionary purpose or advantage.

An analogy might be a tree falling on a road and damaging it. Immediately, cars are parked in front of it so people don't use the road-that is the immediate pain. After the tree is removed, the road has to be repaired, so there are barriers put up and a detour-that is inflammatory pain. Finally, sometimes a barrier is put up even though the road is fine-that would be pathological pain. In many people, the pathological pain is the result of some central administrator insisting on keeping the road closed even when it is just fine, that is, in these people, the brain has changed its processing and thinks there is pain, even when there is not an appropriate signal (maybe just leaves on the road) in the periphery.

How do the neurons that sense pain work?
These neurons need to have sensors for what might be causing the injury. For example, receptors for very cold or very hot. Or a receptor for mechanical injury, like hitting your thumb with a hammer. These sensors will locally change the membrane potential, which leads to activation of Na channels and this activates the whole nerve which then signals the next nerve in the sequence and so on, until the signal is "analyzed" in the brain.

Understanding pain physiology has been tough. As you can imagine, trying to infer when animals have pain is difficult. While you can tell if a human has pain, you can't dissect the tissue to study easily. You certainly can't knockout certain genes. The area of pain physiology research has been helped by genetic studies, studies of rare individuals, like Leslie, Fred, and Emily. At this stage, many studies identify a single gene that seems to be the main contributor to the disease. In these studies, one wants as "pure" a population as possible. That is, you want all the subjects to have as similar symptoms as possible so that it is most likely that they all have the same genetic difference compared to a control population. Thus it is important to define pain and the associated symptoms as precisely as possible. But of course, if the definition and symptoms are too specific, then only one patient will fit the criteria and that is not enough to determine what gene is causing the problem. It takes skill and intuition to work out a definition precise enough to have most patients have a common genetic difference and yet broad enough to have enough patients to be able to study.

In both Fred and Emily's case, the genetic difference is in pain sensing nerve. There are particular Na channels expressed primarily in pain sensing neurons; they are a cousin to the ones in our taste receptors. They both share the property that the initial change in membrane potential causes them to open and allow Na to flow into the cell. They also quickly close the gate, so that the Na stops flowing and the neuron quickly returns to its rest state. In Emily and Fred, the Na channels are mutated so that the nerve becomes extra excitable. The pain-sensing neuron fires even without a signal to the pain sensor protein. So the brain perceives pain even in the absence of an actual stimulus to the pain sensor protein.

Are Emily's disease, erythermalgia, and Fred's disease, paroxysmal extreme pain disorder, dominant or recessive diseases? Remember that you have two genes for every protein you make, one from your father and one from your mother[1]. Typically, you make about the same amount of the protein using your father's genes as your mother's. A dominant inheritance pattern is observed when it only requires one of the two genes to be expressed in order for you to show the trait. A recessive inheritance pattern is observed when both genes have to have the problem in order for you to show the trait.

In some ways, the mutation in the Na channel is a gain-of-function mutant so only 1 copy of the gene per cell is enough to make the nerve cell hyperexcitable.

---

[1] Except for most genes on the X chromosome.

Physiology

If we think about what happens in this case, we can predict the type of inheritance. The basic question is, if the patient has only 1 copy of the mutated sodium channel, would she or he show symptoms? This mutation could be classified as a gain-of-function mutant, because the sodium channel has gained the function of working under conditions where the normal channel does not work. So even one copy of the gene being mutated means that about half of the sodium channels will be open when they shouldn't; this will make it more likely that the muscle will misbehave. This analysis predicts a dominant inheritance pattern.

To test the theory that the gene is dominant, we could look at the patient's relatives and see how many have the trait. As we predict that the gene is dominant, then the patient would only need to have 1 copy of the mutated sodium channel. Thus about half of their offspring will have that copy. Since we predict that the mutation is dominate, the one mutated copy should be enough to see the effect, so about half of his offspring should also show the symptoms. Indeed, that is what is observed, about half of patient's offspring have a similar symptoms.

When I do cases like this, I also like to consider the other case. So what would a recessive sodium channel disease look like? Well, suppose the mutation just made the channel not work. If you have one normal copy and one mutated copy, the nerves are likely to work normally, since half the channels would be normal. But if you had two copies of the mutation, then none of these sodium channels will work and one will see dramatic effects.

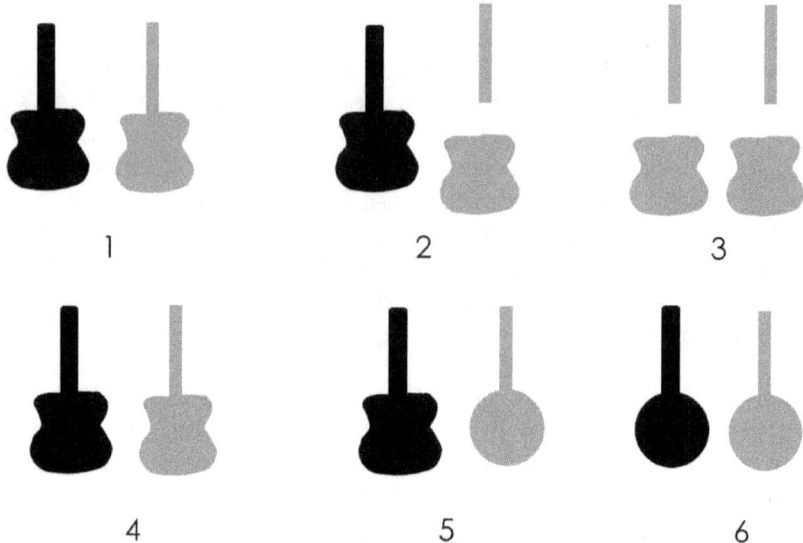

Figure 1. Dominant and recessive genes and loss-of-function vs. gain-of-function. Consider the guitar gene, you have two copies, black and gray, one from Mom and one from Dad. In #1, you have two functional genes and "normal" guitar playing. In #2, you have one functional guitar and one broken guitar. You can still make music. In #3, you inherited two broken guitars and you cannot make guitar music. A broken guitar is loss of function and it usually takes two broken guitars to cause problems.[1] #4-6 illustrate the gain of function case. #4 is just the normal situation, just like #1. In #5, you have inherited one normal guitar gene and one mutated gene, which has now gained the function of a banjo. So #5 will now be "abnormal" with only 1 mutated gene. #6 is the case where one inherits two banjo genes.

We don't fully understand why different mutations in the Na channel lead to different diseases. In the test tube, the mutated channels that cause Emily's disease have a slightly different type of defect than the channels that cause Fred's disease. But for the neuron, the effect is the same; the neurons are hyperexcitable in both cases. In these

---

[1] There are situations where you need to have loud guitar music and having only one functional guitar would be inadequate for "normal" function and therefore #2 would also show symptoms.

diseases, there is pain, but it is from different areas depending upon the disease. So far as we know, the Na channels are expressed in all the pain-sensing neurons, so why mutation A only leads to pain in the feet and mutation B only leads to pain in the anal area is not understood. It might be that, on some neurons, other types of Na channels can compensate.

There is another type of mutation that also results in inappropriate pain sensation, hereditary sensory and autonomic neuropathy Type IV. These people have mutations in their neurotrophin[1] tyrosine kinase[2] receptors. These receptors normally bind nerve growth factor. After binding, the receptor's tyrosine kinase activity is increased. This leads to phosphorylation of tyrosine residues on TRPV1 as well as the Na channels found in pain sensing neurons, thus making these neurons more likely to fire. Nerve growth factor is released after injury to many cell types and thus is important for the brain to sense injury to many cells. It is also released during inflammation and helps to account for some of the chronic pain in inflammatory states. Now that we now that the nerve growth factor receptor (neurotrophin tyrosine kinase receptor) is involved in pain modulation, scientists are working on inhibitors that will block receptor activation and decrease pain. In these patients, the receptor mutations appear to result in impaired development of particular nerve fibers; these fibers are not only important for pain sensation, but also for sweating, so these patients don't sweat normally.

Mice that lack the melanocortin[3] receptor don't feel pain as much. The melanocortin receptor has mutations that account for some cases of red hair and fair skin. However studies in red haired, fair skinned people show conflicting results, so we don't know if the mouse results will apply to humans or not.

---

[1] Neurotrophins are nerve growth factors; trophic is from the Greek for nourishment. These growth factors bind to receptors; upon activation they have tyrosine kinase activity.

[2] Remember kinases from the chapter, Can Viagra™ Kill You? Kinases transfer a phosphate from ATP to another molecule. In this case, the receiving molecule is the amino acid tyrosine in particular proteins, forming phosphotyrosine.

[3] We've discussed two melanocortins before; ACTH and melanocyte-stimulating hormone, see the chapter, Black and White (and Banning Tanning?).

We also have pain-sensing neurons away from the skin that can sense pain in our organs and tissues. We don't understand as much on a molecular level about those processes. There are a number of states were people feel abdominal pain. There is some evidence, but it far from conclusive, that the pain in Irritable Bowel Syndrome might be due to inappropriate behavior of yet another type of sodium channel. If that turns out to be true, finding blockers of that sodium channel might be a way to relieve the pain.

What about Leslie? It turns out that her pain neuron Na channels don't work at all. Thus even though heat activated her heat sensor protein and the nerve membrane potential changed a bit, there was no movement of sodium, because the sodium channel didn't work. So her pain sensing neurons didn't fire and therefore none of the upstream neurons in the spinal cord or brain registered pain. In terms of our alarm analogy, Leslie is a case where the wire from the motion detector to the house alarm box is cut, or doesn't work, so no matter how many potential burglars go past the motion detector, the house alarm box and the police station don't respond. Presumably the Na channel in her other sensory neurons work fine because her other sensory inputs are fine. Leslie's disease is congenital insensitivity or indifference to pain-sodium channels. These people often have more injuries, including breaking their bones and joint deformities. They sometimes require amputations.

One of the injury sensing proteins is one we discussed in the chapter The Good, the Bad, and the Spicy, TRPV1, the capsaicin receptor. This receptor responds to extremes of heat as well as to capsaicin. Interestingly, there is one report of a patient that did not respond to capsaicin and she had about 1/2 the normal amount of TRPV1. She was the only member of her family that could stand eating hot peppers and indeed even at the highest concentrations tested; they didn't give her any pain.

Since inflammation causes pain, one expects that pain-sensing neurons have a mechanism for "detecting" inflammation. When a tissue becomes inflamed, a number of compounds are released including histamine and prostaglandins. Some pain sensing neurons have receptors for either histamine or prostaglandins and this explains the sensation of pain during inflammation. As aspirin reduces the amount of prostaglandin release this explains part of its analgesic effect; another part is aspirin effects on the brain, particularly the hypothalamus. Since aspirin does not affect the cortex, so it does not alter alertness.

Antihistamines are often given to reduce itching.

What is the relationship between itching and pain?
Neurons with TRPV1 do cause itching, but the compounds that cause itching do not activate the TRPV1. Compounds that cause itching activate other receptors. One of these is histamine. Clearly, the signal from this nerve must be a bit different since itching and pain are perceived differently. The exact relationship between pain and itching is still being worked out.

Interestingly, over 50 years ago it was recognized that a tropical plant, cowhage, caused itching. Just recently, how the plant does this has been worked out and this provides additional information about how itching works and solves one of the itching puzzles. The puzzle was why some itches did not respond to antihistamines (like Benadryl). It turns out that cowhage contains a protease. Many of the itch-sensing neurons have a receptor that is activated by a protease, including the cowhage protease. When the receptor is activated, the nerve fires. I don't recall if any of the pain insensitive patients were tested for whether they itch, that might help sort out whether the itch and pain neurons use the same sodium channels.

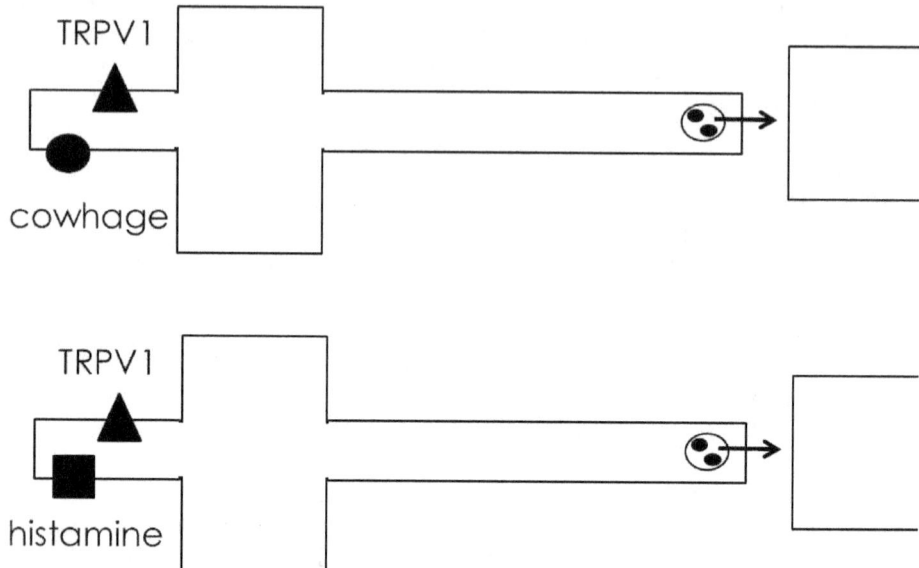

Figure 2. Itch pathways. Both neurons have TRPV1 receptors (triangles). The top neuron has the cowhage receptor (oval) and the bottom neuron a histamine receptor (square). When the neuron is activated, it releases its neurotransmitters (small filled circles) from its vesicles; the neurotransmitter then activates then next neuron.

Are physical and social pain related?

Some scientists wanted to see if the same areas of the brain got extra blood flow in both cases of physical pain as well as social pain. Brain blood flow is measured by functional NMR. The idea is that if a neuron is active, that means sodium moves into the cell. In order to keep being active, the neuron needs to pump the sodium back out of the cell and this uses energy. Thus areas of the brain that are extra active should have high blood flow to supply glucose and oxygen and take away carbon dioxide, just as is true of active areas of muscle. However, NMR measurements require one to be sitting in a NMR tube and you are by yourself. If you are socially isolated, how can one test to for social stresses? The scientists set up a computer game where 3 participants passed a virtual ball. At a set time, the other two "players" stopped passing the ball to the subject, who could play the game on a computer monitor in the NMR tube. When the other "players" stopped passing the ball, the subject would then feel excluded or socially isolated. When this occurred the subject had increased blood flow to one area of the brain that is also active when experiencing physical pain. This suggests some overlap in feeling pain physically and from social isolation.

As you probably know, opiates relieve physical pain. Opiates also seem to relieve social pain in several species of animals. Scientists studied this by separating the animal from its group and count the number of cries and opiates reduced the number of cries. This suggests that there may be some overlap in how animals and humans experience some kinds of pain.

How do pain killers work?
Pain often starts by activating the TRPV1 receptors, so these are an obvious target for pain relief. Research proceeds on this front, but no effective, safe, agent has yet been found. And there is a potential downside. If scientists develop a way to block the TRPV1 channel, a danger will be that the person might burn themselves, as they won't be able to sense heat.

Some people seem not to feel the pain in particular situations, for example, some athletes play through an injury, and many soldiers do not accept pain medication even though they have significant injuries. These observations led to the idea that the brain can activate a system that inhibits the sensation of pain. We now know that this system is also activated by opiates.

This gives us a hint about how opiate pain killers work. Opiates mimic endogenous compounds, such as endorphins and enkephalins, and activate opiate receptors. Some opiate receptors are on the pain inhibition system, so when they are activated, the neurons fire and inhibit the perception of pain. These receptors are also found in the digestive track and account for the constipation that is a side effect of opiate use. The opiate receptors are also found in the reward center and their activation can result in addiction to pain killer medications.

In 2005, almost as many people were dependent on, or abused, pain relievers as were dependent on, or abused cocaine. Most of the people abusing or dependent on pain killer medications started using the medicine to relieve pain, but then found themselves addicted. In fact, some scientists consider than many people are pseudo-addicted to pain killers; they feel that some patients are undertreated for the pain and therefore turn to illicit pain killers.

What is the difference between abuse, dependence, and addiction?
As theories about what happens with continued drug use have changed, the terms have also changed. Many scientists have several categories. The first might be recreational drug use and at this stage, the person presumably is taking the drug primarily for its positive pleasurable effects and could "readily" stop using the drug. There is extreme variability in how long this stage can last; for some people, using the drug once or twice can be enough to move onto the next stage. For others, the use can continue for quite some time before moving to the next stage. The second stage is drug abuse. In this stage, the person is using the drug even though it is definitely having a negative impact on their life-either physiological or social problems. In this state, an unbiased observer would conclude that the damage they are causing by using the drug should be outweighing the pleasure. At this stage, the person is presumed to have the ability to choose to stop using the drug and have the brain circuitry that would allow it. The third stage is drug dependence. While many scientists think this stage is real, there are exceptions; certainly many lay people don't accept this stage, at least for all drugs. In this stage, not only are there severe negative consequences to the drug use, but the drug use has continued so much that some of the brain nerve circuits have changed. While these people can make the decision to stop using drugs, their brain circuits have changed in such a way that they cannot, without outside assistance, actually follow through on this decision.

One of my previous classes (The Science of Sex, Drugs, and Rock'n'Roll) came up with this analogy. You are driving a vehicle through a field. You can chose to go straight ahead and leave the field and get back on the main road. But you are tempted to turn left and go around the field again, because it looks so pretty and what can be the harm? Going straight is the decision not to try a drug initially. Making the first left turn is the decision to try the drug the first time. Let's say you liked seeing the field again, and so you continue to make the left turn. You can do this for a while and then decide it's time to stop seeing the field and drive straight out. Let's say you were supposed to meet a friend for lunch, but you decided to turn left and

Physiology

see the field a few more times. This is the state of abuse. Eventually, you have made the left turn so much that there is a big rut and there is no way you can go straight, the rut is just too big. In order to get out, you not only have to decide you want to stop turning left, you have to call for help-get a tow truck to come and pull you past the rut. This is the state of dependence.

This analogy illustrates several points. The major point is whether or not the state of dependence actually exists. I think the evidence is clear that our brain circuits do change, just like driving a vehicle over and over again in the same path will create ruts. Where the disagreement is, is whether the rut can ever get large enough that outside help is absolutely required to get out. I suspect part of this disagreement is because the data are not completely conclusive on either side and so the interpretation depends upon one's starting point. If one's view of the world is that we always have a choice, it will take a lot of evidence to the contrary. On the other hand, if one goes into the problem thinking that drug dependent people don't have control, then modest evidence supports that idea and it would take a lot of contrary evidence to convince someone that they really had control and a choice.

This analogy also illustrates the interplay between genes and environment and the variability of responses. If the field is the environment and the vehicle the genes, then you can imagine that very muddy (or snowy) fields are likely to develop ruts sooner. You can also see that whether you are driving a Jeep or a VW Beetle or a Ferrari can make a difference.

Most illicit drugs are chemicals originally derived some plants: opiates and marijuana are but 2 examples. In the 1960s and1970s it was discovered that opiate-like compounds (morphine) bound to specific receptors in humans. This was a puzzle: why would the human body have a receptor that bound plant substances? The breakthrough idea was that these plant compounds must mimic an endogenous chemical that goes to the same receptor. This spurred a hunt to find these endogenous chemicals. In order to purify these compounds, the plant mixture was separated, for example into two parts A and B. If A was the active part, then one proceeded to separate it into A1 and A2. If A1 was active then one proceeded to separate A1 into A1a and A1b and so on. The separation was relatively easy. The trick was, how do you know if A or B has the active portion? Giving A and B to different people and seeing who got high was not an option. But even back then, we knew that one common problem that opiate addicts suffered from was constipation. The constipation is a result of inhibiting the contractions of intestinal muscle cells. If a piece of animal intestine is isolated, it spontaneously contracts. If you put morphine on isolated intestine pieces, the contractions stop. This was the bioassay for the endogenous opiates-put fraction A and fraction B on isolated intestine segments and see which one stops the contractions. Eventually, endogenous opiate compounds were isolated from animals, we call them endorphins and enkephalins.

In summary,
- There are specific molecules that respond to injury-mechanical, heat, chemicals, inflammatory mediators.
- These activate appropriate receptors which lead to nerve activation and a signal to the brain, where pain is received.
- The hot pepper channel, TRPV1, is one key pain sensor.
- Gain-of-function mutations in specific sodium channels cause the pain sensing neurons to fire inappropriately and cause several diseases of inappropriate pain sensation.
- Another disease of inappropriate pain sensation is caused by a mutation in the nerve growth factor tyrosine kinase receptor.
- Both of these finds suggest possible new targets for drug development (sodium channel inhibitors and nerve growth factor tyrosine kinase receptor inhibitors).
- The absence of sodium channels in pain sensing neurons leads to a state where pain is not felt and reminds us that pain often serves a useful purpose.
- The main drugs for chronic pain relief are opiates, which activate pain inhibitory pathways in the brain, but these drugs are also addicting.

Physiology

Sleep tight, don't let the bed bugs bite

Some unusual sleep patterns run in families, suggesting a genetic link. David Sedaris, in one of his monologues, talks about growing up with a mother who seldom slept at night and didn't mind her son staying up either. So in his house he often was up and walking around the house at 2 and 3 in the morning. When he was about ten, some friends invited him to a sleepover and he asked about what they would do. All his friends were excited because the big thing to do was to stay up all night. David didn't understand why that was so exciting-of course it was exciting to the others because it was novel and previously forbidden.

Our understanding of the physiology of sleep is still in the early stages. In this chapter we will look at some potential answers to a series of questions about sleep. Because so little is known about sleep, you will also get a sense of how a new area of science tries to grow and some of the obstacles and frustrations along the way.

The sections are:
- Why do we sleep?
- What are the stages of sleep?
- How do sleeping pills work?
- How much sleep do we need? What happens if we don't get it?
- Sleepwalking
- Sleep tight, don't let the bed bugs bite
- Unusual sleepers

Why do we sleep?
The purpose of some physiological processes became obvious in the early stages of modern science. We breathe to supply oxygen. The heart pumps to speed up the delivery of nutrients, and removal of waste products, from the cells. We eat to get nutrients and we excrete to get rid of waste. But why do we sleep? Of course, the short answer is that when we don't sleep, we don't feel as well and we don't perform as well. But what does sleep do that makes us feel refreshed and allows us to perform better? We don't know. There are 3 theories and each may be partially correct. Two are pretty obvious.

The first theory is basically a restatement of how sleep makes us feel-this is the Restorative Theory-that something about sleeping lets our body and brain get reset or restored. Can you think of anything mechanical that needs to rest periodically in order to function well? I can't. But then, mechanical things don't repair themselves. Can you think of any electronics that do better by resting? Maybe this is an analogous situation-many older computers worked better if they were periodically restarted-it allowed them to clear their memory and put away programs that weren't currently being used. Is that what are brain does during sleep?

A second theory for the purpose of sleep is merely that it is energy conservation-by resting and sleeping we are using less calories, so we need to find less calories. We spend about 1/3 of our time sleeping, but it doesn't save us 1/3 of our calories, because we do still burn calories while asleep. But I would guess that it could save us 10 to 20% and in the early days of humans, that could have been a significant advantage.

The third theory is that there was an evolutionary advantage to sleeping during the most dangerous part of the 24 hour day. What was dangerous for early man at night? Was it the possibility of tripping over something? Of getting lost in the dark? These seem unlikely to me, assuming early humans had enough vitamin A as I think we have adequate night vision. Were animals that preyed on humans more active or more successful at night? I don't know enough about the habits of the animals that would have preyed on early humans to answer that question.

Physiology

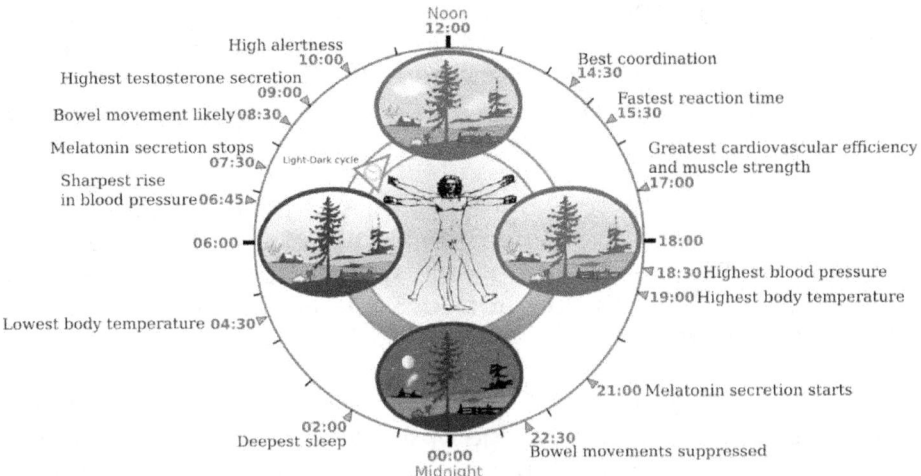

Figure 1. One version of how functions change during the day. Figure from Wikipedia, http://upload.wikimedia.org/wikipedia/commons/5/5f/Biological_clock_human.PNG.

What are the stages of sleep?

Because our understanding of sleep is so rudimentary, the divisions are based on easily observable phenomena. There are two large categories for sleep, rapid eye movement (REM) and non-rapid eye movement (NREM). In addition to eye movements, REM sleep also has brain waves[1] that resemble the brain waves when we are awake, hence a synonym for REM sleep is paradoxical sleep; it is a paradox

---

[1] Brain waves detect electrical activity. Remember that the signal to activate a nerve or muscle cell is the movement of ions, typically sodium (and then calcium). This movement of ions creates a current. Electrodes can detect this current. The heart generates a large current every heart beat-all the ventricular muscle cells move sodium at nearly the same time, so that they contract nearly in unison. The large sodium movement can be detected by placing electrodes on the body surface because the body conducts electricity-we are basically salt water-this is the basis for EKG/ECG measurements (ElektroKardioGramm in German, ElectroCardioGram in English). The nerves in the brain don't fire in such a regular pattern, but we can detect their signals as Electroencephalography, EEG. You probably note cepha- which is from Greek for head.

that the brain appears to be working as much during REM sleep as when we are awake. On the other hand, during sleep, the brain directly inhibits the alpha motor neurons, so all are voluntary muscles are inactive, except, of course, for the muscles controlling eye movements!

REM sleep, one would think, is the opposite of fight or flight. It is true that the parasympathetic system dominates during REM sleep, but during phasic REM sleep, sympathetic systems can be activated just like in fight or flight. This can lead to sudden increases in heart rate and blood pressure.

NREM sleep is divided into a 3 stages. The first stage is when we are falling asleep. Is this really a stage of sleep or a transition to sleep? Is that distinction important? I suppose an analogy might be, when we are chewing food, are we digesting? The second stage is intermediate sleep and the last stage is deep sleep. Note that when I say first, second and third, I am not talking necessarily talking about the chronology, because our sleep at night normally involves about 4 cycles. It is true that the first cycle at night progresses through the 3 stages of NREM sleep in order. The second cycle involves stages 2 and 3 as well as REM sleep. Only stage 2 and REM sleep occur in the third cycle and fourth cycle-isn't it interesting that there is no deep sleep during the second half of the night?

How do sleeping pills work?
What you would like to do is turn off the neurons that are keeping you awake. In some nerves in the brain, there is a chloride channel that is activated by a neurotransmitter called GABA[1]. When GABA is released, it opens the chloride channels and makes it harder (or less likely) for a different neurotransmitter can activate the cell by changing its membrane potential. Thus there is no action potential due to Na entry and the nerve stays quiet. It turns out that these GABA regulated chloride channels are also able to bind other

---

[1] GABA is an abbreviation for gamma-amino-butyric acid. Butyric acid is the chemical name for an acid that has 4 carbons. (2 carbons is acetic acid as in vinegar.) Amino refers to a structure with NH3, a lot like ammonia; Gamma, being the fourth letter of the Greek alphabet is a way to say that the amino group is on the 4th carbon.

chemicals called benzodiazepines. Valium is one example. When benzodiazepines bind, they also open the chloride channels.

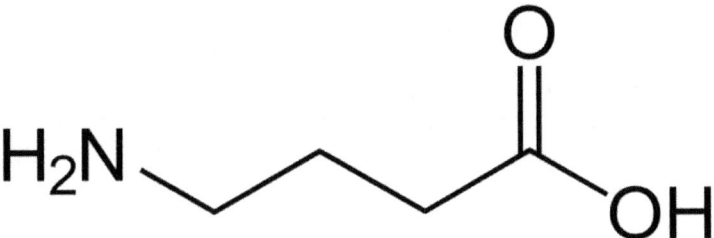

Figure 2. Structure of GABA. Taken from Wikipedia, http://en.wikipedia.org/wiki/Gamma-Aminobutyric_acid.

Benzodiazepine receptors[1] (GABA receptors, GABA activated chloride channels) are found on lots of different neurons and we don't want to inhibit all of them, just the ones that are keeping us awake. Fortunately, there are different types of benzodiazepine receptors, like there are acetylcholine receptors and unlike androgen receptors. The modern drugs target the type 2 receptor and thus more precisely help us sleep rather than have other side effects that could occur if the drug went also to the type 1 receptor.

How much sleep do we need? What happens if we don't get it?
The standard response is that people need about 8 hours of sleep per night. Is there any data to support this? First, of course, we need to remember that everyone is different. There are 2 common ways to determine whether someone is getting enough sleep and it is not clear to me that these two methods necessarily result in the same answer. One way to ask the question is to determine how long someone sleeps when they are not awakened. In the days before alarm clocks, presumably our bodies were designed to wake up when we had had enough sleep. Wild animals don't need alarm clocks-of course, they might use sunlight as a cue.

---

[1] A small point of possible confusion, benzodiazepines refer to a particular chemical structure. The proteins that bind benzodiazepines are called benzodiazepines receptors. But the most popular drugs that bind to benzodiazepines receptors are not actually in the chemical class of benzodiazepines. Benzodiazepines contain benzene and diazepine compounds parts.

The second method to determine if someone is sleep deprived is to see if their performance has decreased. Before I go through the data, I want to point out 1 key result. In these studies, people's self-assessment of their impairment was very inaccurate-they thought they were performing better than they really were. The main affect is on tasks there required sustained attention and on reaction times. Interestingly, there does not appear to be a major effect on executive functioning.

Two labs did paired studies; in one lab people were randomly assigned to groups getting 4, 6 and 8 hours of sleep and the other lab, 5, 7, and 9. The groups went for 2 weeks and were tested on their vigilance. Occasionally during the day, they sat in front of a computer screen that flashed at random times. If they saw a flash, they had to push the space bar. Kind of boring, but the kind of attention that drivers and pilots certainly need.[1] As you might expect, after 2 weeks, those getting 8 and 9 hours sleep performed these tasks well and those getting 4 and 5 hours sleep did poorly. However, after 2 weeks, even the 6 hour sleepers performed poorly. There was substantial variation in the changes in the first few days-some people could go a few days without great impairment, but everyone did poorly after 2 weeks on less than 8 hours sleep.

You might think you are one of the lucky ones that can get away with less sleep. But of the people that think they need less sleep, scientists think that less than 5% actually can. There has been some research to identify the genes that might account for this. It is a trait that does seem to run in families.

Lack of sleep is thought to be the second leading cause of car crashes. Some investigators think that sleep deprivation led to poor judgment by workers at Three Mile Island and Chernobyl. Some even

---

[1] Historically, I guess you'd need it to be aware of a predator sneaking up or some prey close by. It reminds me a bit of the challenge of playing right field in Little League. Very seldom was a ball hit to right field and many youngsters got bored out there and stopped paying attention-then whammo, a ball comes and they aren't ready.

place part of the blame for the Challenger disaster on poor decisions by people who needed more sleep.

Sleep is also important for immune system function. An extreme case was a study done on rats that were sleep deprived for several weeks. I'm actually surprised they were allowed to do this experiment, as the rats were deprived of sleep until they died. Personally, I think that is not an appropriate experiment. But the cause of death in these rats was a systemic bacterial infection because their immune system was not working.

Sleepwalking
There are two behaviors that lay people might term sleepwalking; scientists call one sleepwalking and the other REM Sleep Behavior Disorder. In REM Sleep Behavior Disorder, the person may act out their dreams. They can talk and move and sometimes walk. To understand this behavior, we need to remember a key feature of REM sleep-the brain is very active and normally all the voluntary muscles are inactive. In REM Sleep Behavior Disorder it is thought that the muscle inactivity is misregulated, so that there is some muscle activity.

In contrast, sleepwalking in the scientific sense, seems to occur during deep sleep (the third state of NREM). In this case, parts of the brain do not properly and fully arouse and the person can do things, but without proper control and without any memory.

Both of these disorders are hard to study, even though it is thought that 20% of children and 4% of adults have these problems. A lot of research on sleep in general is done in sleep labs where one has the equipment to monitor many physiological responses while someone sleeps. Unfortunately, those that are prone to sleepwalking seldom do it when they are sleeping in a lab setting. A possible breakthrough in the study of sleepwalking was made in 2008. The investigators examined 10 sleepwalkers (as well as 10 controls). They found that they could get all 10 sleepwalkers to sleepwalk in the lab if they did two things-they sleep deprived them and gave them an auditory tone. The sleep deprivation was done by keeping them awake for 25 hours. The auditory tone was timed to occur during their deep sleep phase of NREM.

If this technique works for other researchers (and other sleepwalkers) then it will allow one to examine what happens during sleepwalking in a laboratory setting and in a reproducible fashion. Then one can not only learn more about sleepwalking but also whether factors such as alcohol or caffeine alter it.

Sleep tight, don't let the bed bugs bite
This is not science, but it may be a piece of history you didn't learn (or forgot). A few hundred years ago, beds did not have box springs. The mattress (often straw stuffed inside a cloth) was kept off the floor by laying it on ropes stretched between the sides of the beds. As you can imagine, after weeks of sleeping like this, the ropes stretched and so the mattress would eventually touch the floor. Bedbugs can't jump, but they can crawl onto the mattress when it is touching the floor. If one tightens the rope, the mattress is raised up from the floor and the bedbugs don't get into the mattress. So they can't bite you while you are sleeping.

Unusual sleepers
Some unusual sleep patterns run in families, suggesting a genetic link. Some scientists examine these families in an attempt to identify a gene that contributes to this behavior. (Note that I say "contribute" not cause.) When it is identified, they can put the gene into mice and fruit flies and see the function of the gene and also whether it alters the sleep behavior of the animal.

One disorder that they have studied is Familial Advanced Sleep Phase Syndrome. These people sleep about the normal amount but seem out of phase to most of us-they are in bed about 7:30 pm and wake up about 4:30 am. They were able to identify a mutant in a gene that is known to influence circadian rhythms in fruit flies that seems to account for this behavior.

There are also a few people who seem to do very well on only a little bit of sleep and a gene for this has also been identified. When they make the same mutations in mice and fruit flies, they claim to also see changes in sleep and activity patterns. After this study was published, many people wrote to them to volunteer for additional studies, claiming that they also needed only a little sleep. The researchers developed a screening test to determine who really needed only a little sleep and only about 5% passed the test. These people sleep

Physiology

about 5 or 6 hours a night and then appear to run full tilt during the day. They have other associated characteristics-they tend to be more upbeat and more tolerance for pain and setbacks. Interestingly, in normal people, sleep deprivation is a risk factor for gaining weight, but these people tend to be thinner than normal.

In this chapter, we provided
- description of current theories on why we sleep
- description of the stages of sleep
- how sleep pills work which involved a review of neurotransmitters and cell charge balance
- theories about how much sleep we need
- observations of what happens if we are sleep-deprived
- description of different types of sleepwalking
- Some people who have unusual sleep patterns

Physiology

Can Viagra™ kill you?

In the movie, *Something's Got to Give* Harry Sanborn (Jack Nicholson) is an older man who is dating a much younger woman, Marin Barry (Amanda Peet). Marin invites Harry to her family's summerhouse for the weekend. Shortly after they arrive, her mother, Erica Barry (Dixon Keaton), also arrives, as she had also planned to stay the weekend at the house. Neither woman had told the other about their plans. Harry is attracted to both women and vice versa. However, at the first dinner, Harry gets chest pains and is rushed to the hospital. As the staff is working on him, the mother and daughter are positioned just outside the room, clearly within hearing distance. At the moment that Harry notices the mother and daughter standing by the door, the doctor asks Harry if he is taking Viagra™. Harry says no. The doctor starts an IV, telling Harry that it contains nitroglycerin and that it is a good thing Harry is not taking Viagra™ because the combination of Viagra™ and nitroglycerin would probably kill him. Harry immediately reaches over and yanks the IV out.

You might discount this story as just being in a movie. But consider this real case, published in a medical journal. A 70-year-old man, George, was given some Viagra™ by his friend. First, even high school students should know better than to take pills from someone else's prescription- but 70-there should be no excuse. Within 2 hours he began to feel some chest pain. Of course, at this point, he should have gone to the hospital. But he didn't. Interestingly, he did tell his mother. His mother gave him some of her nitroglycerin. Nitroglycerin is commonly prescribed to treat chest pain. (If he was 70, you know she has to be around 90, and if he should have known better, she should really have known better than to give him her prescription for nitroglycerin.) Eight hours later he went to see his physician.

The physician diagnosed a heart attack (myocardial infarction) based upon the medical history, the ECG results, and the lab tests. The lab tests involved measuring the level of two specific proteins in blood, creatine kinase and troponin I. One form of creatine kinase is found primarily in heart and skeletal muscle and when these cells are damaged it is released into the blood stream. Other forms are found in the brain and kidney. So elevation of creatine kinase implies one of these organs is damaged and given the history, the heart seems most likely. The creatine kinase assay has been around for a long time and is still routinely done, but the second protein, troponin I, is much more specific for the heart and is starting to be used more now.

Why would the combination of Viagra™ and nitroglycerin kill Harry? We will answer this question in a series of stages, as follows:
- How does dilation of arteries help the heart?
- What is the mechanism by which dilation occurs?
- What is it about nitroglycerin that allows it to alter the smooth muscle contraction process?
- What are the roles for ATP in smooth muscle contraction?
- What is the series of steps involved in relaxation? Who cares?
- What happens if Harry had both nitroglycerin and Viagra™?
- How does Viagra™ work?
- Why isn't Viagra™ marketed to women?
- Another case that raises an additional way for drug-drug interaction.
- Does marijuana lead to arterial dilation?
- Nitric Oxide and Nobel Prize

How does dilation of arteries help the heart?
Both nitroglycerin and Viagra™ dilate arteries. Dilation means the diameter of the artery increases. How does dilation of arteries help the heart? This means there is more blood flow. As an analogy, imagine a particular highway is full of trucks. If you widen the highway, more trucks can flow. Now, if the trucks are bringing food to people, obviously you can feed more people if you have more trucks. What food does the heart need? To supply the energy for contraction, the heart needs oxygen. From your experience with fire you know that oxygen is not enough to get a fire. You also need a fuel. For the heart, glucose or fatty acids work. If blood flow is inadequate, the heart does not get enough oxygen and fuel to supply enough energy to contract properly. The healthy body controls the rate and force of contraction. Just like you don't drive your car at top speed all the time, the heart also adjusts to conditions so that it doesn't waste fuel and have undue stress.

What is the mechanism by which dilation occurs?
The blood vessels can be considered to be cylinder. How can you change the diameter of a cylinder? In the body, smooth muscle cells

Physiology

wrap around the artery. As you might guess from the name, they are able to contract. As these muscle cells contract, the opening in the cylinder gets smaller and the blood vessel constricts. But we don't want to constrict in this case, we want to dilate. In the normal state, blood vessel diameters are a balance between constriction and dilation; this means the vessel can respond in either direction. So if the vessel is partly constricted, we can dilate it if we relax the smooth muscle cells, that it, decrease the amount of their contraction. This can be accomplished by turning off the mechanism that causes contraction. This is similar to thinking about how to stop a rolling car. Not only can you put on the brake, you can take your foot off the gas.

Contraction occurs with 2 proteins, actin and myosin, see Figure 1. Many actins line up together to form a polymer; much like many beads can line up together to form a necklace. One end of the polymer is attached to a wall. Myosin is a bit like a person with a paddle. The myosin paddle attaches to one actin molecule. For contraction, the myosin paddle moves causing the actin molecule to move.

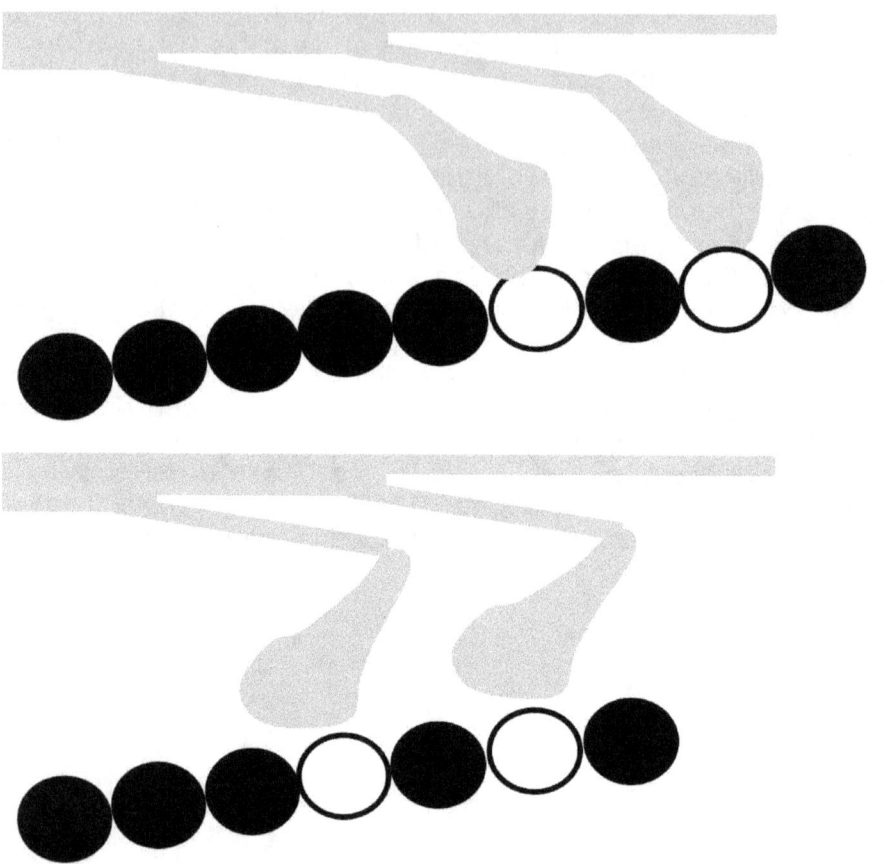

Figure 1. Actin and myosin. Each actin molecule (monomer) is a circle; they are all lined up to form a polymer. The gray structures are the myosins with the head groups (paddle) reaching out to the actin and the tails wound up together. The paddles have rotated from the top to the bottom and this has causes the actin to move.

One way to divide up the myosin molecule is into the head (or paddle), and the tail. Many myosin tails line up so that there is a string of myosins roughly parallel with the actin polymer necklace. However, the geometry is a bit unusual, see Figure 2. The free end of the actin polymer has a gap before the free end of the next actin polymer; it is this gap that gets smaller during contraction, see Figure 3

Physiology

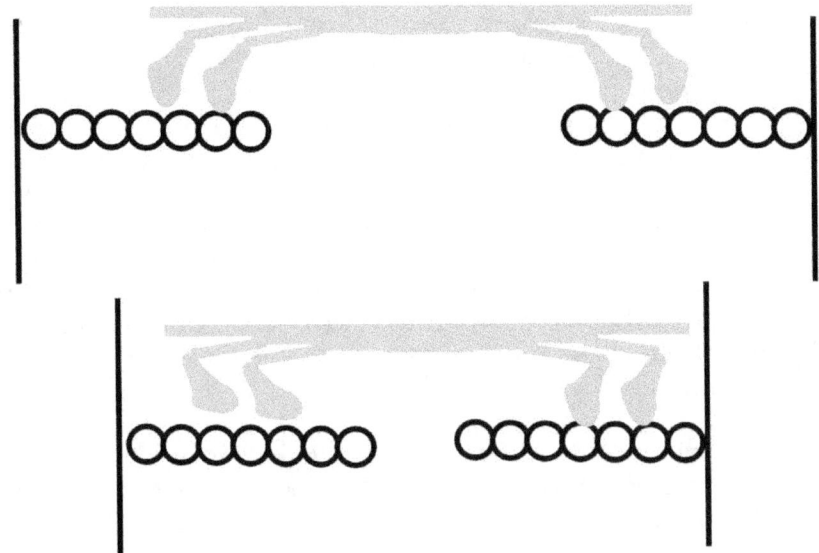

Figure 2. The sliding of the actin necklaces as the myosin paddles rotate, causes the outside lines to get closer, which is a contraction. This is the basic idea behind the sliding filament theory of muscle contraction.

Figure 3. A close up of one the myosin paddles moving. In this part of the cycle, they remain attached to the open circle actins, decreasing the gap between the actin necklaces. To complete the cycle, the myosin paddles detach and reset, then attach to a new actin, much like the stroke of a paddler taking the paddle out of the way, resetting and then putting the paddle into a different place in the water.

What happens when we wind the smooth muscle cells around a cylinder? Now when the smooth muscle cells contract, the cylinder constricts. So what's happening with Harry? Let's say that his cylinder, his blood vessel, is about half way open. So to relax this, we want the actin and myosin to stop their cycling. For myosin to work, it has to have a phosphate on its light chain just like for some people to work, they need to have gloves on their hands. I need to introduce myosin a bit more. As you might guess, another way to divide myosin is into the light chain and the heavy chain. Often the myosin light chain is abbreviated MLC. But I find it hard to follow all the abbreviations, yet myosin light chain also seems awkward, so I'm going to write it as MyosinLightChain.

MyosinLightChain needs a phosphate to work. So we can prevent myosin from holding onto actin by interfering with the process that puts the phosphate on.

What puts the phosphate on? An enzyme, called a kinase, transfers phosphate from ATP to another molecule. Because this enzyme is going to put phosphate onto the MyosinLightChain, you'd be right if you guessed its name was MyosinLightChain-Kinase.

Another way to think of the role of phosphate on MyosinLightChain is to consider the role of the double handle on a push lawn mower. You can only start and run the lawn mower if both handles are close together. Having them close together is not enough to run the lawn mower; you also need to use gas. So for myosin, you need to have phosphate on (the handles close together) and you also need to have ATP (gas) to fuel the myosin.

Physiology

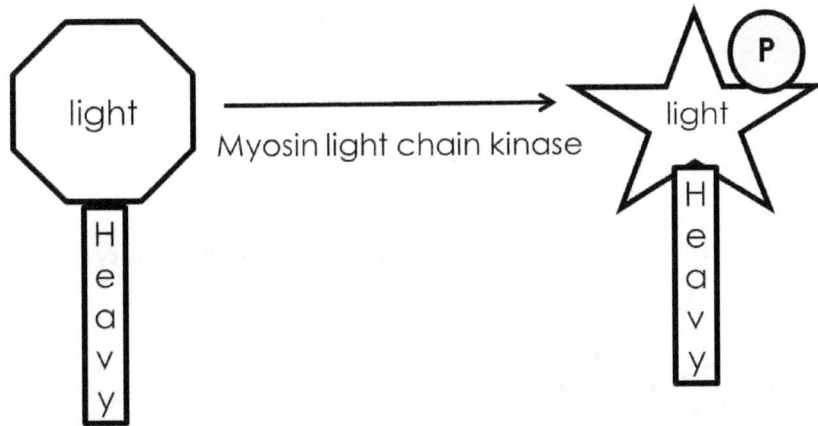

Figure 4. MyosinLightChain becomes active when a phosphate is added to it by MyosinLightChainKinase.
Myosin has a light and a heavy chain. The light chain is inactive (stop size shape). When MyosinLightChainKinase (MLCK) is active, it transfers a phosphate to the light chain, making the light chain active (star shape). Remember kinase just means to transfer a phosphate.

What is it about nitroglycerin that allows it to alter the smooth muscle contraction process? Nitroglycerin is Nitro (nitrogen) and glycerin. Nitroglycerine mimics nitric oxide[1]. Nitric oxide is the component active in the normal body that actually affects the proteins that causes relaxation.

In order to understand how two of the pathways that nitric oxide and nitroglycerin use to cause relaxation, it will help to review the main steps of smooth muscle contraction, see Figure 5
- Cellular calcium goes up. [2]
- Some of the calcium binds to calmodulin and activates it.
- The calcium/calmodulin complex binds to MyosinLightChain-Kinase and activates it.

---

[1] Initially it was thought that nitroglycerin broke down to form nitric oxide, but now there is evidence that that process is too slow to explain the response and a current theory is that nitroglycerin mimics the effect of nitric oxide by binding to the same protein that nitric oxide does.
[2] The calcium inside most cells is very low, so only a little bit of calcium needs to enter the cell cytoplasm to cause a large change in calcium concentration.

- The activated MyosinLightChain-Kinase transfers a phosphate to MyosinLightChain and which allows the myosin paddle to move and contraction to occur.

The amount of contraction is a balance of between active and inactive states. Figures 5, 6 and 7, show 3 different states of the smooth muscle: 1) the reactions are balance, 2) the reactions are poised toward contraction and 3) the reactions are poised toward relaxation.

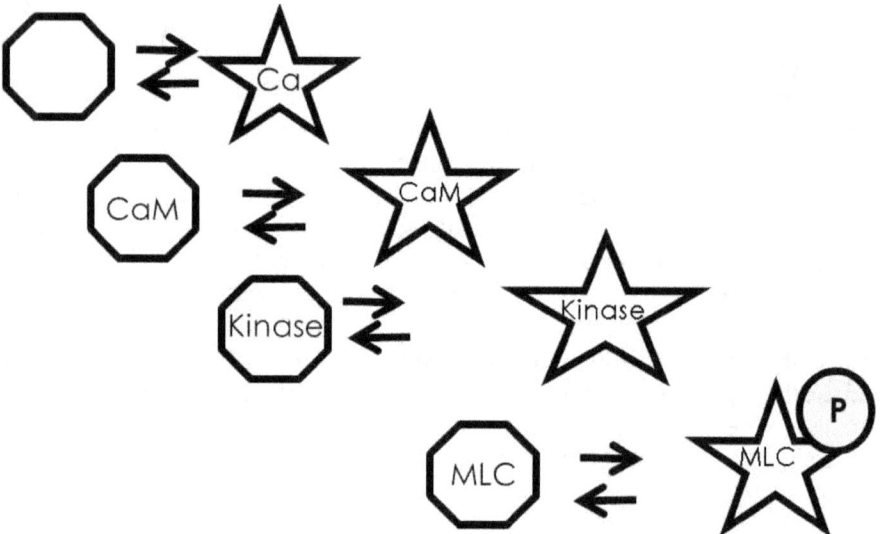

Figure 5. The cascade of events regulating smooth muscle contraction. In this figure the activating and relaxing pathways are balanced so the muscle is partly contracted. The balance is indicated by the two different direction arrows being about the same length, representing that the reaction to the left and the reaction to the right go about the same amount. On the right are inactive states, indicated by the stop sign. On the left are activated states indicated by the star. Each active state activates the reaction below it, for example, activate calcium/calmodulin activates the kinase.

Physiology

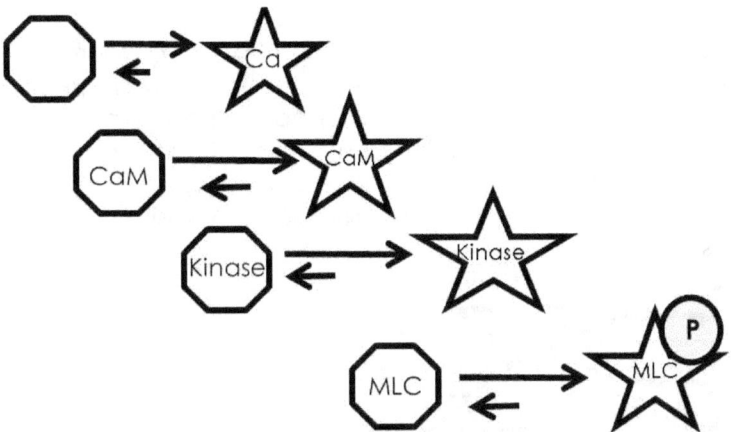

Figure 6. The cascade of events regulating smooth muscle contraction. In this figure the activating and relaxing pathways are poised toward contraction. The top arrow to the right is now much larger toward, calcium the activator. Thus each successive top arrow is also larger to the right, toward the active states.
For details of the symbols, see Figure 5.

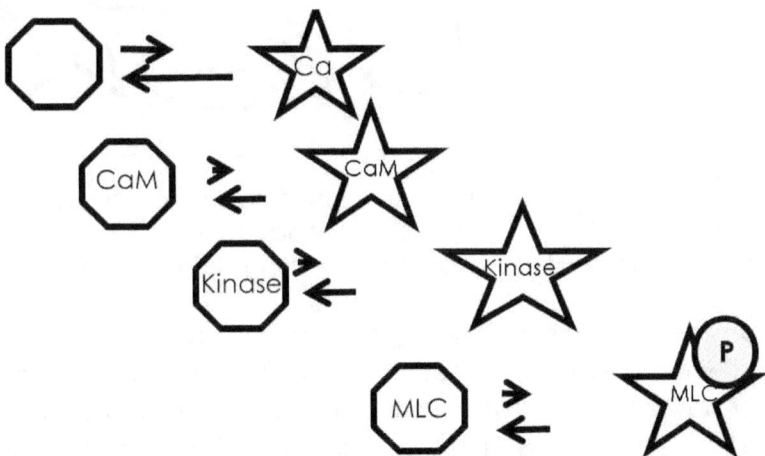

Figure 7. The cascade of events regulating smooth muscle contraction. In this figure the activating and relaxing pathways are poised toward relaxation. The top arrow to the right is now much smaller, so there is less calcium and less successive activation of teach step. For details of the symbols, see Figure 5.

One step where nitric oxide promotes relaxation (that is, a decrease in contraction) is by lowering cell calcium. Another step is by decreasing the amount of MyosinLightChain that has phosphate bound. Nitric oxide could have done this by decreasing MyosinLightChain-Kinase activity, but actually what nitric oxide does is indirectly increase the enzyme that works in the other direction, MyosinLightChain-Phosphatase[1], see Figure 8.

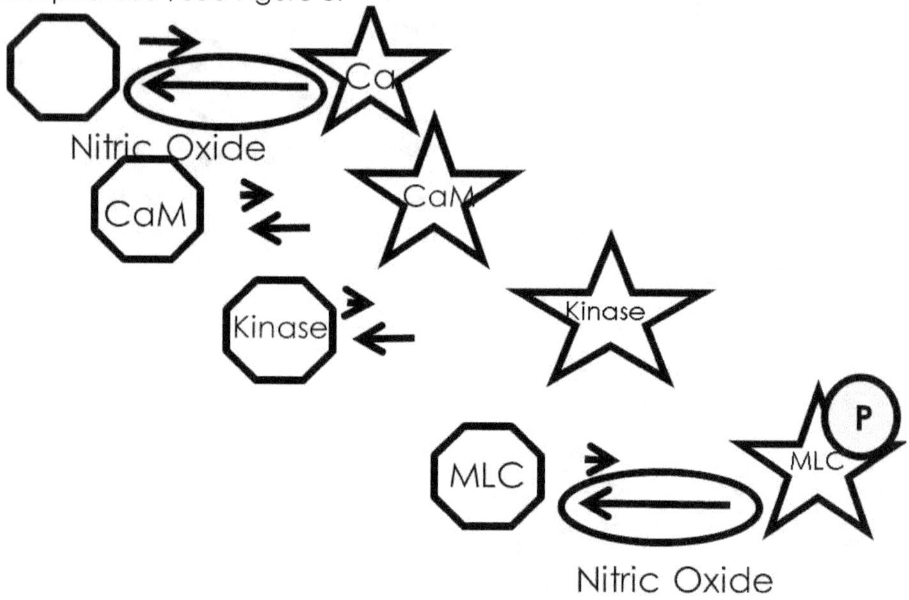

Figure 8. Nitric oxide promotes smooth muscle relaxation by several mechanisms; two are shown. Nitric oxide leads to low cell calcium (top step) and to increased removal of the phosphate from MLC.

As an analogy for this cascade, imagine that the sports figures are enzymes and when they have the ball they are active. Lebron James could pass a basketball to Payton Manning; when Manning catches the basketball, he can pass a football to Abby Wambach; when Wambach catches the football, she kicks a soccer ball to Brittney Griner. When Griner receives the soccer ball, she dunks the basketball.

---

[1] A phosphatase is simply an enzyme that removes a phosphate.

Physiology

What are the roles for ATP in smooth muscle contraction?
One point that often confuses students is what ATP is doing in this system. The confusion is natural, because ATP has two distinct functions. On the one hand, ATP provides the phosphate that is used for regulation. Kinases take one phosphate off ATP and transfer it to a protein. For example, MyosinLightChain gets a regulatory phosphate. The second role for ATP in this system is as a fuel source; as ATP breaks down, it provides the energy that allows myosin to pull the rope and the cell to contract, Figure 9.

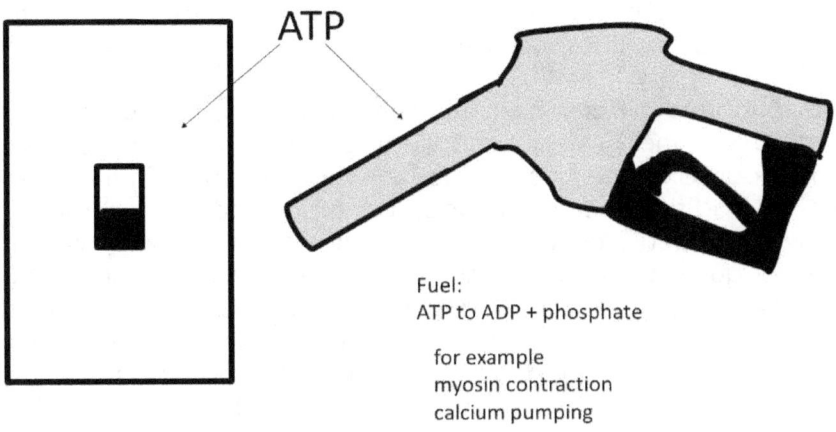

Fuel:
ATP to ADP + phosphate

for example
myosin contraction
calcium pumping

for example
myosin light chain
myosin light chain kinase

Regulation by
transferring phosphate

Figure 9. ATP is need as a switch, by providing a regulatory phosphate to proteins and also as a fuel or energy source.

What is the series of steps involved in relaxation? Who cares? On the surface the details don't seem important as both nitroglycerin and Viagra™ seem to do the same thing, that is, relax smooth muscles, so blood vessels dilate. This is a bit like saying: team A and team B have each scored a touchdown. Who cares how it was scored? But if you are a coach and either want to repeat your touchdown or prevent the other team from scoring again; it makes a difference whether the touchdown was scored on a 90 yard kick return, a 90 yard pass play, or a 15 play, 90 yard drive all involving running plays. Even a fan often wants to know this difference, as it makes following the game more interesting. The following details are as exciting to scientists as this detail about how the touchdown was scored.

We need to know some more details about the steps from nitric oxide to relaxation to find out about the step where Viagra™ works. Nitric oxide binds to an enzyme called guanylate cyclase, which produces cyclic GMP[1]. When nitric oxide binds to GuanylateCyclase, more cGMP is produced. Then cGMP, through a series of steps, leads to relaxation. Viagra™ inhibits the enzyme that breaks down cGMP. One of the functions of cGMP was to activate a protein that activates MyosinLightChain-Phosphatase.

What happens if Harry had both nitroglycerin and Viagra™? Well, what caused the 70 year old patient to have a heart attack? Studies done shortly after the heart attack showed that some of the arteries in his heart had significant blockage, so the parts downstream would not receive as much blood or nutrients. That blockage was presumably there for quite a while and Harry is at an age where he likely also has some blockage. Viagra™ and nitroglycerin both will dilate blood vessels; if this happens too much, and to most of the circulatory system, the overall blood pressure drops. If blood pressure falls too much, then there is less pressure pushing the blood, and the blood flow beyond the blockage is even lower. Thus cells in that part of the heart did not get enough nutrients and some began to die. When many cells die in the heart, we call it a heart attack or myocardial infarction. [2] These cells die because of a lack of oxygen; before they die, the area is usually ischemic, that is, has low blood flow.

---

[1] Note the "cycl" in both the enzyme's name and its product, just like a car factory produces cars. GMP is related to ATP. First, both G and A are nucleosides-also part of DNA (along with T and C). And the P stands for phosphate in both cases. The M and T don't refer to chemicals, the M stands for mono as in monorail so GMP has 1 phosphate and ATP, the T stands for tri as in tricycle, so ATP has 3 phosphates.

[2] Infarction just means an area of dead cells

Physiology

How does Viagra™ work to cause erections? Viagra™ relaxes smooth muscle cells. How do we get from relaxed smooth muscle cells to an erection? Remember, if we constrict blood vessels, we get less blood flow. When your face blood vessels constrict, you get less blood flow to the face and your face is pale. Alternatively, if you get embarrassed, the face blood vessel smooth muscles relax, you get more blood flow, and your face turns red. Viagra™, by relaxing smooth muscle cells, leads to more blood flow. More blood flows into the penis than out of it, so the penis swells up (engorges) which causes the erection.

Why not use nitroglycerin or just increase nitric oxide for erections? As pleasant as erections may be in the short term, there is a clear disadvantage to being always erect. Remember that Viagra™ inhibits the breakdown of cGMP. cGMP causes relaxation. So Viagra™ only works if cGMP is elevated, that is, only if there are other signals to the penis that release nitric oxide and increase cGMP. During sexual arousal, nerves send a message to blood vessels in the penis (but not the rest of the body) which release nitric oxide and increasing cGMP. In the presence of Viagra™ the breakdown of cGMP is prevented. Thus cGMP increases even more.

Here's an analogy for how Viagra™ works on the cell. Imagine that the level of water in your bathtub is analogous to the level of cGMP in the cell. Nitric oxide is like the person that turns on the faucet. The person is the signal. The facet is like the cGMP synthesis enzyme (guanylate cyclase), when it is turned on the water (cGMP) levels increase in the bathtub. But you don't always want a lot of water sitting in the bathtub, so a robot is present that scoops out the water. The robot is analogous to the cGMP breakdown enzyme (a phosphodiesterase). The robot can always be working and keeping the water out of the tube even if the faucet is left dripping. However, when someone comes in and turns on the faucet, too much water comes out for the robot to keep up with. Thus the bathtub fills. (cGMP increases.) Eventually, the person turns off the faucet, but the robot keeps scooping the water and so the water level goes back down. (cGMP returns to normal.) Viagra™ works by stopping the robot. So now, when the person turns on the faucet, the bathtub fills even faster (and higher) and furthermore, when the person turns off the faucet, the water stays high.

One state that leads to blood vessel constriction is stress, that is, the flight or fight response releases epinephrine and norepinephrine, which also usually cause blood vessel constriction. Men that are feeling stressed out have a much harder time having, and maintaining, an erection because the stress is trying to cause blood vessel constriction which counteracts the sexual arousal pathways which are trying to cause blood vessel dilation.

Why isn't Viagra™ marketed to women? There is some evidence that the clitoris has the same enzyme for breaking down cGMP as that found in the penis. Indeed, there is evidence that Viagra™ does lead to blood engorgement in the clitoris. But the female sexual response depends upon much more than clitoral engorgement. There have only been a few studies that have examined whether Viagra™ leads improved sexual satisfaction in women and the results are modest at best. This may be because the women in the studies have other problems besides clitoral engorgement. Viagra™ does not work for all men with impotence, either.

Another case that raises an additional way for drug-drug interaction. A 41 year old man, Marshall, woke up with chest pain. His chest felt tight and the pain radiated down his arms. This combination should make anyone immediately think that a heart attack is a possible explanation. In addition it was reported that he had autonomic upset. The autonomic nervous system controls many physiological responses. An example would be sweating. The pain came and went and after 3 days he finally was seen at a clinic. His creatine kinase levels were elevated and his ECG was also consistent with a myocardial infarction. This man had no standard risk factors for heart disease and no previous history of heart problems, so it would be natural to suspect some recent change. Given the topic of this chapter and what we've learned, it would be natural to ask him if he took Viagra™ recently. And indeed he took Viagra™ the night before he had the initial chest pain. It would be important to know how much Viagra™ he took and it turned out to have been the normal amount. Obviously a high dose of Viagra™ could also lead to massive dilation, lowering of blood pressure, and myocardial infarction. But that doesn't seem to be the case here since he took a normal dose. One could ask if this was his first time taking Viagra™, as perhaps he is extra sensitive to it. The report doesn't mention this, but let's assume this was not his first time taking Viagra™. So why would taking it this time cause a problem? Well, did he also take nitroglycerin? No. Did he take any other similar compounds? No. But it appears this is the first time he's taken Viagra™ at the same time he's smoked marijuana.

Does marijuana lead to arterial dilation?[1]
In fact, it does. In 1967 a study was done observing the dilation of vessels in the eyes. In 8 of the 9 volunteers given a high dose of marijuana, dilation occurred, but in only 1 of the 9 volunteers given the placebo was dilation observed. A common observation of marijuana use is blood shot eyes.

Interestingly, vasodilation by marijuana is not mentioned in the article about this particular case. They focused on another way in which marijuana could interact with Viagra™ and give an increase in blood concentration of Viagra™. The body does not directly excrete most drugs. Rather, the body first modifies the drug. It turns out that compounds in marijuana inhibit the enzyme that modifies Viagra™ for excretion. So the authors of the article presume that, in this man, the Viagra™ concentration stayed higher longer than usual because the marijuana had inhibited the enzyme that would break Viagra™ down. This lead to prolonged hypotension and resulted in his heart attack. It seems to me that the direct effects of marijuana compounds on smooth muscle relaxation could also have contributed to the presumed lowering of blood pressure. But in any case, this is another example of how two drugs can interact and become more dangerous than either one alone.

The first example of drug-drug interactions in this chapter was nitroglycerin and Viagra™. Both dilate arteries, so the combination would tend to be like an overdose of either one. Our second example of drug-drug interactions was marijuana and Viagra™, where one drug alters the breakdown of the other. Another example of a different type of drug-drug interaction we'll discuss in the chapter, Got Cows, where antacids or drugs that modify stomach pH, can alter the amount of the drug that is absorbed.

---

[1] The red eyes that occur after smoking marijuana are one sign of arterial dilation.

Nitric Oxide and Nobel Prize
As mentioned above, the body produces nitric oxide. The discovery of nitric oxide as a cell-to-cell communicating molecule was highly unexpected and resulted in a Nobel Prize award to three of the scientists most involved in the research, Furchogott, Ignarro, and Murad. Nobel Prizes are a bit like MVP awards in team sports; many players contribute to the success but don't get the award and there can be an outstanding player on a losing team that doesn't get the award. And just like sports MVP awards, there can be some politics involved in who actually gets chosen.

To understand why nitric oxide was so unexpected, it helps to understand the conventional wisdom before its discovery. It is a little bit like understanding music history to fully appreciate the impact of the Beatles or Elvis Presley or baseball history to appreciate the significance of the recent Boston Red Sox World Series win.

There are two ways adjacent cells can communicate. The one not involved with nitric oxide is communication via gap junctions. In this situation, the cells' outer membranes are so close to each other, that there is a protein complex that connects them. This complex essentially provides a tunnel between the two cells and small molecules, and electric charge, can easily move between the cells, without needing to go outside. As an analogy, there are some adjoining hotel rooms (adjacent cells), which have a door (gap junction) between the two rooms, so you go from one room to the other without going into the hall. Because electric charge can move between the cells and along the cells quickly, these types of cells are often called electrically coupled. This is of key importance in the heart. It is also important for the cells lining the blood vessels.

The second way two adjacent cells can communicate is that the speaking cell releases some molecules that diffuse across the space between the cells. Typically these communication molecules then bind to receptors on the listening cells and cause a change in the listening cell. So it would be like a kid being sent out from their house to run over to the neighbors. When they get to the neighbors, they push the doorbell (the receptor) and that causes the dogs to bark in the neighbor's house and then all kinds of apparent commotion ensues.

Physiology

Until the discovery of nitric oxide, binding to a receptor on the cell surface was the only way that talking cells communicated with listening cells when the cells were next to each other. I'm pointing this out because nitric oxide, when released from the talking cell, diffuses across the space between the cells. But the receptor on the listening cell is not on the outside, that is, there is no doorbell. The receptor is on the inside of the listening cell, so nitric oxide has to diffuse into the cell and then bind to the receptor.

But this aspect of nitric oxide, that it has a receptor inside the cell and not on the cell surface, is not the most novel aspect. And while my statement that nitric oxide was the first molecule that had an intracellular receptor for adjacent cell-to-cell-communication, is accurate, it might be misleading. There is a whole class of signaling molecules, steroid hormones, (as well as thyroid hormone) which have intracellular receptors. But steroid hormones don't communicate between adjacent cells, they communicate between cells in one part of the body to cells elsewhere. Thus they are released into the blood stream to travel from one place to the next. In our house example, it would be like the kid leaving the talking house, getting into their car, and driving on the road to a house in another part of town. (And the car is actually a good analogy, because most of the steroid hormones are bound to a carrier protein in the blood and only a few are actually walking, that is free, in the blood.) So the novelty is the combination of **adjacent** cells communicating **combined** with the intracellular receptor in the listening cell.

The other novel aspect of nitric oxide is how it is produced. For conventional cell-to-cell communication, the kids are stored inside the cell. I think it is best to change the analogy at this point. But first, it is worthwhile telling you some names for the different terms used to describe the different types of communication. We mentioned steroid hormones above. Hormones are defined as signaling molecules released by one cell into the blood stream that then travel to a distant site and bind to a receptor (either inside or outside) of a different cell. Hormones are released by endocrine cells.[1]

---

[1] Endo-meaning inside, so these secretions stay inside. The opposite type of "crine" cell is exocrine; these cells release their contents "outside" of the body. Outside is used in a topological sense, just like the air in the hole of the donut is

For cells communicating to neighboring cells, the term used is paracrine; "para" is Latin for beside or next to. Conceptually, I would classify communication between 2 adjacent nerve cells as paracrine, but that communication is specialized and most scientists consider it special enough not to put it in the category of paracrine, even though it fits there conceptually.

Let's get back to the other novelty of nitric oxide. Until its discovery, all the paracrine (and neuronal) communication involved molecules that were made in advance and stored in the talking cell. When the talking cell needed to speak, these signaling molecules were released in packets, to diffuse across the space between the cells[1]. Nitric oxide doesn't last long so the cell cannot store it. Thus the cell makes the nitric oxide at the moment it decides to talk; it is a bit like the difference between food that is prepared and food that is made only after the order.

I'd like to share one of the experiments that gave the first hint that there was a molecule that was a signal from the endothelial cell to the smooth muscle cell. I need to explain those terms and this involves a bit of the structure of a blood vessel.

Smooth muscle cells don't actually touch the blood. There is a layer of endothelial cells that form the inner lining[2] of the blood vessel and that actually touch the blood. The smooth muscle cells wrap around the outside of this endothelial layer.

It had been known for some time that acetylcholine added to blood caused relaxation of smooth muscles (and you remember that if you relax the smooth muscles surrounding the blood vessel, the vessel dilates.) The assumption was that the acetylcholine bound to a

---

outside of the donut. All the stuff inside your digestive tract is considered outside of the body, so saliva, for example is an exocrine secretion. So is stomach acid secretion.

[1] The discovery of these packets is a terrific story of scientific inference; the packets were deduced before anyone actually identified that the talking cell had packets.

[2] Remember that "endo" means inside so endothelial would be the inside cells.

receptor on the smooth muscle cell and it controlled relaxation. However, if one removed the endothelial cells, acetylcholine usually caused a constriction. This was a puzzle. Some scientists thought that maybe acetylcholine also bound to a receptor on the endothelial cells and the endothelial cells released a factor, X, that then bound to a receptor on (or in) the smooth muscle cell and factor X's effect trumped any direct effect of acetylcholine binding to its smooth muscle cell receptor. A cute idea, but how could one test this?

Endothelial cells are right next to smooth muscle cells and we didn't know what factor X was. So the scientists did an experiment similar to one that was often used to determine if a hormone might exist. For hormones, e.g., insulin, one can manipulate the donor animal so that one expects its blood to have high levels of hormone. In the case of insulin, one can feed the donor animal sugar. This should lead to a rise in insulin. If you now take some of the donor animal's blood and give it to the recipient animal, you should be adding insulin. Insulin makes blood sugar fall, so the recipient animal should have a drop in sugar. This indeed happens. If the actual hormone molecule has not been identified, then one can separate the blood into different fractions and see which fraction has the effect of lowering blood sugar in the recipient animal. This is a bit like solving the problem that you think there is one disruptive person at school A, but you don't know who it is. This person becomes disruptive when exams are given. So you give an exam at the school A, take a sample of school A students (blood) and put them into school B. If school B now responds as if it has a disruptive student, you can try to identify the disruptive student by dividing the school A students into categories: all wearing the same brand of shoes. For each shoe brand, divide by the brand of pants. And so on.

To find out factor X, an experiment similar to the blood transfusion experiment between animals was tried. But since the blood vessels themselves were enough to see the effect, the experimental design was simpler. In one bath was a complete blood vessel with endothelial cells and smooth muscle cells. Acetylcholine was added. Presumably the endothelial cells secreted factor X. If you take some of this donor solution and add it to blood vessels that have had their endothelial cells removed, one would expect factor X to cause the recipient smooth muscle cells to relax. But they did NOT. Oops. So the experiment was repeated with various changes. It didn't work. At this point, most scientists decided this theory did not work, that is, that endothelial cells did not release a factor X. But a few scientists were convinced that endothelial cells did release a factor X. This is a very common dilemma in science: when do you decide the theory is wrong and it is time to move on and when do you decide to keep trying?

The only way to keep the theory and be consistent with the negative results was to say that factor X didn't last very long, by the time you took the donor solution and got it to the recipient solution, factor X was gone. The breakthrough came when one laboratory group decided to flow the solution over the donor blood vessel and then directly onto the recipient blood vessel. If the flow is fast enough, then some factor X should last long enough to arrive at the recipient vessel and relax it. This worked! And factor X was given the descriptive name, endothelial derived relaxing factor, EDRF, for short. Hooray. But now came the chemical nightmare. How can one purify this compound since it only seems to last 50 milliseconds? It took a lot of work, but finally, nitric oxide was determined to be EDRF.

There are now thought to be several molecules that communicate between adjacent cells that are like nitric oxide in that they are gas molecules and they are made on demand, for example carbon monoxide, discussed a bit in the chapter, From Anemia to Vampires. When the molecules have a short life time, like nitric oxide, they are hard to identify.

Physiology

In summary, in this chapter we have discussed
- the effect of blood volume on blood pressure
- the effect of blood pressure and dilation on blood flow
- how actin and myosin filaments slide
- that enzyme activity can be regulated by adding a phosphate; kinases are enzymes that transfer phosphates
- that calcium activates calmodulin and the calcium/calmodulin complex activates MyosinLightChain-Kinase
- that MyosinLightChain-Kinase transfers a phosphate to MyosinLightChain which activates it
- that nitric oxide causes an increase in cGMP
- increases in cGMP promote smooth muscle relaxation
- Viagra™ inhibits cGMP breakdown
- ATP is important as both the source for phosphate which is a regulatory switch and as an energy source
- one mechanism of drug-drug interactions occurs when two drugs affect the same process, as both nitroglycerin and Viagra™ affect smooth muscle relaxation
- another mechanism of drug-drug interactions occurs when two drugs are broken down by the same enzyme and this kind of multitasking slows down each process
- the discovery on nitric oxide
- nitric oxide is made on demand, not stored

Physiology

In A Heart Beat

A friend of a friend of mine, John, had his wallet picked. John immediately turned and started to chase the pickpocket. For the first 3 blocks, the pickpocket kept increasing his lead from 5 yards to 10 yards to 15 yards. But John kept after him. John started chanting, "I'm going to get you, I'm going to get you,...." After a mile, the pickpocketer couldn't believe he was still being chased. John finally caught him at 5 miles; John wasn't even winded because he was a marathon runner.

What about John's heart allows him to run so much longer than most people? Most of us pump a bit more than 1 gallon of blood per minute (5 liters per minute). The amount of blood we pump per minute is called our cardiac output. When we exercise, our cardiac output increases about 4 fold to over 4 gallons per minute; if you have done some training, you can increase your output about 5 ½ fold to about 6 gallons per minute. But John and other elite endurance athletes can increase their cardiac output about 7 fold. Interesting, trout and pigeons increase their output 8 to 10 fold when they are swimming or flying rapidly.

This chapter will discuss how the body changes cardiac output, heart rate, and stroke volume. We will discuss why our heart beats fast when we are excited or have a lot of caffeine. We will talk about the pressures and volumes during a cardiac cycle. This will allow us to understand the short term responses to high blood pressure and how elite endurance athletes' heart function differs from ours.

In this chapter we will cover:
- What influences cardiac output?
- What regulates heart rate?
- What else besides heart rate influences cardiac output?
- How do pressure and volume change in the cardiac cycle?
- How do Pressure Volume Curves work?
- Is the elite athletes' and trout's ability to increase stroke volume due to their hearts having better contraction strength?
- What is the mechanism for elite endurance athletes' and trout's ability to increase stroke volume?

What influences cardiac output?

Cardiac output is the amount of blood the heart pumps per minute. It should be obvious that this is the amount pumped per beat times the number of beats per minute, just as your weekly income is the amount earned per hour times the number of hours worked per week. The amount pumped per beat is called the stroke volume. My analogy with weekly income does not include a complication that occurs with the heart. The complication occurs because the faster the heart beats, the less time there is between beats. (duh!) Less time between beats means there is less time to fill the heart. There are conditions where an increase in heart rate actually leads to a decrease in cardiac output, because stroke volume and heart rate are not independent variables, whereas the number of hours you work per week and your hourly pay are independent (except for overtime).

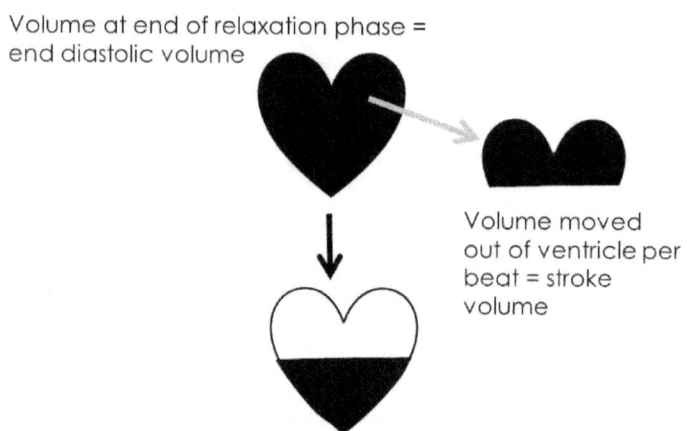

Volume at end of relaxation phase = end diastolic volume

Volume moved out of ventricle per beat = stroke volume

Volume at end of contraction phase = end systolic volume

> Figure 1. The volume of the heart at the end of the relaxation phase (that is, just before contraction) is the end diastolic volume. As the ventricle contracts, about ½ of the volume is ejected into the blood stream. This amount is the stroke volume. The amount of blood remaining in the heart at the end of the contraction phase is the end systolic volume. Another way to phrase this is that the stroke volume is the difference between the end diastolic volume and the end systolic volume.

Physiology

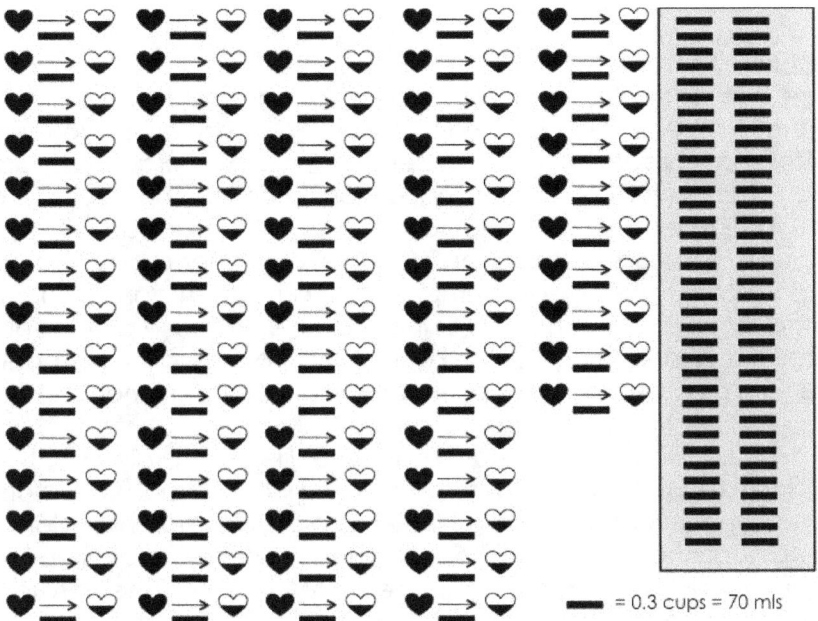

Figure 2 illustrating the calculation of cardiac output. The amount of blood pumped per minute is the number of heart beats per minute (70 bpm) times the amount of blood pumped per beat (stroke volume, 0.3 cups or 70 mls). Thus the cardiac output in this case is 70 x 0.3 = 21 cups per minute or 70 bpm x 70 mls = 4900 mls = 4.9 liters per minute.

An increase in cardiac output means there is an increase in blood delivered to the tissues; since blood carries oxygen, this means that tissues (especially muscles when exercising) get more oxygen. More oxygen delivery to tissues is a positive result of an increase in cardiac output. An increase in cardiac output, all else held constant, leads to an increase in blood pressure. Since cardiac output is related to heart rate, if stroke volume is constant, an increase in heart rate also increases blood pressure. Another factor that influences blood pressure is the diameter of the blood vessels, as we discussed in the chapter, Can Viagra(TM) kill you? There are many times when acute increases in blood pressure are appropriate physiological responses. However, long term elevated blood pressure leads to damage of blood vessels and the heart. Later in this chapter, we'll discuss one of the effects of blood pressure on heart.

What regulates heart rate?

There are two rhythmic activities that are critical to life. If you think about the First Aid ABCs, airway, breathing, circulation you will recognize that breathing and pulses (or heart contractions) are rhythmic. The rhythm for breathing is controlled by the brain; if a person is brain dead, they need to be kept on a ventilator if their organs are to be used for donation. But they don't need mechanical assistance to keep their heart pumping, because the heart rhythm is generated by cells within the heart. Indeed, one can take a heart out of the animal and so long as it is keep in an appropriate liquid and supplied with glucose and oxygen, it can beat for a long time.

As you well know, the heart rate can change; the brain has two systems that regulate the heart rate as well as hormonal factors. When you get excited reading a textbook, your adrenal glands release epinephrine[1] (also called adrenaline). Epinephrine acts on the heart to increase the heart rate. Another way to increase heart rate is to activate nerves that release norepinephrine (also called noradrenaline) onto the pacemaker cells.

If you have not had caffeine for several weeks, you might find that your heart rate increases the first time you have a large amount of caffeinated beverage. How does this happen? Caffeine probably has 2 effects; one effect is to block adenosine receptors. When adenosine receptors are activated, they slow the heart rate. When caffeine binds to the adenosine receptors, the receptors are not activated and the caffeine prevents the adenosine from binding. Thus caffeine blocks the signal that normally slows the heart rate, thus the heart rate increases.

The second way caffeine might increase the heart rate involves cyclic adenosine monophosphate (cAMP). This explanation will be reminiscent of the mechanism of Viagra™ and also relates to how epinephrine/norepinephrine increase heart rate.

---

[1] Adrenaline" comes from adrenal, so this compound was first isolated from the adrenal glands. Where are the adrenals located? "Ad-renal" that is, above the renal, or kidney. Epinephrine is the same compound as "epi" means "above" and "nephrin" is another word for kidney.

Physiology

The heart has two general kinds of receptors for epinephrine and norepinephrine. The two main receptor types have the very clever names, alpha and beta. When epinephrine or norepinephrine bind to the beta receptors in the heart, the receptors change shape. The fancy name for a shape change is a conformational change. When the receptors change shape to the activated form, that causes the enzyme that makes cAMP to become more active. You will recall from the Viagra™ chapter that the enzyme that made cGMP was called guanylate cyclase and similarly the enzyme that makes cAMP is called adenylate cyclase. An increase in cAMP alters proteins that regulate the rate of the pacemaker cells in such a way that the rate increases.

One of the effects of caffeine is to inhibit the enzyme that breaks down cAMP. Both enzymes that break down cAMP and cGMP are called phosphodiesterases. Rather than call them alpha and beta phosphodiesterases, the different types have different numbers. For example, Viagra™ inhibits primarily PDE5 at therapeutic concentrations. By inhibiting a phosphodiesterase that breaks down cAMP, caffeine leads to an increase in cAMP. In contrast epinephrine increases cAMP by activating adenylate cyclase. Thus both caffeine and epinephrine leads to an increase in cAMP. It is this increase in cAMP that leads to an increase in heart rate.

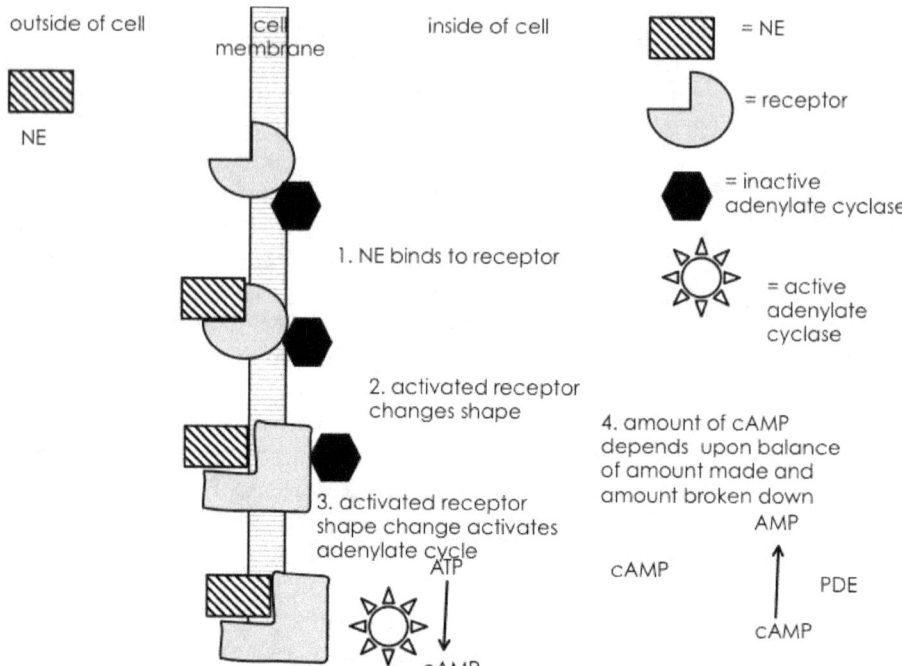

Figure 3 summarizing the actions of epinephrine, norepinephrine (NE), and caffeine on heart muscle. Blood epinephrine or nerve released norepinephrine bind to the adrenergic receptor in the cell membrane. When the receptor has bound epinephrine or norepinephrine, the receptor changes shape. This receptor shape change indirectly activates the enzyme adenylate cyclase. Active adenylate cyclase converts AMP to cAMP. cAMP is broken down by phosphodiesterase (PDE). Caffeine inhibits the PDEs that break down cAMP. As far as I know, it has no effect on the PDEs that break down cGMP. (Adrenaline is a synonym for epinephrine, so adrenergic receptors bind adrenaline = epinephrine and noradrenaline=norepinephrine. Few people use the term epinephrinergic.)

If only epinephrine altered heart rate, that would be like having a car with only a gas pedal. You can regulate the rate, but not as well as with a car that has both a gas pedal and a brake pedal. The brake pedals for heart rate are the nerves that release acetylcholine onto the heart.

Physiology

The acetylcholine receptors on the heart are different from the acetylcholine receptors on skeletal muscle cells. Remember from the chapter, Blood, Sweat and Tears, that the skeletal muscle acetylcholine receptors, when activated, open an ion channel. They are called nicotinic because they bind nicotine better than the other receptors. The other type of acetylcholine receptors are called muscarinic because they bind muscarine better than nicotine, not because physiologist wanted a trick question since muscarinic receptors are NOT on skeletal muscle cells. There are several types of muscarinic receptors; rather than call them alpha and beta, like adrenergic receptors, different muscarinic receptors are given numbers, M1, M2, M3, M4, and M5.

The muscarinic acetylcholine receptor (M2) on heart cells, when activated, decreases the activity of adenylate cyclase. Remember that adenylate cyclase converts AMP to cAMP. So activation of the muscarinic acetylcholine receptor (M2) will lead to a decrease in cAMP and cause the heart rate to decrease.

I want to emphasize the regulation of the heart rate involves two different balances. First there is the balance of how much acetylcholine versus norepinephrine is being released. Just like a race car's speed is a balance of how much the brake pedal vs. the gas pedal is being pressed. So there are 2 ways to speed up the heart; you can increase the amount of norepinephrine, or you can decrease the amount of acetylcholine, activating their respective receptors on the heart. This balance is summarized in Figure 4. The second balance is the amount of cAMP inside the cell: it is a balance of how much is made (the activity of adenylate cyclase) and how much is broken down (the activity of some phosphodiesterases).

Many therapies for treating the heart involve altering either or both of these balances. Can you predict what will happen to heart rate if you add an acetylcholinesterase inhibitor? Remember that acetylcholinesterase breaks down acetylcholine thus decreasing acetylcholine concentration, thus allowing acetylcholine to come off its receptor and allowing the receptor to change shape back to the inactive shape.

As we learned in the chapter, Blood, Sweat, and Tears, acetylcholinesterase is the enzyme that "stops" the acetylcholine signal. So is "nor-adren-al-in-ase" the enzyme that stops the norepinephrine signal? Unfortunately, no. A different mechanism is used to reduce the norepinephrine signal. The norepinephrine in the synapse decreases because there are proteins that transport norepinephrine from the synapse back into the cell. You may have heard that cocaine also increases the heart rate. How does this happen? One of the effects of cocaine is to inhibit the protein that transports norepinephrine from the synapse back into the cell. So cocaine effectively increases the amount of norepinephrine available to activate the adrenergic receptors.

What else besides heart rate influences cardiac output?
As stated earlier, if all else is constant, an increase in heart rate leads to an increase in cardiac output. When we exercise and when pigeons fly, it is the increase in heart rate that accounts for most of our increase in cardiac output. But trout only increase heart rate a little bit; the heart rate change does not account for their increase in cardiac output. And our maximal heart rate is about the same as for John and other elite endurance runners, so there is something else that changes to account for their higher cardiac output.

Both trout and elite athletes have a larger stroke volume (amount of blood pumped per beat). To understand the influences on stroke volume, we need to review the steps in the pressure and volume changes in the cardiac cycle.

How do pressure and volume change in the cardiac cycle?
Many students have a hard time following the volume and pressure changes in the heart during a pumping cycle and end up memorizing what happens and the terms. I'd like you to understand the cycle and I think the difficulty some students have has several causes, including
- the odd words
- forgetting some basic principles
    - water and blood flows from high pressure to low pressure
    - water and blood only flows when there is a pathway
    - one-way valves only allow water and blood to flow one direction

Physiology

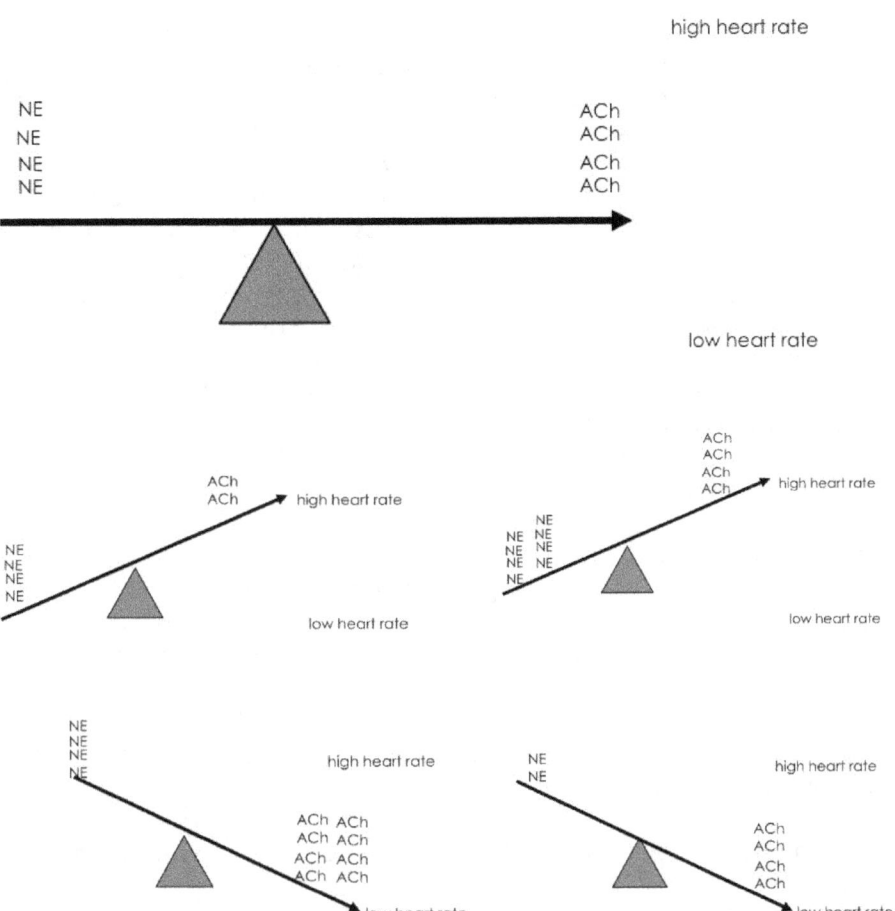

Figure 4. The heart rate is a balance of the amount of acetylcholine (ACh) and the amount of norepinephrine (NE) (or epinephrine) that binds to cardiac receptors. In the top panel, the effects of NE and ACh balance and so the heart rate is between high and low. The middle two panels illustrate two ways that the heart rate can increase. The obvious way is to increase the amount of NE (right) but another way is to decrease the amount of ACh (left). Just as there are 2 ways to speed up a race car since before the start, the driver has their feet on both the brake and the gas. Releasing the brake or increasing the gas can both speed up the car.

First a little anatomy:

> There is a one-way valve between the atria and the ventricle that only allows flow from atria to ventricle.
> The atria are the small chambers of the heart that receive blood from the systemic veins or from the pulmonary veins.
> There is a one-way valve between the left ventricle and the aorta that only allow flow from the ventricle to the aorta.
> The aorta is the major blood vessels that receives blood from the left ventricle and distributes it to the first set of arteries.
> I hate it that atria and aorta are such similar words and that both are "connected" to a ventricle. But you have dealt with this situation if you've ever read a story about Alice and Alex or Marc and Marcia.... Now you should be alert to pay close attention in the following to whether I use aorta or atria. (And I sure hope I didn't make any inadvertent mistakes.)

Since flow requires a pressure difference (gradient) and a pathway, when there is no flow, you know that either there is no pressure difference or there is no pathway (or neither is present).

To get around some of the difficulties with the standard presentation of the cardiac cycle, I will discuss the cardiac cycle for the pulmonary vein, the left atria, the left ventricle and the aorta and provide two analogous cycles, one for a Gatorade™ water bottle and one for a hallway with two rooms.

The Gatorade™ water bottle analogy only works for ventricular function, so let's start with that.
In this analogy,
- the Gatorade™ water bottle is like the ventricle
- the water in the water bottle is like blood in the ventricle
- the one-way valve at the top of the water bottle is like the one-way valve from ventricle to aorta (aortic valve)

For this section, we will have the valve from atria to ventricle closed at all times. The Gatorade™ water bottle does not have an analogy for this valve.

As you squeeze the Gatorade™ water bottle, water is forced out of the top. Just as when you squeeze the ventricle, blood is forced out of the ventricle into the aorta[1]. The Gatorade™ water bottle needs

Physiology

an addition to make the analogy more complete. When the ventricle squeezes, it doesn't squirt blood into the air (I hope!) but rather into the aorta. When the ventricle is not squeezing, there is still blood in the aorta. So let's imagine that the Gatorade™ water bottle has a straw from the valve standing up and that the straw is filled with water. You don't want the straw water to go into the Gatorade™ water bottle and fortunately the Gatorade™ water bottle and the ventricle both have one-way valves so that water (or blood) can only move from bottle, ventricle to straw, aorta.

With this modified Gatorade™ water bottle, as you first start to squeeze, no water squirts out of the bottle. It only starts to squirt when the pressure inside the bottle exceeds the pressure in the straw: at the point, the one-way valve opens and then the pressure gradient can cause movement. The same is true for the ventricle: as the ventricle first starts to contract, or squeeze, the pressure inside goes up, but the blood does not go anywhere. At the point where the pressure inside the ventricle exceeds aortic pressure, the valve opens and blood rushes out of the ventricle into the aorta.

The amount of water that the Gatorade™ water bottle squirts out per squeeze is the Gatorade™ water bottle stroke volume, just at the amount of blood the left ventricle squirts out per squeeze is the ventricular stroke volume. When you stop squeezing the Gatorade™ water bottle, you are at the end of systole. The volume that is left in the Gatorade™ water bottle or the ventricle when you stop squeezing is called the end systolic volume. (I wish we could just call it end squeeze volume or end contracting volume.)

The ventricle has the nifty property that I want to introduce here[1]. In a "standard" person at rest, the ventricle fills to about 120 mls (24

---

[1] The squeezing part of the cycle is called systole (conveniently, both start with S; it is also when the ventricles contract). The relaxing part of the cycle is called diastole, maybe this is the part of the cycle when the ventricles calm down, so we have d in calm down and in diastole.

[1] Frank was one of the first scientists to describe this nifty property and Starling did some follow up work. This phenomena is called the Frank Law, the Starling Law or the Frank-Starling Law. Politics and culture likely influence which term

teaspoons, roughly ½ cup). The ventricle then pumps out just over half, about 70 mls (14 teaspoons, 1/3 cup), leaving about 50 mls (10 teaspoons). The nifty property is that if you fill the ventricle more, say an additional 10 mls (2 teaspoons) to 130 mls (26 teaspoons), then heart pumps out an additional 10 mls (so 80 = 70+10). If the ventricles fill an additional 30 mls over normal (so 150 mls), then the heart pumps out an additional 30 mls (so 100 = 70+30). Or to put it another way, the more the heart fills, the more it pumps out. Of course, there is a limit to this. We'll come back to this nifty property later, but first I want to talk about the rest of the cycle and I think it gets too confusing with the Gatorade™ water bottle.

In this new analogy, a building is analogous to our body (we are going to ignore the lungs and the pulmonary circulation and the right side of the heart). The systemic circulation is analogous to the halls in the building. People are analogous to blood. When it is more crowded, that is analogous to higher pressure. The one-way valves are like one-way doors; you can only go through them one-way (duh!). The left side of the heart is like two connecting rooms that separate. Room A connects to room V. Room V connects to the aorta hallway. The pulmonary vein hallway connects to room A.
There is a one-way door between the pulmonary vein hallway and room A so you can only go from hall to room A.
There is a one-way door from room A to room V so you can only go from room A to room V.
There is a one-way door from room V to the aorta hallway so you can only go from room V to aorta hallway.
People moving is analogous to blood moving.

Let's start the cycle:
People move from vein hallway into room A because there are more people in vein hallway and room A is fairly empty.
People move on from room A to room V because room V is fairly empty.
Once in room V, people can't get out. Why not? Well, going to aorta hall doesn't work because there are more people in aorta hall so the pressure is the wrong direction AND the door is closed. And going

---

is used; Frank's article was in German and Starling's in English.

Physiology

back to room A isn't an option because the door is one-way the other way.

Now room A is made smaller which increases the crowding which pushes more people into room V. As room A gets smaller and smaller, it gets more and more crowded, forcing more and more people into room V.

Now there is a signal that causes room V to start to get smaller. This quickly makes room V more crowded than room A and this shuts the one-way door between room A and room V.

As room V continues to get smaller, room V gets more and more crowded. Eventually room V is more crowded than aorta hall. Thus the door is pushed open and people move into aorta hall.

Room V now stops getting smaller and begins to get larger. At the point where aorta hall is more crowded than room V, the one-way door closes and no more people move from room V to aorta hall.

Room V becomes less crowded because
- people were leaving room V into aorta hall
- people could not enter room V from room A because room V was more crowded
- room V was starting to get larger and return to its former size

Eventually room V is large enough that it is less crowded than room A, so the one-way door opens and people move from room A to room V, starting the cycle again.

Now we have the tools in place to understand what happens during exercise, in endurance trained athletes, in patients with hypertension, in patients with heart failure, and in trout.
The main parameter is cardiac output.

I think from the analogies, it is obvious what happens to blood pumped by the ventricle when the patient has high blood pressure (in their aorta and arteries). This is like having a higher volume of water in the straw placed in the Gatorade™ water bottle or more crowding of people in the aorta hallway. The one-way valve between the room and the hallway opens at a higher pressure than usual and thus less blood is pumped out, there is a smaller stroke volume. In order to maintain cardiac output, either the heart rate has to increase or the heart has to fill to a greater volume. If the heart fills to a greater volume that stretches it more and, over the long term, creates problems.

How do Pressure Volume Curves work?
Many scientists like to use a pressure volume plot to easily illustrate what happens. It is not so essential for what happens with high blood pressure, but it really helps understand what happens when the ventricle's stiffness changes and when the ventricle can contract stronger or more weakly. The hearts from trout and endurance trained athletes are less stiff than sedentary individuals. Furthermore, when we exercise, our heart contractility improves. These ideas can be difficult to explain with just words, but I hope will be clearer with the pressure volume plot.

First let's review the steps in the heart. We will start at the point where the ventricle is full and it starts to contract. As the ventricle starts to contract, very quickly the pressure in the ventricle is greater than the pressure in the atria, which closes the one-way valve, so even though there is now a pressure gradient, no flow goes back from ventricle to atria. The one-way valve between the ventricle and the aorta is also closed because the aortic blood pressure is about 80 mm Hg at this point and the ventricle blood pressure is much less. As the ventricle contracts, the pressure inside increases (just like the pressure inside a Gatorade™ water bottle) but because both valves are closed, no blood moves and so the volume in the ventricle is constant at this point.

At the point where the ventricle pressure is 81 mm Hg, the one-way valve opens and since there is a pressure gradient, blood starts to move into the aorta. But the ventricles are still squeezing and the pressure in the ventricles will increase further. About half of the blood in the ventricle goes into the blood stream. As the ventricle volume decreases, the ventricle pressure starts to fall; as it falls, blood is still moving from ventricle to aorta until the ventricle pressure reaches 99 mm Hg. At that point, aortic pressure is greater than ventricle pressure so the one-way valve closes. If there were a hole in the valve, blood would move from aorta to ventricle because of the pressure difference. But without a hole in the valve, the amount of blood in the ventricles at this point does not change because none is entering and none is leaving. The ventricle continues to relax. As the pressure drops from 99 mm Hg to 10 mm Hg, both one-way valves are closed and so while the pressure decreases, there can be no volume changes because there is no way for blood to enter or leave the ventricle. Eventually the ventricle pressure is less than the atrial pressure (especially when the atria start to contract). When this happens, the one-way valve between atria and ventricle opens and blood flows from atria to ventricle and starts to fill it.

Now, let's graph the pressure and volume in the ventricle as a function of time during the cardiac cycle. We start when the ventricle starts to contract. As we move from point w to point x the pressure increases rapidly, but the volume does not change (the line is essentially flat). The volume does not change because there is no flow into or out of the ventricle at this point. Remember, that, in general, no flow means either no gradient or no pathway or both. In the heart from point w to point x, both the one-way valves are closed, so there is no pathway for blood to flow, even though there are gradients. (Do you understand why both one-way valves are closed? The pressure in the ventricle is higher than in the atria, so that closes the first valve. The pressure in the aorta is higher than in the ventricle, so that closes the second valve.)

At point x, the pressure in the ventricle starts to exceed the pressure in the aorta. This opens the second one-way valve and because there is a pressure gradient, blood flows from ventricle to aorta. From x to y the graph indicates blood is flowing by the decrease in volume. Just after halfway between points x and y the ventricle starts to relax, so the pressure starts to fall. At point y, the pressure in the aorta exceeds the pressure in the ventricle, so the second one-way valve closes which means there can be no more flow and no volume change. From y to z, then, the volume is constant, but note that the volume is not zero-the heart did not pump out all the blood. From y to z, the ventricle continues to relax, so the pressure drops a lot. At point z, the pressure in the atria exceeds the pressure in the ventricle, so the first one-way valve opens which means there is a pathway and a gradient for blood to flow and the cycle starts again.

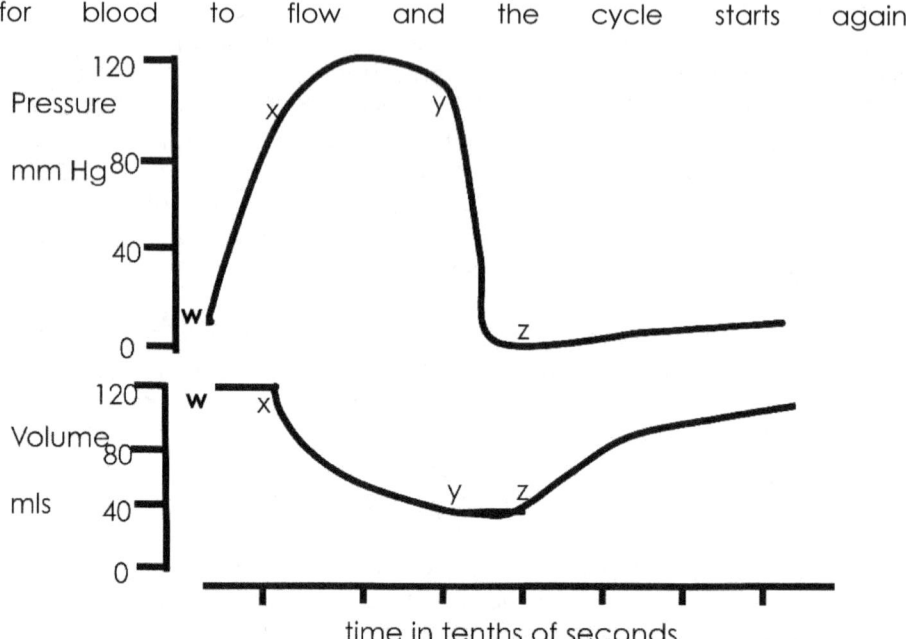

Figure 5 Changes of pressure (top) and volume (bottom) vs. time in the cardiac cycle, starting at w, which is when the ventricle is full of blood and just starting to contract. See text for details. Let's see how thinking about our main principles helps us interpret this plot. From w to x, the volume does not change. That means that either there is no pressure gradient or no pathway. But clearly there is a pressure gradient before we get to x. So it must be that between w and x both valves are closed. At x, the volume starts to change, which means there is both a pressure gradient and a pathway. Where does the blood flow when the ventricle is contracting? The blood flows from left ventricle to aorta, so it must be that the aortic valve opened. From y to z, there is also no volume change, so we can apply the same reasoning. At y, the open valve, the aortic valve, must close. At z flow begins, so at this point a valve must open. The pressure is too low to allow flow to the aorta, so at z, the valve that opens must be the atrial to ventricular valve and it allows flow from the atria to the ventricles. That also tells you that the atrial pressure must be higher than ventricular pressure at times after point z.

Physiology

Over the years, physiologists have found it useful to plot this in a different way and I hope by the end of this section you see the value of this new plot. The plot we just did had 2 lines, one for pressure vs. time and one for volume vs. time. The new plot combines this into 1 lined cycle by plotting pressure vs. volume. We can connect these lines (and arrows will help us keep track of the time).

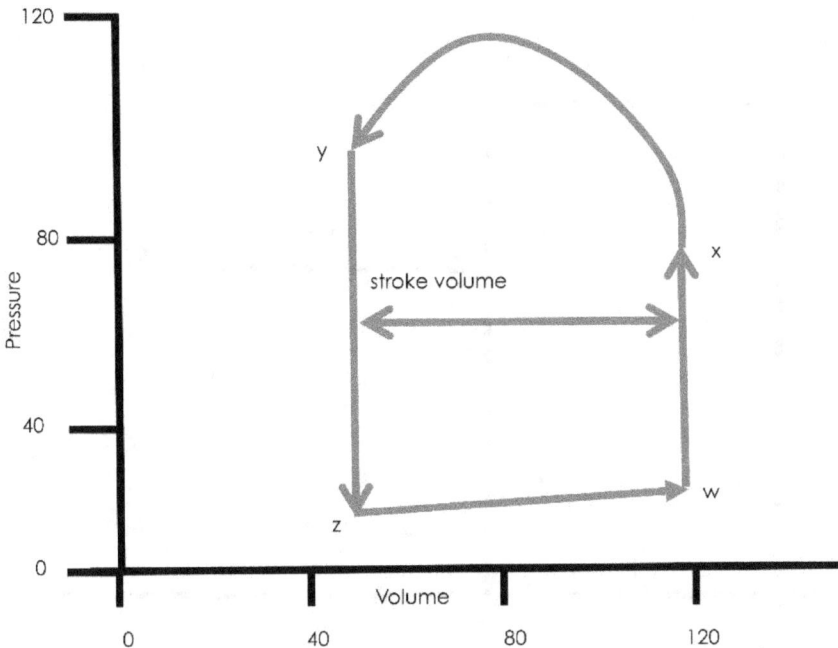

Figure 6. I think it is important to read graphs, just like you read an equation. For example when you see $E=mc^2$ you should say to yourself, eee equals em cee squared. It's harder to read a graph because one often doesn't know where to start. On this graph, we could start at w. So here's how I would read the graph. As we go from w to x, the volume is 120 mls and does not change. The pressure increases from 20 to 80 mm Hg. From x to y, the volume changes from 120 to 70 and the pressure from 80 to 120 and then back down to 100. From y to z, the volume stays fixed at 70 and the pressure drops from 100 to 10.

On this new plot, the distance from x to y is the amount of blood that was pumped out of the heart (i.e., the left ventricle in this case) is called the stroke volume.

Let's first look review what happens with high blood pressure on this plot. Stroke volume on this plot is the distance between the two parallel, vertical lines. If blood pressure increases (Figure 7), then the one-way valve from ventricle to aorta opens at a higher pressure and closes at that same higher pressure. Therefore, all else staying the same, the stroke volume decreases.

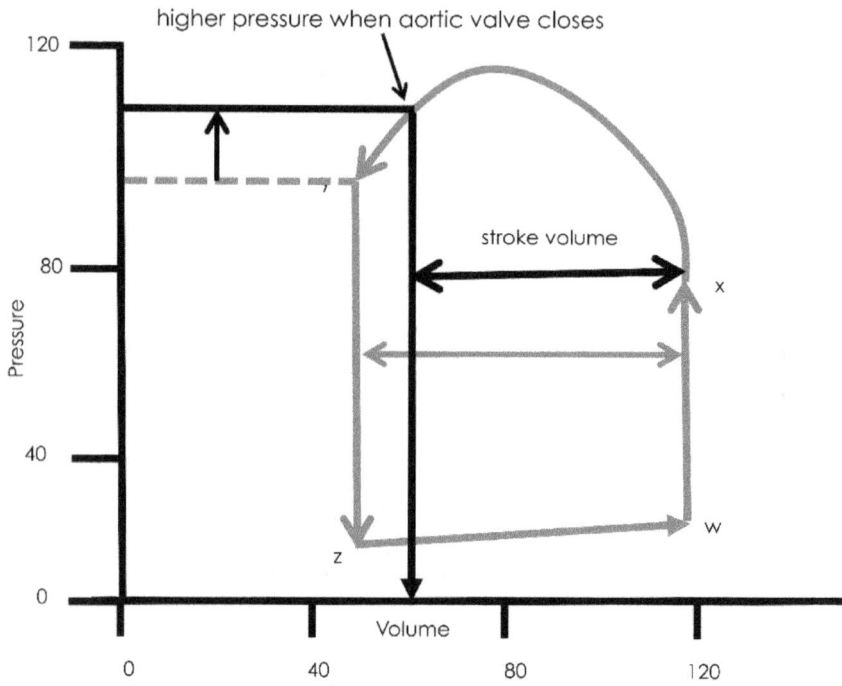

Figure 7.

One-way to test one's understanding of a concept or graph is to change it and see if it still makes sense. If you really understand the layout of campus, can you turn the map so that North is down and still make sense? In this particular case, I happen to like to put volume on the y axis because I am more used to examining changes in vertical distances. The plot turned on its side is shown below. One trick for doing this is to take the P vs. V plot and turn the paper over and then turn it clockwise by 90 and then redraw it.

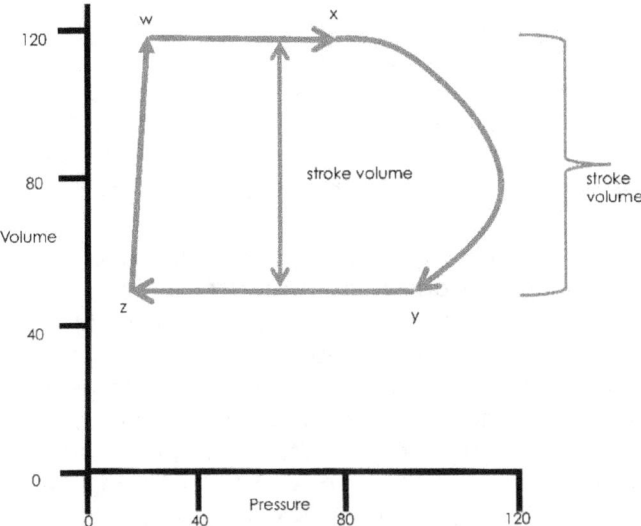

Figure 8.

We can also illustrate the nifty heart property (the Frank-Starling Law) by changing the amount of filling and then plotting the stroke volume vs. the filling volume. Over a range of filling volumes, the stroke volume increases, which is a mathematical synonym for the statement that the heart pumps out all the blood that enters it, as in Figure 9.

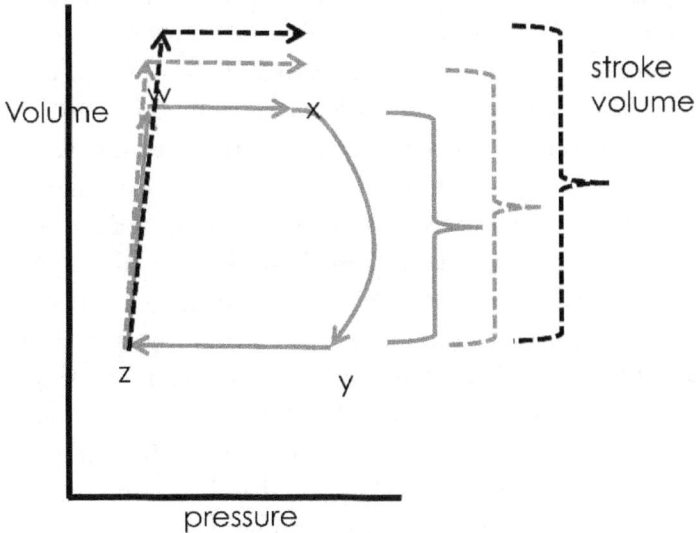

Figure 9.

Trout increase their cardiac output mostly by increasing stroke volume with little or no change in heart rate. Elite endurance athletes have about the same maximal heart rate as other people doing exercise so this does not account for their larger cardiac output. The major way that elite endurance athletes' hearts achieve higher cardiac output than our hearts is that their hearts can obtain larger stroke volumes.

Is the elite athletes' and trout's ability to increase stroke volume due to their hearts having better contraction strength?
First, let's discuss how better contraction strength influences stroke volume. Stroke volume depends upon the difference in the volume of the ventricle between the end of the relaxation stage (end diastolic volume) and the volume at the end of the contraction stage (end systolic volume). The contracting ability of the heart will influence the contracting phase, so it is the volume at the end of the contraction stage that is important. This volume is set by two parameters; we've already mentioned one, which is the pressure between the aorta and the ventricle that closes the one-way aortic valve. The other factor is how much pressure the muscle can generate at a particular volume.[1] On our pressure vs. volume curve, I have drawn the normal force-length curve in Figure 10 (see footnote 1). You will note that the line

---

[1] In skeletal muscles it should be obvious that you can hang up a strip of muscle vertically and measure the force it generates. You can then stretch the muscle to change its length and measure the new force. The new force can be measured when the muscle is not stimulated (cell calcium remains low) and this is the passive force or when the muscle is stimulated (cell calcium is high) and this is the active force. It is the active force that is important for the contracting phase. In recent years, scientists have been able to dissect single cardiac myocytes and do similar experiments on them. But historically, scientists studied isolated hearts to generate these data. Since the muscle cells wrap around the outside of the heart chamber, one indirect measure of how much the fibers are stretched is how much volume is in the heart, just like if the larger the volume of air in the balloon the more the balloon molecules are stretched. The force the heart generates creates the pressure that moves the blood, so pressure can be used as a surrogate for force. At rest, the left ventricular pressure reaches 120 mm Hg, which means that the ventricle has enough pressure to squirt blood about 5 feet straight up.

starts on the x axis at the point where the muscle has no force (generates no pressure). For the heart, this occurs at the volume of blood that would remain in the heart if the heart were maximally contracted, about 30 mls. For the moment, we will assume that the tension vs. length (or pressure vs. volume) curve is linear. The volume at the end of the contraction stage (end systolic volume) is given by the intersection of the contraction curve and the systolic pressure, the point where the aortic valve closes and the ventricle volume does not change.

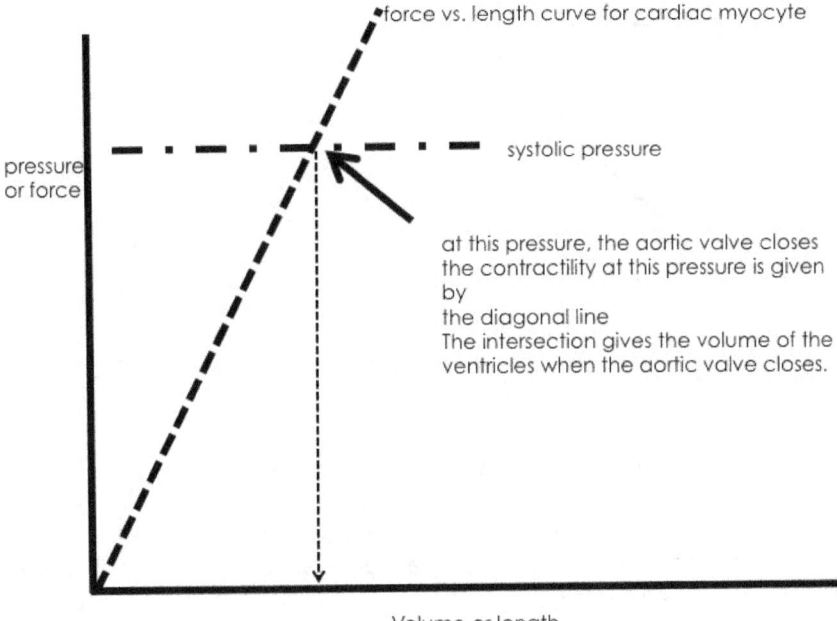

Figure 10.
When we say that the heart is "stronger" we mean that, for a given muscle fiber length (or ventricle volume), the heart can generate more force.

In Figure 11, is line 2 or line b the line for a stronger heart muscle?

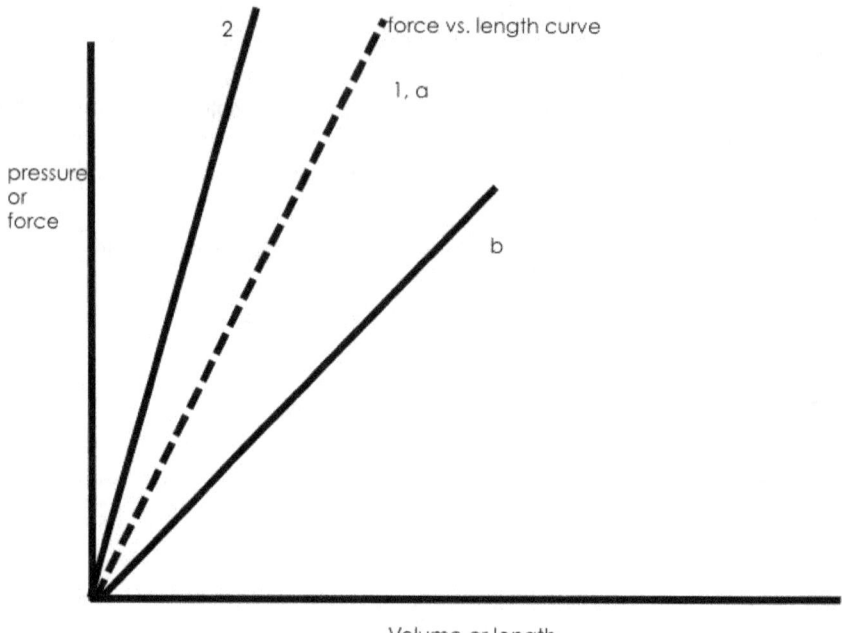

Figure 11. The steeper line is for the stronger heart muscle.

When we get excited and release epinephrine and stimulate the norepinephrine nerves that act on the heart; this leads to an increase in cell calcium. We already pointed out that activating the beta receptors leads to an increase in cAMP. The increase in cAMP not only speeds up the heart rate, it also leads to the cell reaching higher levels of calcium. The higher levels of calcium means more activity of the actin and myosin and more contraction.

If we get excited and increase cardiac muscle calcium we shift the tension vs. length (pressure vs. volume curve) from 1a to 2. If all else remains the same, what happens to end diastolic volume? As you can tell from the plot, if blood pressure stays the same, the volume remaining in the heart at the end of the contracting phase (end systolic volume) is less than before we got excited. If the volume at the end of the relaxing phase (end diastolic volume, volume just before contraction starts) stays the same, then we have pumped more blood in that contraction, that is, the stroke volume has increased.

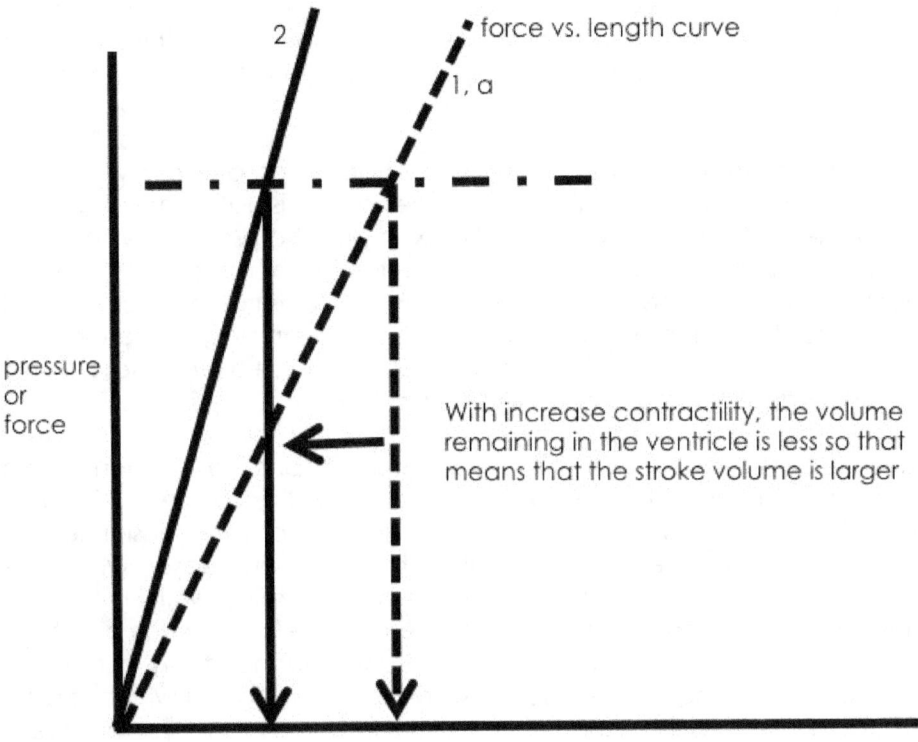

Figure 12.

Elite endurance athletes respond to going from rest to exercise by shifting the contraction curve from 1a to 2. The surprising thing is that we have about the same shift!! So it is not that elite endurance athletes have "stronger" hearts that accounts for their different cardiac performance.

What is the mechanism for elite endurance athletes' and trout's ability to increase stroke volume?

Stroke volume is just the difference in volume between when the ventricle is full and when it has finished ejecting blood. In other words, the difference between the volume at the end of the relaxation stage (end diastolic volume) and the volume at the end of the contraction phase (end systolic volume). We just said that the change in the volume at the end of the contraction phase (end systolic volume) is the same for us and the elite athletes. So the only other thing we can change is the volume at the end of the relaxation stage (end diastolic volume).

The volume at the end of the relaxation stage (end diastolic volume) is influenced by factors that are properties of the isolated heart and by properties external to the heart. An example of a property external to the heart is how much blood returns to the heart (enters the left ventricle). The property of the isolated heart (its intrinsic property) that influences the volume at the end of the relaxation stage (end diastolic volume) is the muscle fiber properties when relaxed, the passive tension-length (pressure-volume) relationship.

Elite athletes and trout have a different passive tension-length (pressure-volume) relationship. It turns out that the difference is not in the slope; remember the difference in our active tension-length (pressure-volume) relationship is that the slope increases when we release epinephrine or the nerves communicating with the heart release norepinephrine. For the passive curve, elite athletes and trout have hearts that are able to be stretched further than our hearts during the relaxing phase. Scientists are still working out which factors account for these changes. This research may have some great benefits as patients in heart failure often have reduced stroke volumes. It may be that if we can turn on the genes that elite athletes turn on to have the ability to stretch their hearts further, that we may find a way to turn on these genes, or activate the proteins that are these gene products, in patients with heart failure and so help their hearts obtain a larger stroke volume.

Physiology

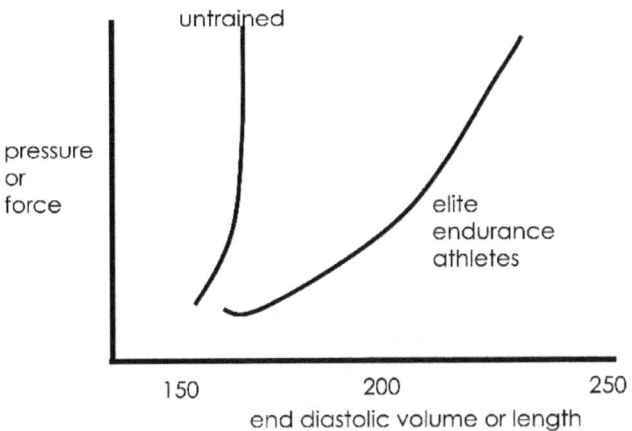

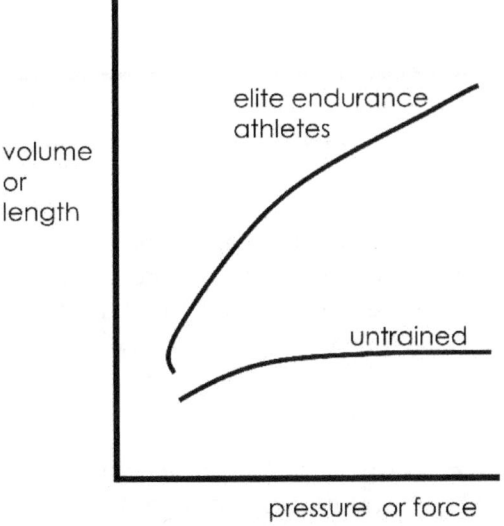

Figure 13. These two figures summarize one set of data on untrained humans and elite endurance athletes. The top graph is what is normally plotted in textbooks, pressure vs. volume. The bottom graph has the same data, but it plots it as volume vs. pressure to show that elite athletes are able to obtain a higher volume.

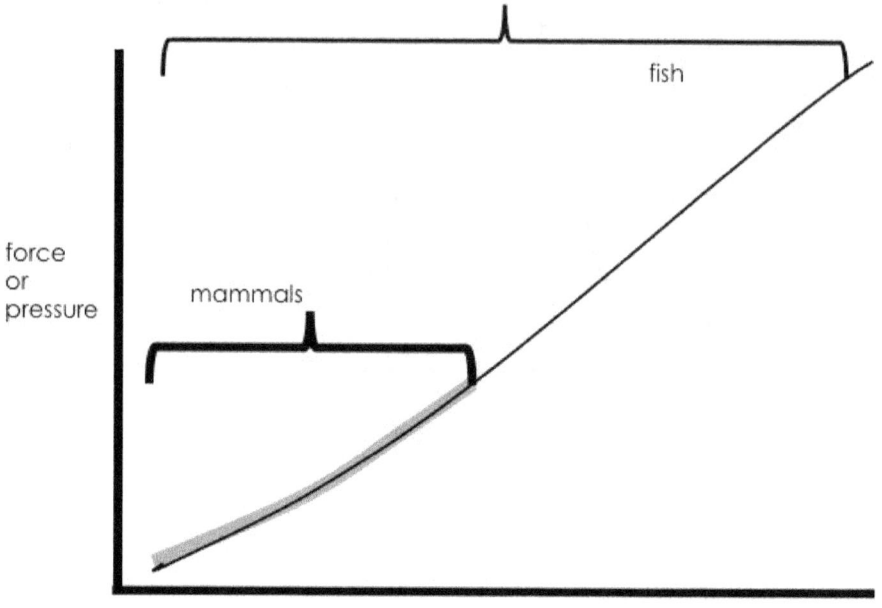

Figure 14.

These are some summary data for studies on single cardiac myocytes. The black line is data obtained from trout and the gray line are data obtained on different mammalian hearts. The point is that the slope of the line is similar for the different species but that it is possible to stretch the fish myocytes much further.

Major Concepts
The major concepts we have covered in this chapter are:
- The amount of blood the heart pumps per minute, the cardiac output, is the product of the amount pumped per beat (stroke volume) times the heart rate.
- An increase in cardiac output, all else held constant, leads to an increase in blood pressure
- Heart rate is a balance of the influence of acetylcholine and norepinephrine.
- Acetylcholine decreases cAMP in the heart and slows the heart rate

Physiology

- Norepinephrine increases cAMP in the heart and increases the heart rate as well as the force of contraction because of the increased calcium levels.
- The flow of blood is governed by the pressure gradient, always flowing from high pressure to low pressure.
- In order for flow to occur, there must be a pathway (open valve).
- The pressure volume curve for the heart summarizes the changes during the cardiac cycle.
- Stroke volume, the amount of blood pumped per heart beat, is the difference between the volume of blood in the heart at the end of the ventricular relaxation (end diastolic volume) and the volume of blood in the heart at the end of ventricular contraction (end systolic volume).
- The amount of blood in the heart at the end of the ventricular relaxation (end diastolic volume) depends upon the amount of blood returning from the venous system, the amount of time to fill, and the ability of the ventricle to stretch.
- The amount of blood in the heart at the end of the ventricular contraction (end systolic volume) depends upon the aortic blood pressure and the relation between cardiac muscle fiber length and force.
- Ventricular volume is a surrogate measure for the length of cardiac muscle fibers.
- Ventricular pressure is a surrogate measure for the amount of force cardiac fibers can generate.
- When we exercise, and pigeons fly, cardiac output increases primarily because of an increase in heart rate. Stroke volume also increases because of an increase in the force vs. length relationship, allowing the volume of blood remaining in the heart at the end of the contraction phase, the end systolic volume, to be less.
- When trout swim and elite endurance athletes run, there is an additional increase in stroke volume because the volume of blood remaining in the heart at the end of the relaxation phase, the end diastolic volume, is greater, because their hearts are able to stretch to greater lengths.

Physiology

Simon Says "Move"

In the 1800s and early 1900s, there were 6 day races. In one of the earlier races, the winner ran almost 500 miles which is like walking/running from Boston to Philadelphia and back. Or from St. Louis to KC and back. Or Portland to San Francisco. In 6 days! That's an average speed of more than 3 miles an hour including rest time. One estimate suggests that this requires about 9000 calories a day. On the Lewis and Clark expedition, each of the men ate about 9000 calories a day when they were going upstream on the Missouri River.

Some animals can outrun humans at short distances, for example, for a ½ mile run, the best sprinters run about 10 meters per second, but horses run 19 meters per second, lions 30 meters per second and cheetahs 35 meters per second. I've also heard that Grizzly bears are fast. In fact the advice is, if you are on a horse being chased by a Grizzly, to get off the horse, as the Grizzly can outrun the horse and will chase it and not you because it is a bigger meal. I think this is mostly a tall tale, but interesting. But once you talk about running for hours, or traveling for longer distances, humans are pretty close to the top.

Some dogs are thought to travel about 6 miles per day and hyenas about 12. But humans may well have traveled more per day before cars and bikes and even horses. It has been suggested that the top two animals for long term endurance are the Husky dog and humans. In the Iditarod races, Huskies can easily go 1000 miles in 8 days. The ability to sweat gives humans an advantage in a hot environment so we can probably go further than a Husky in warm climates.

There are a number of programs encouraging people to exercise more. Many exercisers take various supplements. What is the theory behind the supplements? Do they actually work? What exercise program is best? What about exercise improves health? There are several opinions on the answers for all these questions. In this chapter we will examine the basic science foundation for some of these claims as well as examine some of the population studies that examine the actual outcome in patients.

The sections in this chapter are:
- Limitations and strengths of different types of studies
- Basic science for exercise
- Supplement 1: Creatine
- Supplement 2: Carb loading
- Supplement 3: HMB
- Supplement 4: Whey protein hydrolysates and carbohydrates, for example, maltodextrin
- Response to exercise, oxygen, and Antarctic fish
- Types of exercise and physical activity

Limitations and strengths of different types of studies

It is important to remember the different types of limitations and strengths of different types of studies. Thus studies on molecules, cells, and tissues tend to be more reproducible, often many different labs obtain similar results. In contrast, studies on animals and humans are more variable, often different labs get different results.[1] The lab experiments that provide reproducible results are very appealing. However, extrapolating from the behavior of an isolated muscle cell being fed creatine in a lab dish to what happens to a human taking a creatine supplement is exactly that: an extrapolation. Such an extrapolation requires making assumptions that may not be justified or supported by new data.

Basic science for exercise

We will now go through some of the basic science background of muscles and their energy. Then we will discuss creatine, carb loading, HMB, and whey hydrolysates. Depending upon your interest and learning style, you might prefer to read about the supplements first and then come back and read this basic science overview. Then we will examine some of the broader issues about what type of exercise is good and what type of inactivity is bad.

Muscle contraction uses the energy released by breaking down ATP to ADP[2]. In a very active muscle, the ATP can be burned up in less than

---

[1] Two common reasons for these differences are the genetic backgrounds of the animals or people or the different environmental history before the experiment started.

[2] Remember **ATP** stands for adenosine triphosphate and **ADP** for adenosine diphosphate-so the only difference is 1 phosphate.

10 seconds. A short-term supply of energy in muscle is stored as creatine phosphate[1]. This lasts about another 10 seconds. All the other sources involve converting glucose to ATP. In the absence of oxygen, glucose to ATP only produces 2 ATP. The product of the reaction is lactate (or lactic acid). This is not a very efficient process but it does allow your muscles to do some work. In the presence of oxygen, glucose to ATP produces about 10 times more ATP! Muscles have saved some glucose energy by having stored excess glucose as glycogen. Thus they can break down the glycogen to glucose and make lots of ATP. Of course, glucose also arrives in the blood[2]. During exercise, this glucose is transported into the cell and burned. As one uses up the glucose in the blood, the liver makes more. The liver breaks down its own glycogen. In addition it makes glucose out of other raw materials. Another important source for energy for the muscle is fatty acids. Fatty acids are produced by breaking down the fat stored in fat cells.

$$\text{Creatine \& ATP} \longleftrightarrow \text{Creatine-P \& ADP}$$

$$\text{ATP} = \text{ADP} + \text{Pi} = \text{Adenosine-P-P-P}$$

Figure 1. Creatine-phosphate can generate ATP when the muscle is active; at rest, the reaction runs in the other direction, to increase creatine phosphate in preparation for the next bout of exercise.

---

[1] **Creatine** is a positively charged compound related to the amino acid arginine. Creatine phosphate is made from ATP by creatine kinase. Remember that a kinase transfers a phosphate from ATP to an acceptor, in this case creatine. The reaction from ATP to creatine phosphate is close to equilibrium under resting muscle conditions; the enzyme speeds up the reaction in both directions so at rest, one can build up creatine phosphate and when exercising one can break it down to form ATP.

[2] How much glucose is there in blood at any given time? The standard body has 5 liters of blood containing 5 mMoles/liter of glucose, which is 25 mmoles of blood glucose per body. The molecular weight of glucose is 180 daltons (grams/mole), so 25 mmoles is 4.5 grams of glucose at 4 calories per gram gives us about 20 calories-not a whole lot.

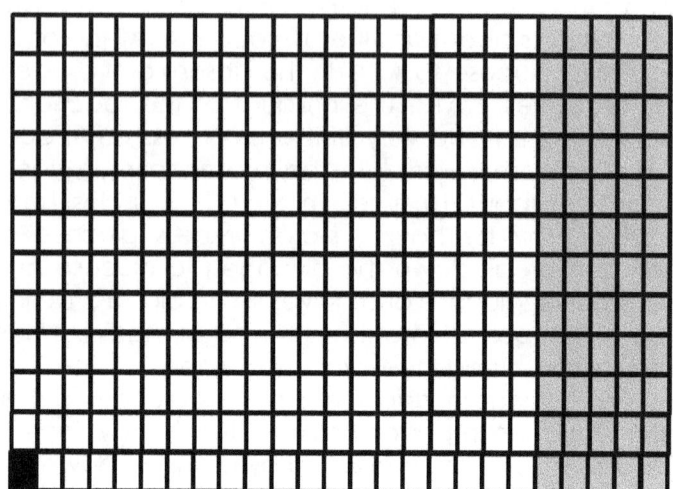

Figure 2. How much of one's weight is stored as fat and glycogen. Each square represents ½ pound for a 150 pound person. About 30 pounds is stored as fat, the gray shaded boxes. About ½ pound is stored as glycogen, the lone black filled box.

Figure 3. How many calories stored as fat vs. glycogen. The open boxes represent the amount of calories stored as triglycerides, about 120,000 calories; each of the 300 squares represents 400 calories. Only about 1000 calories, (the 2.5 black filled boxes) are stored as glycogen.

Muscle cells can also get larger. This occurs by making more proteins. Proteins are built out of amino acids. Both exercise and an increase in insulin led to an increase in glucose transporters in the muscle cell membrane. Insulin also increases the number of amino acid transporters in the muscle cell membrane. This makes me wonder if the amino acid and glucose transporters are stored in the same vesicles.

In addition to insulin other hormones also regulate the amount of protein that is made and broken down in muscle. Two important hormones that regulate muscle mass are the steroids cortisol and testosterone; high cortisol tends to lead to muscle wasting and high testosterone tends to lead to increases in muscle mass. We will cover more of the effects of steroids in the chapter Steroids make the World go Round.

Supplement 1: Creatine
Why might creatine help? I think the original idea of taking creatine supplements arose from an understanding of muscle energetics. The energy to fuel muscle contraction comes from ATP. As ATP's energy is released, ATP is broken down to ADP. During intense exercise, say a sprint or weightlifting, the muscle can burn up a lot of ATP. It takes time to convert glucose to more ATP. In muscle cells, some of the energy from burning glucose is stored as creatine phosphate instead of ATP. So being able to use creatine phosphate gives the muscle cell a bit more energy for the short term.

It was natural to think that if one could increase the creatine concentration in muscle, then the creatine phosphate concentration might also increase and therefore one could have a bit more energy. So people started to use creatine supplements.

Studies do suggest that creatine supplementation can increase the creatine phosphate content of muscle a little bit, about 10%. Surprisingly, creatine supplementation seems to have an even bigger effect on some athletic performances. It is thought that many of these effects do not related to extra creatine phosphate in muscle but to creatine effects on other aspects of muscle cells.

You need to keep in mind that there is probably an inherent bias in science towards finding that drugs, supplements, and treatments have an effect. Suppose you are in a lab and you are testing whether drug X has effect Y. What result would interest you most? What result would interest your boss? People get excited with new effects. If you do a few experiments and see no effect, you might decide to look at another drug and not complete the study on drug X. But if you found that drug Z had an effect, you would be very likely to continue the work and publish the result. Once someone has published that drug Z has an effect, if a new lab sees no effect, even if they complete the study, it may be hard to publish. The reviewers are likely to have a bias towards the first result. (A bit like many fans and sports poll voters are biased by the pre-season polls.)

In squid and some other invertebrates, their muscles store energy as arginine phosphate instead of creatine phosphate. The chemical structures are similar. I think creatine phosphate is better because of the equilibrium for the reaction. Since both creatine and arginine are similar cations and both are phosphorylated, I wonder if the 2 enzymes are evolutionarily related and whether one amino acid results in the switch for an enzyme that phosphorylates arginine to one that phosphorylates creatine or did it take several steps.

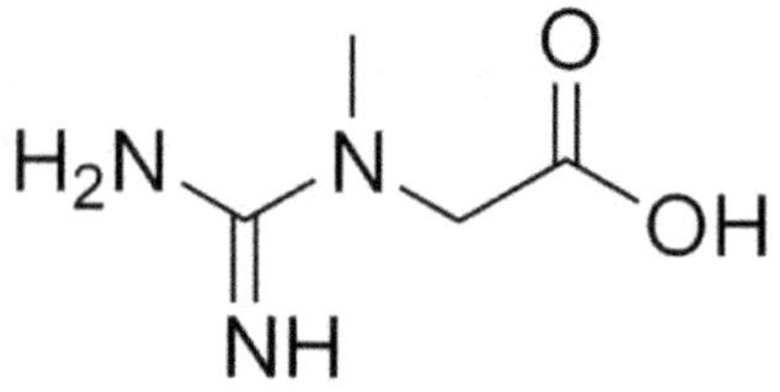

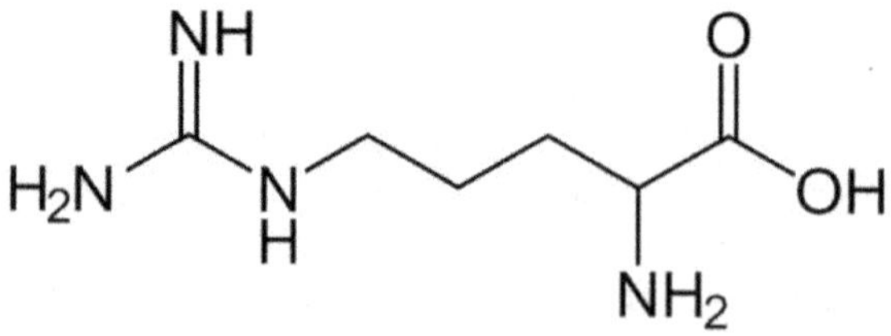

Figure 4. The structures of creatine and arginine, from Wikipedia.
http://en.wikipedia.org/wiki/File:Creatine_neutral.png
http://en.wikipedia.org/wiki/File:Arginin_-_Arginine.svg

Physiology

The current evidence supports the idea that creatine supplements can provide a modest boost in performance; it is important that these studies include a placebo[1] control, because obviously, one's motivation can also influence performance. There has been some concern about whether the high amounts used in creatine supplements could be harmful, but so far the human data tends to suggest no problems. The animal data are mixed. Here is a concluding paragraph from a paper commenting on another paper that found that giving creatine damaged the liver in mice but not in rats.

> "These findings are important because they indicate that different species may respond uniquely to creatine supplementation. Therefore, care must be taken to consider species-specific responses when 1) designing studies to assess the potential therapeutic role of creatine monohydrate supplementation on various human medical disorders; 2) assessing the pharmacological effects of various medications that may involve creatine metabolism; and 3) attempting to extrapolate results from long-term creatine feeding studies in animals to humans. Additionally, because neither the mice nor rats experienced renal or other tissue pathological changes after long-term creatine supplementation, these findings provide additional evidence that creatine supplementation does not appear to adversely affect renal function or health outcomes supporting recent clinical trials in humans. From Kreider RB. Species-specific responses to creatine supplementation. Am J Physiol Regul Integr Comp Physiol. 2003 Oct;285(4):R725-6.

---

[1] A **placebo** pill is when you give someone a sugar pill instead of a drug in order to control for any psychological response a person has when they are given something. So likewise a placebo treatment would be to give the control group athletes a powder that has no effect. But you want to be sure the athlete, their coach, and the person measuring the performance don't know which group has the active powders and which the inactive, because if they did know, they might unconsciously work more with the athlete (coach) or be biased in their observations (the researcher).

Personally, I think this paragraph shows an inconsistency. In the first half, they talk about different species effects of creatine, yet in the second half, they extrapolate from the safety of the studies in rats and mice to suggest that supplements will not affect renal function in humans which is precisely, to my mind, what they objected to in the first half.

Supplement 2: Carb loading
What is the point of carbohydrate loading? This is also based on some basic science. In the chapter How sweet it is, we will talk about storing the energy from glucose as glycogen in liver and skeletal muscles. Having glycogen available right in the muscle means it can provide an extra boost of energy without needing to increase blood flow or mobilizing glucose from the liver. Which came first, expression of glycogen enzymes in liver or muscle? And which enzymes came first, those that make glycogen or those that break it down? I find these are interesting questions and have not yet seen if scientists have found the answer. In order to make glycogen, one wants lots of glucose and insulin, so a high carbohydrate diet can do this. Such a diet maxes out the stores of glycogen, since glycogen stores, unlike fat stores, have a maximum.

Carbs probably have more effect in men than women because women seem not to use glycogen so much according to a review article. I was not able to track down the original data. But I did come across a study looking at metabolism in males versus females[1]. The data include the sleeping metabolic rate normalized to weight (but removing the fat component of the weight). If we look at the effect of the phase of the menstrual cycle on metabolism, they observed 88 +/- 5 for the pre-ovulation phase and 95 +/- 7 for the post-ovulation phase. The ranges overlap. (88 + 5 = 93 vs. 95-7 = 88[2]) so they don't appear different. But the researchers very nicely show us what

---

[1] Meijer GA, Westerterp KR, Saris WH, ten Hoor F. Sleeping metabolic rate in relation to body composition and the menstrual cycle. Am J Clin Nutr. 1992 Mar;55(3):637-40. PubMed PMID: 1550036.

[2] Since this is only 1 standard deviation, it is not even different at the 66% level let alone the 95% level. For a discussion about statistical significance, see the last section of the chapter, Is there proof?

happened to each woman. I would say that 14 of the 16 had an increase, though I will admit that for 2 of those 14, the increase wasn't very much. That data would support an effect of the menstrual cycle on sleep metabolism. I would feel more confident of that interpretation if they had also measured the sleep metabolic rate in men at 2 different times, 2 weeks apart. The men don't have a cycle, but this would get at how variable the data is. In addition, this study was done on college students and I can imagine several cyclic changes in college student lives that might affect metabolism. Exam schedules might be monthly, so 14 days apart could reflect the difference of an exam the next day vs. 2 weeks away from an exam. I can imagine some dining halls menus repeat monthly.

Supplement 3: HMB
Hydroxy-methylbutyrate, HMB is a compound related to the amino acid leucine. Leucine concentration is thought to regulate how fast muscle cells make proteins. So the idea is that HMB would lead to more leucine and one would get more muscle mass.

Many studies have examined whether HMB is effective. Not all the studies showed an effect. An analysis of many of these studies (called a meta-analysis) was recently completed. The investigators decided on the criteria for which studies where done "carefully enough" to be included. They then sorted through all the studies they could find and only used data from the studies that met their criteria for "high quality". Their conclusion was that HMB was effective. However, other scientists wrote in to comment on the study and I found their comments interesting.

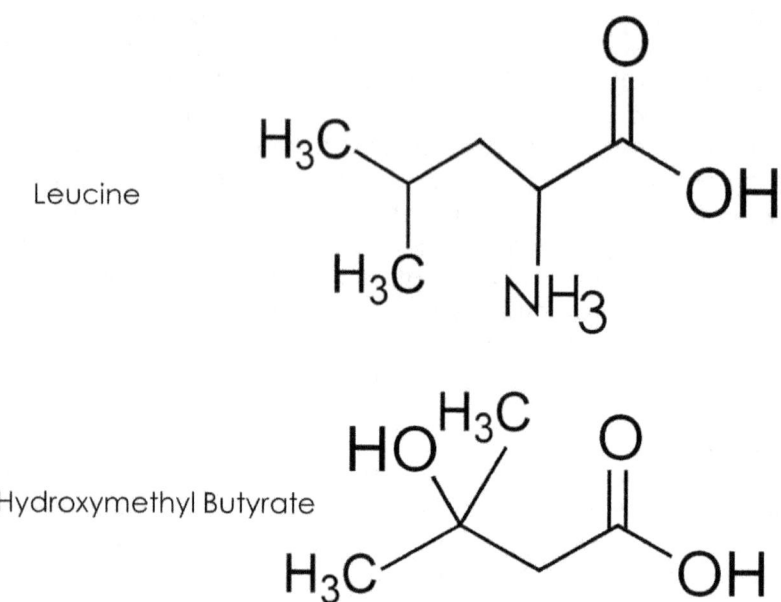

Figure 5. The structures of leucine and hydroxy-methylbutyrate, Modified from Wikipedia.
http://en.wikipedia.org/wiki/File:Hydroxymethylbutyric_acid.png

First the commenting scientists praised the analysis scientists for their high quality criteria, "Only randomized, placebo-controlled clinical studies published in peer-reviewed journals were selected." So far, so good. But the commenting scientists then observed that of the 9 studies for HMB were only from 3 independent groups. That is, some of the studies had the overlapping authors or were from the same institution. The commenting authors then make the following comment, which I personally fully agree with: "Demonstrations of efficacy produced by only a small number of separate sources are intuitively less persuasive than those from a large number of independent laboratories. Among reasons for this, we believe that investigations led by connected authors or institutions are likely to share similar approaches to a given question and use a smaller variety of methods than truly independent ones." [1]

---

[1] From Décombaz J, Bury A, Hager C. HMB meta-analysis and the clustering of data sources. J Appl Physiol. 2003 Nov;95(5):2180-2; author reply 2182.

Physiology

Supplement 4: Whey protein hydrolysates and carbohydrates, for example, maltodextrin

During exercise, you want high blood glucose, so having high glucagon and cortisol makes sense. Insulin is actually low during exercise. From our readings in the chapter How Sweet It Is, you might think this is strange as insulin led to an increase in glucose transport in the muscle. But exercise also results in the movement of glucose transporter vesicles to the cell membrane, so that muscle glucose transport is high during exercise.

For reasons that I don't yet understand, immediately after exercise is the optimal time for the muscle to increase its mass. One of the effects of insulin is to promote muscle growth. One of the ways is to increase amino acid and glucose transport into muscles. So increasing insulin just after exercise would be a help. Apparently some athletics inject themselves with insulin. This is dangerous, because with too much insulin, blood glucose can drop too far and one can become unconscious or die. As an alternative, many athletes eat carbohydrates. As the sugar from the carbs is absorbed, the blood glucose levels go up, and the pancreatic beta cell releases insulin. The feedback systems in the body ensure that the insulin does not get too high and the glucose does not fall too low.

The muscle cell also needs protein to grow. Some athletes eat whey protein hydrolysates. Hydrosylates means the protein has been pretreated so that it is partly broken down. The amino acid content of the protein and its hydrolysate are the same. But it is thought that the hydrolysate increases blood levels of amino acids faster because the stomach and intestine does not have to break down the full protein[1]. Which protein is best is a matter of debate. Personally, I would have thought you want the same amino acid mix as found in muscle, so predigested meat (muscle) would make the most sense.

PubMed PMID: 14555678.

[1] I am not convinced that this is true and I plan to check the literature some more and maybe even do some experiments on this.

The timing, and decision of what amino acids are needed, reminds me of advice that vegetarians used to be given. They were advised to be sure that every meal was complete in the amount of essential amino acids. So commonly the advice was to eat rice and beans as their proteins were complementary. I think this advice is no longer given. But it does raise to me the question of what happens inside a cell when it runs out of an amino acid.

When the cell is making a protein, it is assembling amino acids in order. Just like when you are following a recipe, you add ingredients in order. What happens when you get to an ingredient you don't have? Do you put the dish aside and finish it after your next shopping trip? Do you stop and go to the store to get it? In the same way, does the cell stop making the protein and have it just wait until the next meal? Does it go to get the amino acid from an amino acid store? You know we store excess glucose as glycogen. But I'm not aware of an amino acid store, but it is possible that the body "considers" some proteins more important than others and "raids" the unimportant ones to get amino acids for the important ones. Of course a good cook probably checks their pantry before they start making the recipe; I wonder if there is there some way that a cell checks its supply of amino acids before making a protein.

To summarize this section, ATP is the fuel used by your muscles to contract. There are several different sources to replenish the ATP and some are available quickly and used up quickly, while others last longer, but also take longer to be used. Perhaps an analogy might help with the overall response to exercise.

ATP is the fuel, lighter fluid, say, in a spray bottle or refillable Zippo lighter. As you use up the spray bottle, you can refill it from a jar, which is analogous to creatine phosphate being used by the muscle to regenerate ATP. You have the jar right next to you, so you can refill quickly, but the jar only has a limited capacity. The next two sources of energy to make ATP are glucose and glycogen. Glycogen can be broken down to make glucose maybe like a case of canisters. It takes a bit of time and effort to open the case, take out a canister, and open it up. Just as it takes time to break the glycogen back down to glucose. Finally, the muscle can rely on fatty acids; the fatty acids are stored as fat in fat cells, which is like having to go to the gas station to get fuel, gasoline.

Physiology

Response to exercise, oxygen, and Antarctic fish
One can divide exercise into 2 extremes. Not surprisingly the adaptations are different for each type. At one extreme is relatively low intensity, long duration exercise. If this is done repeatedly, the muscle responds by an increase in mitochondrial[1] density in the muscle fibers and an increase in capillary density. In this way, more oxygen can be delivered (capillaries) and the muscle can make ATP faster (mitochondria).

If you start to run, your body needs to get more oxygen. The response is to increase your respiratory rate; it can increase 20 fold. The amount of blood your heart pumps per minute also increases, but only about 7 fold. Most of the increase is due to an increase in heart rate, but the amount of blood per beat also increases a bit. This results in what is almost a paradox. When the heart rate increases, there is less time for the heart to fill before it pumps. So without compensation, as one increases heart rate, there would be less to pump per beat, not really a big help. During exercise the body does several things to increase return of blood to the heart so that the heart fills faster with blood between beats.

In contrast, when there is exercise of short duration and high intensity, like weight lifting, the response to doing it a few times a week is to increase muscle strength. In men, this means the muscle fibers get larger. In women, other adaptations can take place that also increase strength, but don't necessarily require larger muscle fibers. It should also be noted that during weight lifting there is a point where the pressure in the muscle exceeds the blood pressure. This means that the blood vessels in that working muscle are squeezed closed and for a short time there is no blood delivery[2].

---

[1] **Mitochondria** are the power houses of the cell. They take the energy from glucose and oxygen to produce ATP

[2] This is another example of our concept that movement requires a gradient and a pathway. Or conversely that lack of movement (of blood) in this case means that there is either no pathway (artery is closed) or no gradient (but blood pressure is high, so there is a gradient).

Many of the effects of exercise are stimulated by the need to deliver oxygen. In researching the topics of respiratory and exercise physiology, I came across the Antarctic ice fish. Some of these fish lack hemoglobin so their blood is not red. We need hemoglobin to transport oxygen, what do they do? Remember that oxygen has the odd property that it is more soluble in cold water than warm water. And these fish live in the Antarctic where the water temperature is below 0 C. Thus their blood has about 10% dissolved oxygen; this is less oxygen content than we have if you count what is on our hemoglobin but a lot more than there is in our plasma[1] at 37°C. The current fish have larger hearts, larger blood vessels, and many more capillaries. While no one has mentioned this that I can find, I find it striking that this is what happens to trained athletes: they have bigger hearts and more capillaries. It makes me think that part of the signal for the change relates to having low oxygen during exercise.

Types of exercise and physical activity
Finally, I want to conclude with some ideas about the type of exercise that is needed. I first need to point out that careful studies of the long-term health benefits of exercise are hard. Long term means 10 to 20 years or more. Everyone has to remain consistent in their diet and exercise for the length of the study; this includes those in the no exercise group as well as those in the exercise group. And once we have the data, can we apply it to the next generation? That is, studies done from 1970 to 1990, say on people your parents' age, can they apply to you? The genes should be similar, but the environment and diet choices are quite different. To avoid some of these problems, scientists have come up with surrogate markers for health, for example, low cholesterol, normal fasting glucose, etc. But whether these surrogates are really accurate is hard to determine.

Americans today are certainly not as active as Lewis and Clark or the 6 day racers mentioned in the start of the chapter. And I certainly agree that the advice to get 30 to 60 minutes of vigorous exercise at least 3 times a week is good. But a number of studies have prompted some researchers to point out that sitting for long periods of time is hazardous.

---

[1] Remember that plasma is the non-cell component of blood, see the chapter From Anemia to Vampires for more details.

Physiology

Part of the advice to not sit and to be at least lightly active is merely a numbers game. You have 112 or so waking hours per week, which is a lot more than the 3 hours per week of suggested vigorous exercise. I think people spent more time on housework and yard work in the 50's than is spent today. These activities, such as vacuuming, standing up to do dishes, cooking and stirring, mowing the lawn, raking leaves and washing cars took more energy and time in the 1950s than today. I also think there were many fewer riding lawn mowers in the 50s. And even power steering was rare. So we have become much less active in some ways.

I will now discuss some recent studies, but first want to make it clear the paradigm I'm following. A number of researchers (led by some at the University of Missouri) argue that the "normal" state for humans is one of constant activity. For example, think about living on a farm in Missouri before the Civil War. I'm guessing they probably burned about 5000 or more calories per day. Clothes were washed by hand. Wood was cut with an axe. Even recreation was active. Do you remember square dancing or country dancing in school? By the 1970s, we had shifted to a much less physically active society and hence the recommendation for vigorous activity for about ½ to 1 hour three times a week as a way to increase health. But now, researchers are finding that people sit a lot more than they did even 20 years ago and that going from light activity to sitting probably also increases one's unhealthiness.

Historically, I think the implicit attitude that going from inactivity to moderate activity to high activity had a monotonic and possibility linear relation with health, a bit like going from sea level to 1,000 feet to 2,000 feet. But now some scientists think that going from moderate activity to little activity is quantitatively different from going from high activity to moderate activity, more like going from sea level to -1000 feet is quantitatively different (one is under water) than going from 1000 feet to sea level (you still have air to breathe).
One set of data that supports this concern was done in England in the 1950's. The researchers presumed that bus conductors and bus drivers were of the same socioeconomic class and ate similar foods and had similar behaviors outside of work. But the conductor is walking around the double-decker busses all day, collecting money and giving tickets whereas the driver is sitting. It is reasonable to ask whether other

factors, besides sitting, might account for this difference. The researchers themselves thought about whether the bus drivers had more stress and that it was their stress, not their sitting that increased their risk. But in a survey of bus drivers and conductors, both groups thought that the bus conductors had the more stressful job; I guess more than a few bus riders can be pretty rude!

Many recent articles suggest that sitting, or being very sedentary, creates increased risks rather than standing and light work. I am so impressed by the data, that I have arranged to do all my computer work standing up and I try to sit for less than 30 minutes at a stretch, and less than a few hours per day. I tell you this so you know my bias.

Some studies, at least as reported in the popular press, suggest that sitting all day and exercising 3 times a week for 1 hour each, is LESS healthy than not sitting most of the day and not getting any "extra" exercise. That interpretation appeals to me, but it does not stop me from looking critically at the data.

Here is a quote from an article in the NY Times in 2010[1]:
> Regular workout sessions do not appear to fully undo the effects of prolonged sitting. "There seem to be different pathways" involved in the beneficial physiological effects of exercising and the deleterious impacts of sitting, says Tatiana Warren, a graduate student in exercise science at the University of South Carolina and the lead author of the study of men who sat too much. "One does not undo the other," she says.

Here is a quote from the original study article[2]:

---

[1] Second to last paragraph in: Phys Ed: The Men Who Stare at Screens By Gretchen Reynolds. July 14, 2010. The New York Times. http://well.blogs.nytimes.com/2010/07/14/phys-ed-the-men-who-stare-at-screens/?scp=1&sq=tatiana%20warren&st=cse

[2] p. 883, end of first (continued) paragraph in Sedentary Behaviors Increase Risk of Cardiovascular Disease Mortality in Men Tatiana Y. Warren, Vaughn Barry, Steven P. Hooker, Xuemei Sui, Timothy S. Church, And Steven N. Blair Med Sci Sports Exerc. 2010 May;42(5):879-85. PMID 19996993.

Physiology

> Results further showed that, regardless of time spent in riding in a car or in combined sedentary behavior, being older, having normal weight, being normotensive, and *being physically active* were associated with a lower risk of CVD mortality in this cohort. (italics added)

A bit later they state[1]:

> Another major finding of our study was that, for any given amount of time spent riding in a car, men who were physically active (Fig. 1C) maintained lower CVD mortality rates than men who were classified as physically inactive.

This does not seem to exactly agree with the quote in the newspaper article.

Research into what type(s) of exercise and activity most promote health as well as what type(s) of inactivity are most unhealthy continue.

---

[1] Second column of p. 883, first full paragraph

In summary,
- ATP is the fuel used by your muscles to contract. There are several different sources to replenish the ATP and some are available quickly and used up quickly, while others last longer, but also take longer to be used.
- Extra energy is stored in muscles as creatine phosphate
- Another source of local energy is glycogen
- An increase in muscle cell size occurs when the cells make more proteins than they break down. Insulin and anabolic steroids increase protein production; cortisol decreases protein production in muscles.
- Studies are ongoing to determine what types of exercise and activities promote good health; it may be that too much time sitting can lead to poor health outcomes even in those that regularly exercise a few times per week.
- Studies that have interesting or novel results are more likely to be pursued, and published, than experiments that show no effect, which can bias the literature towards effects.
- Several studies done by independent groups that reach the same conclusion are more convincing then many studies done by related groups.

Physiology

Blowing in the Wind

Imagine that a group of college students goes to Colorado for a spring break ski trip. On their flight from Kansas City to Denver, one of the older passengers on the plane, Harriet, becomes ill. When the students arrive at their ski lodge at 7,000 feet, Betty and Paul get sick. Roger goes off to stay at a family cabin nearby. When he does not show up in the morning, Mike goes to check on him. He can't wake him up and Roger looks very red. Mike immediately takes him out of the cabin and rushes him to the local hospital, where they are able to treat Roger and release him in a day. On the second day, Pat looks up and way above them are some birds soaring; as all the humans are still breathing more heavily than usual, he wonders how the birds can do it. On the third day, Sally catches a cold and spends a lot of time coughing and sneezing. On the return trip, they meet up with Sandy, who says she also spent her time sneezing and coughing due to the pollution in Denver, where she had stayed with her grandmother, Harriet. On the fourth day, Emily starts to hiccup and can't stop for 8 hours. On the fifth day, a stranger collides into Bob, and he has the wind knocked out of him. After what seems like a long time, but was probably only a few minutes, he is finally able to breathe normally again. While at the resort, the students run into several Olympic team members, but they compete in the summer Olympics, why are they training at high altitude?

In this chapter, we will talk about the physiology that occurred on this trip, including these sections:
- Wind knocked out
- Acute mountain sickness
- Oxygen saturation
- Special cases: alligators, birds, and fetuses
- More red blood cells: Athletes, Andeans, Tibetans
- Red Roger
- Cough, sneeze (and taste?)
- Hiccup

Wind knocked out
When Bob had his wind knocked out of him, it was because the stranger had run into his diaphragm. This made his diaphragm spasm. The diaphragm is a set of muscles at the bottom of your lung. In order

to breathe in, the diaphragm muscles contract. This increases the volume of the lung, which makes the pressure inside the lung (and in the alveoli) less than atmospheric. If you do this with your mouth open, you have a pressure gradient and a path. By now, you know this means air flows into the lung. At rest, exhaling is simply due to relaxing the diaphragm. This reduces the volume of the lung and now the pressure inside the lung is greater than outside so air flows out.

When the diaphragm is in spasm, it can't contract and relax, so Bob cannot change his lung volumes, and therefore he can't move air. Fortunately, the spasm ends pretty quickly and he can catch his breath and go back to skiing.

Acute mountain sickness
Betty and Paul both complain of headaches. They also feel tired and listless. These are symptoms of Acute Mountain Sickness. Acute Mountain Sickness occurs to about half of the people that go to high altitude too quickly. At the moment, no one knows how to identify in advance, who will suffer from high altitude sickness, but if you have suffered before, you are likely to suffer again. A common symptom for almost everyone suffering from high altitude sickness is a headache. In fact, it is believed that a Chinese official, Too-Kin, when traveling over the Kilik Pass in what is now Afghanistan was the first to describe high altitude sickness about 30 years BCE. He called the mountains he was traveling over, "Headache Mountains". Other Acute Mountain Sickness symptoms sometimes include anorexia, nausea, vomiting, fatigue, dizziness, and the inability to sleep well.

When you go up in altitude, the air gets thinner, that is, there are less molecules of air per volume. A key gas for us is oxygen. At high altitude there are fewer oxygen molecules going into our lung for each breathe. In addition, there is a smaller gradient from the lungs to the blood. The result is that we don't load quite as much oxygen onto our blood in the lungs. Oxygen sensors along the blood vessels in our neck sense this lower oxygen and direct the brain to increase the rate and depth of our respiration. This means we are able to get more total oxygen into our blood.

Physiology

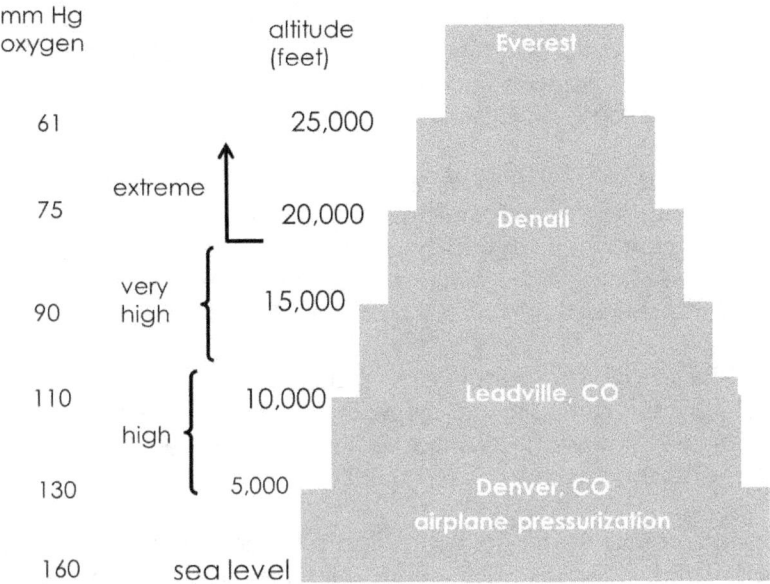

Figure 1. An increase in altitude leads to a decrease in the amount of oxygen available as air pressure decreases. Air remains about 20% oxygen, but there is less air at high altitude.

However, the faster rate of respiration also means that we are exhaling more carbon dioxide (CO_2). That might seem like a good thing, since carbon dioxide is the waste product of converting glucose and oxygen to ATP and water. However, carbon dioxide also serves another function. Carbon dioxide reacts with water and forms bicarbonate, HCO_3^- (and a proton, H^+).

$$CO_2 \text{ \& } H_2O \longleftrightarrow HCO_3^- \text{ \& } H^+$$

If we blow off CO_2, then the reaction goes in reverse, bicarbonate and protons to carbon dioxide and water.

This means there are fewer protons in the blood. This changes the blood pH[1]; because the change is in the alkaline direction, it is called

---

[1] pH is defined as the negative of the log of the proton concentration*. Because of the negative sign, as pH increases, the proton concentration decreases. Because pH is a log scale, a change of pH of 1 unit, say from pH 7 to pH 8, is a

alkalemia. This contributes to the out-of-sorts feeling that Paul and Betty feel. It might also be one contributing factor to their headache. Over a few days, the kidneys should compensate and retain more protons by excreting more bicarbonate.

In order to prevent Acute Mountain Sickness, it is best to go up gradually. If the students had spent a day or two in Denver before going to the ski lodge, they might have avoided suffering from Acute Mountain Sickness. Once they have it, the best thing is to relax and not go higher for a few days.

It is also healthier to avoid alcohol when going up in altitude. Alcohol, among other things, is a respiratory depressant, that is, it tends to slow ones breathing. But you want to breathe faster at altitude, not slower, in order to get enough oxygen.

An inappropriate response to going to high altitude can result in rare, but potentially fatal, diseases. The person can develop severe lung or brain problems or both. Recently, the explanations for these problems have been criticized, but the evidence remains unclear. Therefore, we will call them by their current names, but the names may not be accurate descriptions of the physiology, high altitude cerebral edema (HACE) and high altitude pulmonary edema (HAPE). At the first signs of HACE or HAPE, a person should try to descend or get in a tent that has higher oxygen levels. Poor coordination and altered consciousness are two key symptoms of HACE. HAPE patients often

---

10 fold change in the proton concentration, from 100 nM to 10 nM, where nM is nanomolar, 10^{-9} Molar. *Concentration is easy to measure as it is just the ratio of the amount of a substance (for example protons or salt) in a known volume. However, when as the concentration gets high, the added molecules interact in different ways than when the solution was dilute; the term "activity" is a measure of the actual behavior. A rough analogy is that concentration is the number of people in a building. (Amount per volume.) The more people, the high the concentration. When there are only a few people, they behave independently and a description of their behavior is straightforward, but when the number of people gets pretty high, they start interacting with each other and the simple way of describing their behavior doesn't work. How the people actually behavior is their chemical "activity"; in this particular example, activity is the lay word for how the people behave, as well.

have a dry cough and then higher heart rates and breathing rates at rest. Sometimes they even turn blue. The common term, edema, indicates that for both these diseases, there is swelling of the tissue. In the lungs, there is increased blood pressure, this causes the lung capillaries to leak fluid, and this can be extremely dangerous. In the brain, the extra leakiness expands the brain volume and this is also not a good thing.

Oxygen saturation
Harriet has a respiratory illness, Diffuse Interstitial Lung Disease. Because of this, her blood saturation with oxygen is less than normal. On the plane trip, Harriet also suffered some of the same sort of acute symptoms that Betty and Paul did, but she was on an airplane, not high up in a mountain. You probably realize that airplane cabins are pressurized, but they are not pressurized to sea level, but rather to about being at 2000 m (~6000 feet). This is fine for most people; they get slightly less oxygen saturation, but not enough to cause them any problems. However, for Harriet it means her blood saturation gets low enough that she begins to have problems with delivering enough oxygen to her tissues.

What is this oxygen saturation we are talking about? Oxygen dissolves pretty poorly in water. The air has about 200 mls of oxygen per liter. But the water in a tank below the air only has about 3 mls of oxygen per liter. Not very much. Our blood has close to 200 mls of oxygen per liter because our blood contains hemoglobin. Each hemoglobin can bind 4 oxygen molecules.

One interesting property about oxygen is that it is more soluble in cold water than warm water. That explains why trout like cold water and why there is a lot of sea life in tidal basins of cold areas. Some fish in the Antarctic manage without hemoglobin.

At sea level in a healthy person, about 98% of the hemoglobin molecules have 4 oxygen molecules bound, so our blood is essentially saturated. (100% of the hemoglobin with 4 oxygen molecules bound is completely saturated.) However, the curve of oxygen molecules bound to hemoglobin vs. oxygen molecules dissolved in the blood is not linear, see Figure 2. This means one can decrease the oxygen levels in the air substantially and still have the hemoglobin nearly saturated.

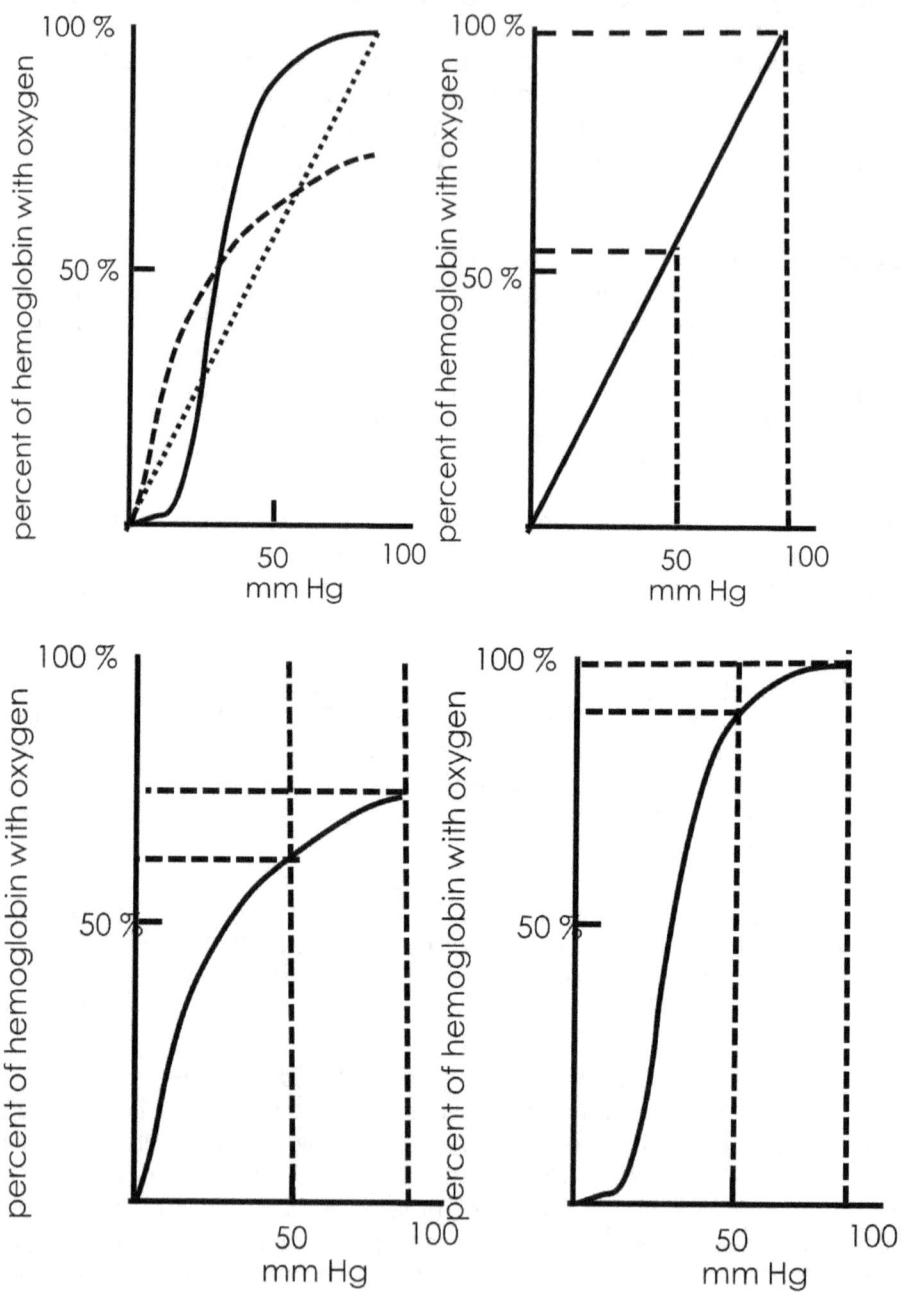

Physiology

Figure 2 The advantages of hemoglobin's oxygen curve.
Upper left graph shows three different oxygen binding curves.
The solid line is for hemoglobin which cooperatively binds 4 oxygens.
The dashed line is for a protein that only binds 1 oxygen.
The dotted line is a linear response.
Upper right graph shows that changing oxygen from 90 to 50 mm Hg decreases the oxygen saturation from 98% to 55%.
Bottom left graph shows that changing oxygen from 90 to 50 mm Hg decreases the oxygen saturation from 74% to 62%.
Bottom right graph shows that changing oxygen from 90 to 50 mm Hg decreases the oxygen saturation from 98% to 88%.
Clearly, hemoglobin has the highest percentage of oxygen bound even at 50 mm Hg.

In this chapter, an understanding of saturation and of affinity will be important. An analogy will illustrate the idea of saturation. The hemoglobin of blood in the lung is nearly saturated with oxygen at sea level. While the oxygen concentration decreases with altitude, even at 1000 m (~3000 feet) most of the hemoglobin is saturated. In the analogy, hemoglobin is like a spoon. Oxygen is sticky cookie dough. The lungs are a bowl of sticky cookie dough. The cookie pan is the tissue that accepts the cookie dough. When the bowl is full of cookie dough (the lungs have plenty of oxygen), then the spoon is almost full of cookie dough (the hemoglobin is almost full of oxygen). When the spoon gets to the cookie pan (when the hemoglobin gets to the tissue) about half of the cookie dough comes off the spoon and lands on the pan. About half of the oxygen comes off hemoglobin in the tissue.

A pressurized airplane is like having a bowl only partially full with cookie dough. As long as the transfer from bowl to spoon is still pretty good, the spoon is still almost full of cookie dough and so the cookie pan gets about as much cookie dough. However, if the transfer from bowl to spoon is not so good, as in Harriet's case, then the cookie pan is going to get less cookie dough and then the cookie eater (the tissue) is going to be grumpy. But even if the transfer from bowl to spoon is good, there is a level of cookie dough in the bowl where the spoon will not be as full. This is like high altitude where the oxygen in the air is too little to fully saturate the hemoglobin.

If one stays at altitude for a long time, then the affinity of oxygen for hemoglobin decreases. In contrast, the affinity for oxygen is higher in the fetus than the mother. And this is also true of the Bar Headed Goose that can fly in the Himalayas. We need to discuss the implications of this shift in affinity and also just a bit about the mechanism.

A decrease in affinity of oxygen for hemoglobin is a bit like putting some butter on the cookie dough spoon. Even with the butter on, you can usually get a full spoonful of dough out of the bowl, but now, more of the dough should fall off onto the pan. So you have delivered more dough to pan. The particular way that hemoglobin works is that even for large changes in the amount of cookie dough in the bowl, the spoon is usually close to full. So butter doesn't really affect how much dough is on the spoon. But butter greatly helps the dough to come off on the pan.

For the change in the other direction, that is higher affinity we make the spoon more sticky. The Bar Headed Goose lives at high altitude where the oxygen concentration is about 1/3 of sea level, so that it like having a bowl with very little cookie dough, it needs to have the spoon be as sticky as possible. The fetus is like another spoon (baby spoon?) that needs to be extra sticky; if it were only as sticky as the big spoon, it would not get enough dough.

What is the equivalent of butter for hemoglobin? In most mammals, it is a compound called diphosphoglycerate[1]. Diphosphoglycerate binds to hemoglobin and reduces its affinity for oxygen, so it acts like butter on the spoon.

Special cases: alligators, birds, and fetuses
The fetus needs to have hemoglobin that can bind oxygen better than the mother's hemoglobin so that it can get enough oxygen. You

---

[1] Diphosphoglycerate can be understood by breaking the word up: di-phospho-glycerate. Glycerate is a 3 carbon compound, roughly half of glucose. Di-phosphate means 2 phosphates. Diphosphoglycerate is an intermediate in glucose metabolism and is found in low amounts in most cells. But it is high in red blood cells. Also, note that phosphate is an anion so 2 phosphates means diphosphoglycerate has 2 negative charges.

might think that fetal hemoglobin has a mutation to make it bind oxygen with higher affinity than maternal hemoglobin and while that would work, that is not what actually happened in nature. Rather, fetal hemoglobin has a mutation that decreases diphosphoglycerate binding. Since diphosphoglycerate binding made the mother's hemoglobin have lower oxygen affinity, the fetus now has higher affinity! Or with our analogy, Mom's spoon has the butter, but not the baby spoon.

In birds, it is not diphosphoglycerate that is used to modulate hemoglobin, but inositol pentaphosphate[1].

The Bar headed goose does have a mutation that allows it to bind oxygen slightly better than its relatives, that is, it has a slightly stickier spoon. But, unlike its relatives, inositol pentaphosphate has little effect, so in the red blood cell, the affinity for oxygen is much greater in the Bar Headed Goose than its relatives.

What about the alligator? Alligators sometimes drown their prey. Why don't the alligators drown too? While both are under water, they are burning oxygen and producing CO_2. Some of the CO_2 will react with water and form bicarbonate (and a proton, which we'll ignore). Remember that bicarbonate is an anion and so is diphosphoglycerate; bicarbonate has 1 negative charge and diphosphoglycerate has 2. Well alligators have a mutation in their hemoglobin so that instead of binding diphosphoglycerate, it binds 2 bicarbonates. This also lowers the affinity of hemoglobin for oxygen-- thus more oxygen comes off the hemoglobin and alligator tissues can use this "extra" oxygen and stay alive.

One of the effects of Harriet's disease is to increase the distance that oxygen has to diffuse to get from the small sacs of air in the lungs (the alveoli) to the blood in the capillaries. Diffusion is a relatively slow process and it one of the limiting steps in getting oxygen to the tissues. The distance between the air in the lung and the blood in the capillary is only 0.2 microns[2]. This is extremely thin from our perspective, about

---

[1] Inositol is a molecule a lot like glucose but it is not sweet. It is part of the signaling molecule group called inositol phosphates. Pentaphosphate just means the inositol has 5 phosphate bound.

1/250th as thick as a page of paper, but still a long ways for oxygen to move, about 1000 oxygen diameters. The membrane between air and blood is pretty amazing. In our lungs, if we spread out this layer it would be about 100 m², that is, about the size of a tennis court. It has to expand and contract about 20 times a minute or more than 20,000 times per day. Pretty tough membrane for being only 0.2 microns thick!

Why doesn't the staff at the ski lodge feel as tired as the healthy college students? They have been at high altitude for quite some time. The body begins to compensate in many ways. One compensation is to increase the number of red blood cells; so even though their oxygen saturation is low, they have more hemoglobin per volume of blood so they can carry more oxygen.

More red blood cells: Athletes, Andeans, Tibetans

One other way the body can increase the fraction of blood that is red blood cells[1] in the short term is just to decrease the amount of fluid in the blood; since hematocrit is the ratio of red cells to fluid, the ratio also increases if fluid goes down, as well as if red cells go up. A decrease in fluid is easy to accomplish, the kidneys just excrete more water.

While higher hematocrit is good in the short term, it may not be so good in the longer term. Two human populations have lived at high altitude for thousands of years, those living in the Andes and Himalayan mountain ranges. Interestingly, they have somewhat different adaptations. The people in the Andes have higher hematocrits than those living in the Himalayans. The higher hematocrit is the expected response; most people that move to high altitude have this response. However, the higher hematocrit creates a risk for Chronic Mountain Disease. The Andeans have a high risk for Chronic Mountain Disease; the incidence is about 13% for older teenagers and increases to 36% in middle age. Interestingly, some rare individuals living at sea level have many symptoms in common with Chronic Mountain Disease. They also have too high a hematocrit. Some of these individuals have a mutation that activates the hypoxia

---

[2] Micron is abbreviated uM; 0.2 uM is the same as 200 nanometers.
[1] The term for the fraction of blood that is red blood cells is hematocrit.

inducible factor gene. The name of this factor indicates that it was discovered to increase when tissues were give low oxygen (hypoxia). This gene is thought to be one factor that increases red blood cell production[1].

In contrast to the Andean response, the high altitude Tibetans do not have high hematocrit. Many of the Tibetans have a mutation in their gene for hypoxia inducible factor. The Tibetans mutation results in a gene that works less well than normal, thus blunting the response to hypoxia inducible factor. Only 1% of the Tibetans get chronic mountain disease.

Some Andeans offer coca leaves to tourists climbing in the mountains. These leaves contain small amounts of cocaine. Cocaine is known to increase respiratory rate and it also makes one feel more energized, but that is presumably a perception as there is no evidence that cocaine increases muscle strength or exercise performance.

Red Roger
In my made up story, I imagined that Roger was sleeping in a cabin with a defective wood stove and no ventilation. The stove was producing carbon monoxide CO as well as heat. While CO_2 is important for us, CO is deadly. The molecule oxygen is composed of 2 oxygen atoms, O_2. For this paragraph, I'll denote it OO. CO (carbon monoxide) is similar enough to OO that CO can also bind to hemoglobin. But CO is different enough from OO, that once it is bound it takes a long time to come off. If most of your hemoglobin has CO bound instead of oxygen, then your blood is carrying much less oxygen. Too little oxygen can mean death. Mike's decision to immediately move Roger outside was important as that should immediately stop Roger from breathing any more CO. Getting Roger to the hospital quickly is also a good idea, because they were able to keep Roger alive until the CO came off his hemoglobin; when the CO came off the hemoglobin in the lungs, he was able exhale the CO. A bit of interesting chemistry: While CO binds to hemoglobin like OO, when it is bound the hemoglobin has a slightly different shape and is

---

[1] Another factor, apparently used by some athletes, is erythropoietin. Erythropoietin is made by the kidneys and some patients with severe kidney disease need to be given erythropoietin to prevent anemia.

bright red. That is why Roger looked so red. What grocery store products are sometimes treated with CO to make them stay red longer? Beef products.

Cough, sneeze (and taste?)

On this trip, both Sally and Sandy were coughing and sneezing, but for different reasons[1]. Sandy's was in response to the pollution in Denver. Sally had a cold. One important cause of coughing is breathing the wrong chemicals. Some of the nerves that line the lung and sense these toxic chemicals contain the TRPV1 channel, which you remember can be activated by capsaicin. The TRPV1 channel also responds to some other chemicals and activates the nerve, which signals the brain to begin the cough reflex. But in mice in which the TRPV1 channel is knocked out, while the mice don't cough with capsaicin, they do respond to other chemicals. Many of these sensory neurons contain a cousin to TRPV1 called TRPA1.[2] TRPA1 does not bind capsaicin, but it does bind other chemicals, many of which are produced in fires such as acrolein. TRPA1 also binds to chemicals found in mustard, wasabi, cinnamon, and garlic.

The cough reflex involves vagus nerve[3] control of the glottis and larynx as well the diaphragm. Overstimulation of the system, for example, by inspiring too much capsaicin, can overactivate the system. This can lead to respiratory arrest and even death.

When a respiratory virus infects the nasopharynx, the resulting inflammation leads to bradykinin and prostaglandin release[4]. It is

---

[1] **How can you tell if you have a cold or the flu?** There is substantial overlap in the symptoms. In adults, colds seldom lead to fevers. About 80% of patients with a cough and a fever have the flu and not a cold.

[2] In science, like in Chinese, the family name or initials often are given first.

[3] The vagus nerve is also called the pneumogastric nerve or cranial nerve X. Pneumogastric nerve is actually a very helpful name as the vagus nerve communicates to the lungs (pneumo-) and the stomach. A branch of the vagus also goes to the heart. In seems like the vagus goes all over the place and that is where its name comes from as vagus is Latin for wandering.

[4] Bradykinin is from the Greek, "brady" meaning slow and "kinin" meaning to move. Roche e Silva gave the compound the name based on the fact that it slowed gut contractions; bradykinin also is important in the inflammatory and

Physiology

thought that the bradykinin is the cause of the sore throat; if you spray the nose and throat of healthy volunteers with bradykinin they develop a sore throat. TRPA1 can be activated when sensory nerve cells bind bradykinin. Scientists and drug companies are now working on targeting the TRPA1 and TRPV1 channels as way to develop anti-coughing medicines.

Scientific studies have seen little effect of over the counter cough medicines, but the studies have been of small size and have not optimally designed. So the original authors tend to say that there is no clear evidence for or against cough medicines. But later authors quote this as saying that there is no scientific evidence that over the counter cough medicines are effective which seems to me to be a partial truth and misleading.

In the nasal cavities, the swelling of blood vessels leads to part of the feeling of congestion. Also the swollen blood vessels impede the drainage of fluid and that also increases the congestion. In the first few days of a cold, the fluid is more watery, and represents fluid secretion by the nasal glands[1]. The molecule that the nerve releases to signal the gland is acetylcholine[2]. Thus anti-acetylcholine drugs can reduce the amount of fluid secreted early in the cold. But after about 4 days, the fluid comes from leaky capillaries. It is much richer in protein because blood plasma has lots of proteins. The leakiness is not controlled by acetylcholine. Thus anti-acetylcholine drugs won't work at this stage.

In order to decrease nasal congestion, some people spray nasal decongestants up their nose. The active ingredient in some of these mimics the effect of norepinephrine. In the nose, norepinephrine leads to blood vessel constriction. This reduces blood flow and so the

---

pain responses. Prostaglandins were initially isolated from seminal fluid which is partly generated by the prostate gland. Prostaglandins are now known to be made by a variety of cells and to have a variety of effects.

[1] A runny nose is called rhinorrhea. Rhino referring to nose and -rrhea is the same suffix as in diarrhea and refers to flow.

[2] Remember acetylcholine from Blood, Sweat, and Tears and from In a Heart Beat?

epithelial cells are able to secrete less fluid. In addition, the smaller volume of the blood vessels when constricted might also contribute to the relief of congestion. We discussed vessel constriction in the chapter, Can Viagra™ Kill You, so you might guess, correctly, that norepinephrine, or these drugs that mimic it, lead to a rise in smooth muscle calcium, triggering vascular smooth muscle contraction. Do you think Viagra™ sprayed in the nose would be useful for decongestion? No, the effect of Viagra™ is the opposite-to dilate vessels!

Often in response to a cold, the eyes get teary. Figure 3 shows the a schematic of the tear duct anatomy.

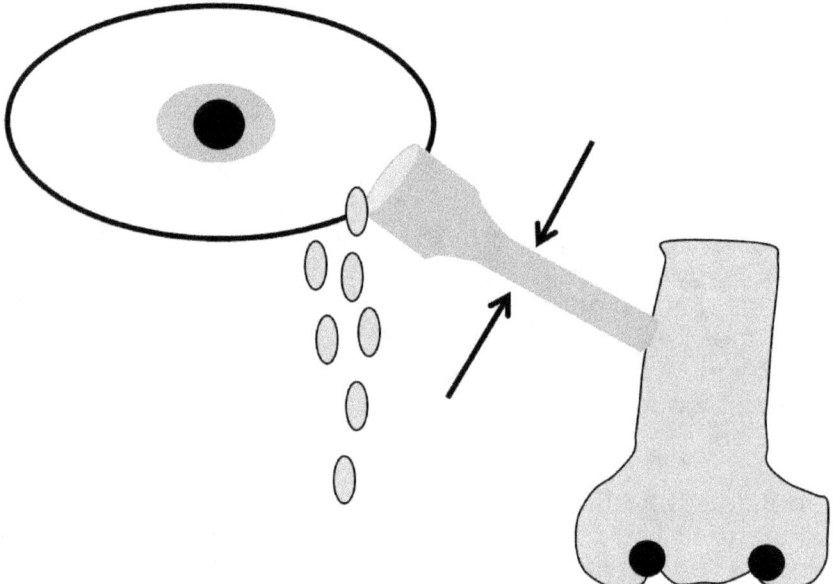

Figure 3. Blockage of the tear duct leads to tears going down the face and not draining into the nasal cavity. Normally, the fluid generated by the lacrimal (tear) gland is drained away from the eye by the lacrimal duct to the nose. When blood vessels near the duct dilate (at arrows), the lacrimal duct narrows and cannot accommodate all the fluid generated by the lacrimal gland during a cold. The excess fluid runs down the face.

Physiology

Sneezing is activated by nerves in the nose; this makes sense as sneezing is designed to clean out the nasal passages. These nerves signal the brain; the brain interprets this irritation or even pain and starts the sneezing reflex.

The sneeze reflex involves the brain triggering the nose and tear glands to secrete fluid. The face muscles contract so that eyes close and one grimaces. There is then an explosive exhalation which hopefully dislodges whatever triggered the reflex.

Some people sneeze when they look at the sun. This has not been studied in great detail, but about ¼ of people have this response and it appears to be inherited in a dominant fashion. Somehow the optical signal gets crossed with the sensory signal for sneezing. This could occur when the optical and nasal nerves pass close to each other or it could occur higher up in the brain where the signals get crossed. Usually it is a nuisance, but it can be deadly in fighter pilots or even in car drivers emerging from a dark tunnel to a sun-drenched highway. (For an amusing description, see http://www.straightdope.com/columns/read/527/why-do-some-people-sneeze-when-going-out-into-bright-light).

Hiccup
Emily had the hiccups. What causes hiccups? The current theory is that the vagus nerve gets activated; it leads to convulsion of the diaphragm muscles, causing you to gulp air. The vagus also leads to the closing of the glottis, but with a slight delay. This gives the "hic" sound. The vagus nerve goes to lots of places in the body, so irritants in lots of places can set off hiccupping. The most common cause of sustained hiccupping is a digestive system problem; even distending the stomach too much can set it off in some people. Another interesting case was of a girl who was hit in the jaw and starting to hiccup. The physicians' figured out that the blow made one of her blood vessels enlarge and in squeezed the adjacent nerve, causing the hiccups. They put in a plastic spacer and the hiccupping immediately stopped. When the spacer fell out, she started again. The vagus even has nerve endings in the ear and maybe the strangest ear irritation to cause hiccupping was a person who had an ant crawling around in his ear.

Recently a new suggestion has been made for how to stop hiccups. One is to rub a moistened cotton bud for 1 minute on border between soft and hard palate. This stimulates vagus so there is more vagal tone and interrupts the hiccup loop.

In summary,
- Movement of air in the respiratory system requires a pathway and a gradient
- Inhalation occurs when the diaphragm contracts, increasing lung volume, decreasing lung pressure. If there is a pathway open, then air moves in
- We have discussed some of the physiology of coughs, sneezed and hiccups, and related concepts from the chapter The Good, the Bad and the Spicy as well as the chapter Can Viagra Kill you?
- We have used what happens at high altitude and low oxygen to explore some of the physiology of the respiratory system
- Changes in hemoglobin-oxyen affinity can be modulated by anions such as diphosphoglycerate, inositol phosphate and bicarbonate and alter how much oxygen binds to hemoglobin in the lungs and how much is released in the tissues
- Carbon dioxide functions as a ph buffer
- Changes in carbon dioxide levels alter ph
- Oxygen binding to hemoglobin is not a linear response, but shows saturation
- More hemoglobin and more red cells increases the oxygen carrying capacity of blood but also increases blood viscosity

Physiology

Soy Sauce and Licorice as Medicine

Literature Cases

1. A middle-aged man, Solomon, complained of general weakness. For two years he had noticed that he urinated and drank a lot more than usual. His systolic blood pressure was 70 mm Hg, much lower than the usual 120 mm Hg. The main cause of this man's low blood pressure turns out to be excess loss of fluid; his fluid losses were greater than his fluid intake.

2) A Vietnam Vet, Victor, had gone to his dentist's office because he had broken a tooth. A student helper who escorted him to the dental chair noted that the man had to stop 3 times on the way because he was feeling dizzy. Upon questioning, he indicated that he had been feeling dizzy for a few days and had blurred vision. The clinicians measured his blood pressure, 75/33 mm Hg, which was much lower than normal, 120/80 mm Hg. He was taken by ambulance to the hospital and treated. The Emergency Department physicians commented that if he had just stayed home because he felt sick he probably would have died within 24 hours. So the broken tooth probably saved his life.

3) Alex, an eighteen-year-old man, complains to his health care provider that he has to drink a lot per day; if he doesn't get enough water he gets nervous and irritable. His blood pressure was normal, but his daily urine output was 10 liters! When he was deprived of water, his urine osmolarity (number of particles) did not change much, even though, without water, he should have concentrated his urine. So far, his symptoms are similar to that expected from patients that do not make enough of the pituitary hormone, antidiuretic hormone, see Diana Inspired in the chapter, "How Sweet It Is". Alex's levels of antidiuretic hormone were above normal, suggesting that his pituitary is responding appropriately. What else go be going wrong?

4) Sue is an interesting case because the physicians eventually worked out that she had inadvertently been self-medicating. After about one month of not feeling well, Sue went to her physician. Besides just not feeling energetic, she would often feel dizzy when she stood up. Over the years, she had increased her intake of soy sauce to the point where she was consuming about 1/3 of a bottle of soy sauce each

day-that's a lot of salt! It would be like using up a standard 26 ounce salt box every 2 months. Sue was also eating a lot of real licorice. It turns out that soy sauce and licorice were actually important for "treating" her disease.

5) Lily had recently moved from the Netherlands to Canada; many of her relatives still lived in the Netherlands. In January, she presented with high blood pressure, which was very unusual for her. A blood test also indicated that she had low blood potassium levels. What could have happened around the winter holidays to cause high blood pressure and low blood potassium levels?

Have you ever drunk a nice cold glass of water? Or eaten a whole bag of potato chips? These put very different demands on your body. How does the body maintain the correct amount of water? How does the body maintain the correct amount of salt? And what happens when these amounts are not maintained?

In this chapter we will discuss
- Fluid losses
- Water regulation
- Salt regulation
- What happens if one becomes dehydrated?
- Can you get too much water?
- Fish and aldosterone
- Aquaporins and the Nobel Prize
- More on Solomon

In addition to keeping sodium and water in balance, the **ratio** of salt to water in the body is important to balance. The number of particles in a given volume of water is called osmolarity; molarity is the number of a specific type of molecule per volume. If you put sucrose into water, each sucrose molecule dissolves as one particle. So the osmolarity would be the same as the molarity. On the other hand, if you put table salt (sodium chloride) into water, each salt molecule dissolves into two particles, sodium and chloride. Thus the osmolarity would be twice as much as the molarity. And if you put both sodium chloride and sucrose into water, the osmolarity will be the molarity of sucrose plus 2 times the molarity of sodium chloride since both sucrose and sodium chloride contribute particles. Most membranes in the body allow water to pass through freely. Thus the water will move so

Physiology

that the ratio of salt to water is the same on both sides of the membrane. The primary exception is the kidney and this is what allows the kidney to produce either very concentrated urine or very dilute urine.

Fluid losses
Both Solomon and Victor presented with symptoms consistent with losing too much water. The main cause of their low blood pressure turns out to be fluid loss; their fluid losses were greater than their fluid intakes. The health care providers had to determine why the greater water loss. How many ways can you think of for water loss? The most obvious one is to urinate too often. This turns out to be a rare cause of excess volume loss, unless one is taking diuretics, drugs that are designed to cause extra urination. The 2 year history of excess urination for Solomon and his thirst should make you wonder about type II diabetes which we will discuss in the chapter How sweet it is. Certainly a possibility, but it is not the cause in this case. In addition to excess urination, the other 3 obvious ways are sweating, vomiting, and diarrhea. The final way to lose water is through your breath; your airways and lungs add water to the air your breath. The air you exhale has more moisture than the air you inhale; this is visible when you breathe in cold air and the excess water condenses[1].

Water Loss Possibilities

Pee
Perspire
Poop
Puff out water with breathe
Puke

---

[1] Panting can be a major cause of dehydration in dogs; dogs use panting for cooling as they can't sweat.

Solomon turns out to have a genetic defect that makes his kidney not function properly. Two quick lab tests would have immediately suggested to the clinicians a kidney problem. The amount of salt (and osmolarity) in his urine were in the moderate range yet his body should have been trying to conserve water. Thus if his kidneys were functioning properly, he should have had high salt urine. In addition, his blood levels of potassium were very low. Yet his urine has high levels of potassium. So his kidneys are not responding properly as they should try to retain potassium. The fact that the urine values are unexpected for his blood values is a key clue to a clinician. Either his kidneys or the signal to the kidneys are part of the problem. This would be true even if his urine values were in the normal range, because the normal response of a healthy kidney would be abnormal urine values for abnormal blood values! Solomon's problem is consistent with a kidney issue and he has him a disease in the Bartter/Gitelman spectrum. For example, his genetic defect could be in a protein that transports sodium and chloride. When it is defective, the kidneys lose sodium and potassium and water.

Once the clinicians gave Solomon enough intravenous fluids, his urine output was 7 to 10 liters per day. He can avoid problems by remaining in water and potassium balance. So all he has to do is drink a lot of water and take adequate potassium. Six months after this episode, he was doing fine, though his potassium was still a bit low, even though he was taking over 10 grams of potassium per day, about the equivalent of 25 bananas per day.

Victor's urine output was very low which is exactly as expected if he needed to retain fluid. So his kidneys are ok. So we think about the other ways to lose fluid. He was not sweating abnormally and he had not been vomiting. But he had taken too high a dose of a colon cleansing regimen. Colon cleansing induces diarrhea and so involves losing a lot of fluid. The excess fluid loss had led to the low blood volume and consequently low blood pressure. The low blood pressure accounted for his dizziness.

Water regulation
In order to understand what has gone wrong with Alex, we need to review how the healthy person is able to reduce urine output, as Alex seems unable to reduce his urine output. When the body needs to retain water, the pituitary secretes antidiuretic hormone. Antidiuretic

hormone acts on specific ADH receptors[1] in particular cells in the kidney and raises the cAMP levels in those cells. These special cells have vesicles. In their membranes, there are water channels (aquaporins). When the cell is activated, the vesicles fuse with the surface, plasma membrane, thereby putting water channels in the membrane. These cells are located in the last portion of the kidney, where the blood side of the cells has high osmolarity[2]. The high osmolarity provides a gradient to move water. If the water channels are in the surface membrane then water can move from the tubular fluid to the blood side, thus reducing the volume of fluid that becomes urine.

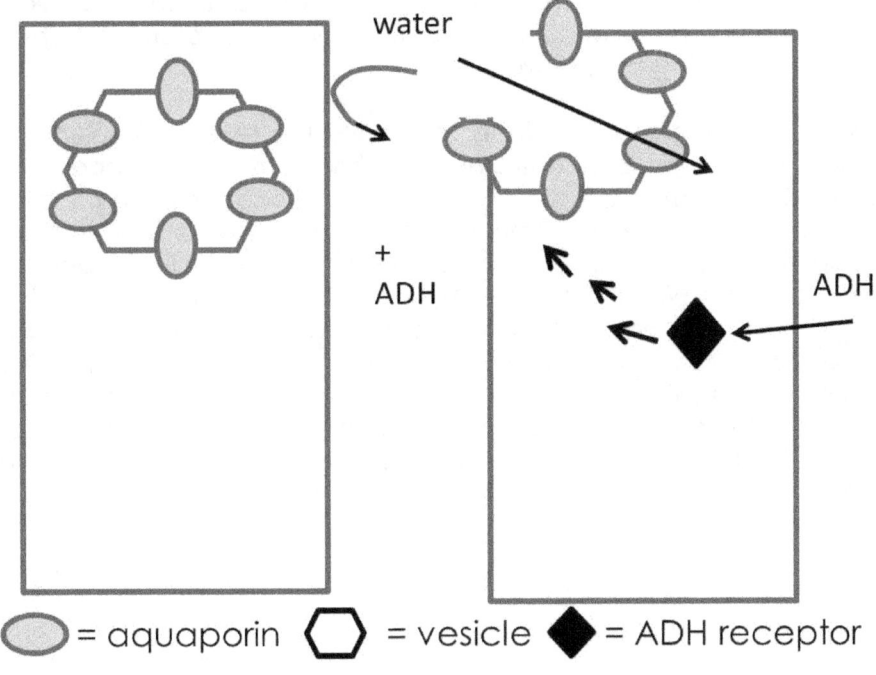

◯ = aquaporin ⬡ = vesicle ◆ = ADH receptor

---

[1] By the way, the ADH receptors are in the G coupled protein receptor family, the same as some of the taste receptors in The Good, the Bad, and the Spicy.
[2] Remember from Got Cows? that Osmolarity is the measure of the number of particles in a solution.

Figure 1. Inserting more aquaporins into the surface membrane provides a path for water to move into the cell in response to the osmotic gradient.

There are 3 primary places where ADH regulated water absorption can go wrong. In Diana Inspired's case, her pituitary didn't secrete ADH. In other cases, the patient has a mutation so that the ADH receptor either does not bind ADH or it does not change shape (conformational change) properly, and the cell does not get activated. These patients, like Alex, would be expected to have high AVP. Alex, though, had a defect in the third place: his cells were not able to have the aquaporin vesicles move to the surface membrane. Hence, even though his kidneys had an osmotic gradient, they did not have a pathway, and so water could not be reabsorbed from the tubule and he lost a lot of water.

Interestingly, there are patients that have the reverse problem; their urine is always too concentrated. You should realize 3 places where their ADH system might have gone wrong. Since their urine is always concentrated, the ADH system is behaving as if it were always on. This could be because ADH is always being secreted; this happens with some tumors. What about the ADH receptors, what could go wrong with them? In some patients there is a mutation so that the receptor is always active. So even at low or no ADH, their special kidney cells are activated, thus the aquaporins are in the surface membrane. So they always have a pathway and a gradient and water is always being reabsorbed. Finally, patients could also have a mutation so that their aquaporins are always in the surface membrane.

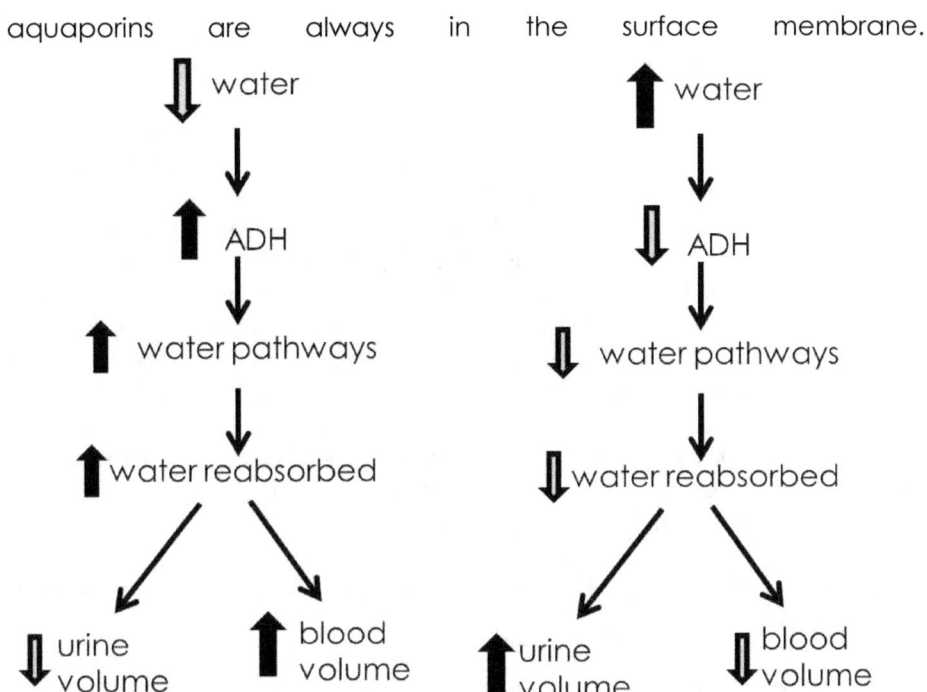

Figure 2. Response to low and high blood water (volume).

Salt regulation

Sue is an interesting case because the physicians eventually worked out that she had inadvertently been self-medicating. A blood test showed that she had antibodies to her adrenal gland, which is consistent with the beginning destruction of parts of her adrenal gland. The outermost part of the adrenal glands produces two hormones, cortisol and aldosterone. Cortisol is released in times of stress and aldosterone is involved in salt and water balance. To test for adrenal function, she was given an injection of another hormone, ACTH. In a normal person, an injection of ACTH[1] would lead to a

---

[1] ACTH is adrenocorticotropic hormone and it is the hormone that regulates cortisol release. Adrenal, of course, refers to the adrenal glands. They get their name because they are adjacent to (ad) the kidneys. (Renal is the Latin word for kidneys; kidneys are apparently an old English word. Interestingly, the term

doubling of blood cortisol levels. In Sue's case, her blood levels of cortisol went up less than 10%.

While not measured, her aldosterone levels would also have been very low. Aldosterone is a hormone which regulates sodium reabsorption in the kidney. Patients with insufficiency in the outermost part of the adrenal gland have Addison's disease and typically lose a lot of salt. Sue's antibodies, poor response to ACTH, and dizziness are consistent with Addison's disease. Typically, these patients have low blood pressure, but Sue's blood pressure at the clinic visit was normal. So her physicians were puzzled: did she have a disease that overlapped with Addison's but was something different? Then they realized that Sue seemed to have compensated for her Addison's disease by altering her diet. The 2 cups of soy sauce she consumed each week made up for the salt she lost in her urine because her aldosterone was low. The licorice also helped her.

To understand how licorice helps Sue, we need to explain how aldosterone works.

Aldosterone[1] is the major hormone that regulates salt balance. It is secreted by the outer part of the adrenal glands. Aldosterone probably has 2 classes of receptors, one extracellular and one intracellular. The intracellular one is better understood and we will focus on that. One class of intracellular receptors is transcription factors that alter the rate of the transcription from DNA to mRNA and so regulate how much protein will be made. (Steroid hormones, including cortisol, estrogen, and testosterone, as well as Vitamins A

---

kidney bean is used because the bean has the same shape as the organ.) Cortico has the same root as cortisol and is named for the location from which the hormone was isolated. Cortex meaning the outside part; in this case the adrenal cortex, the outside part of the adrenal gland. Tropic has the same root has tropism and comes from the Latin word, to turn to. As a scientific suffix, it now means to have an affinity for. Thus adrenalcortico-tropic means this compound has an affinity for the cortex of the adrenal gland.

[1] In contrast to cortisol, which is named for the location of where it is made, aldosterone is named for its chemical properties. It is in the class of compounds chemists call aldehydes and also in the class of compounds called steroids.

Physiology

and D, all have intracellular receptors and all regulate transcription of various genes.) The aldosterone receptor also regulates transcription and when the receptor is active it leads to an increase in the production of transporters required for sodium reabsorption. When the body has low salt, aldosterone is increased and this leads to more sodium reabsorption

What does licorice have to do with the case? It turns out that licorice inhibits an enzyme that metabolizes cortisol. The enzyme is called hydroxysteroid dehydrogenase. This enzyme is present in the kidney and its function is important for aldosterone action. Confusing? Let me explain.

Aldosterone binds to, and activates, a receptor in the kidney. When the aldosterone receptor is activated, it causes the kidneys to retain more salt in the body. The aldosterone receptor also binds cortisol very well. Blood levels of cortisol are about 1000 times greater than aldosterone. If the enzyme hydroxysteroid dehydrogenase were not present, then cortisol would always be bound to this receptor and it would always be active. So the kidney would always be retaining too much salt and there would be no chance for aldosterone to regulate the salt.

But because the kidney has the enzyme hydroxysteroid dehydrogenase that metabolizes cortisol before it can get to the aldosterone receptor, changes in aldosterone do regulate this receptor in normal people. [1]

---

[1] Here is an analogy for the cortisol and aldosterone system where cortisol is coffee and aldosterone is tea. One country (USA) drinks only coffee and not tea. The other country (England) could drink both tea and coffee, but there is a machine (the cortisol breakdown enzyme) that destroys all the coffee in England before it can be used. Ships can take coffee all over the world and other ships take tea. The coffee activates the Americans, but the coffee does not activate the English, because it is destroyed before it gets to them. But the English can be activated by tea.
Licorice is like having a stick that blocks the coffee destroying machine. So now, even if there are no ships carrying tea, the British can still be activated because now they can drink some coffee.

Sue does not make much aldosterone so her receptors are not activated and so her kidneys are not reabsorbing much salt and she is losing a lot of salt.

But with the licorice, she is inhibiting the enzyme. Her cortisol is high enough in the kidney that it can activate the aldosterone receptor when it avoids being broken down by the enzyme. So her kidney is reabsorbing a bit more salt than it would if she skipped the licorice. In fact, before the days of synthetic aldosterone, licorice was part of the treatment for Addison's disease! Since she inhibited hydroxysteroid dehydrogenase then that would mean that cortisol would be bound to the aldosterone receptor, which would activate the receptor.

To summarize, Sue has Addison's disease, which is the loss of function of the adrenal cortex (the outer part of the adrenal gland). In her case, it was due to an autoimmune attack on her adrenal cortex. The low ability to secrete cortisol explains her low energy levels. Her low ability to secrete aldosterone explains why she loses so much salt and water. The fact that she ate licorice meant that her cortisol, even though low, could activate her aldosterone receptor and thus her kidneys work to retain a bit of sodium. The rest of the sodium she lost in her urine didn't cause a problem because she had compensated by eating 2 moles or ¼ pound of sodium per week!

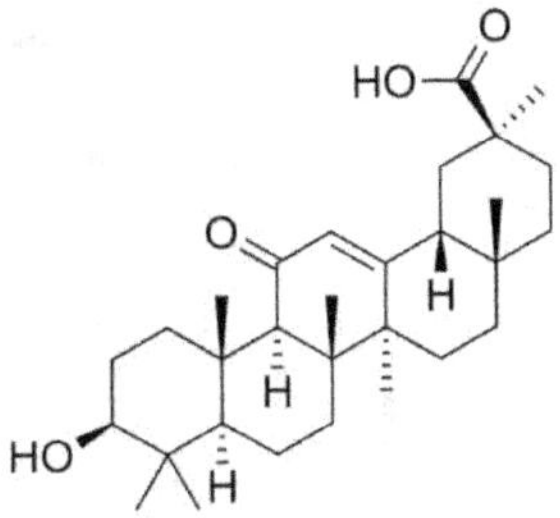

Glycyrrhetinic acid

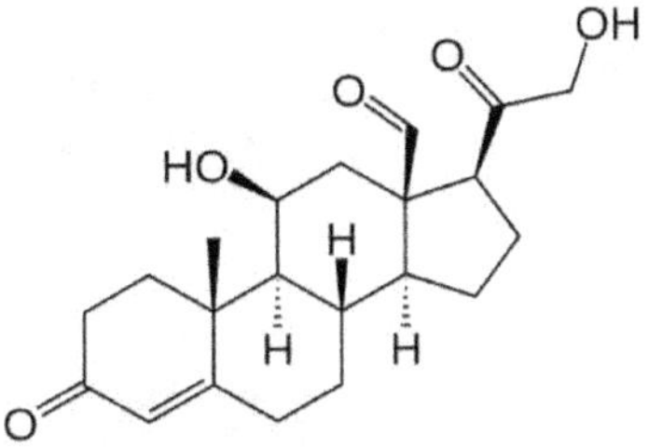

aldosterone

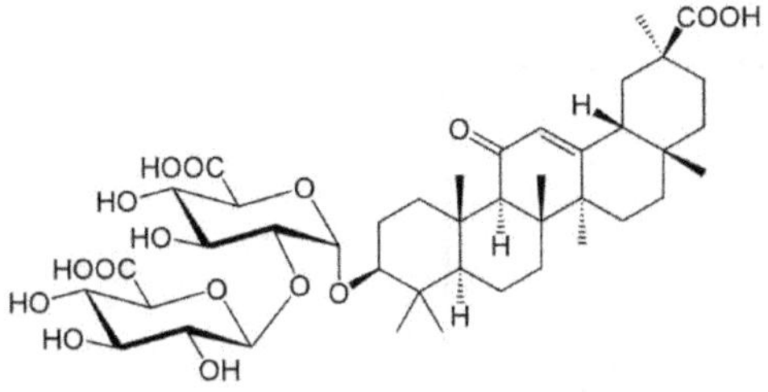

Glycyrrhizin

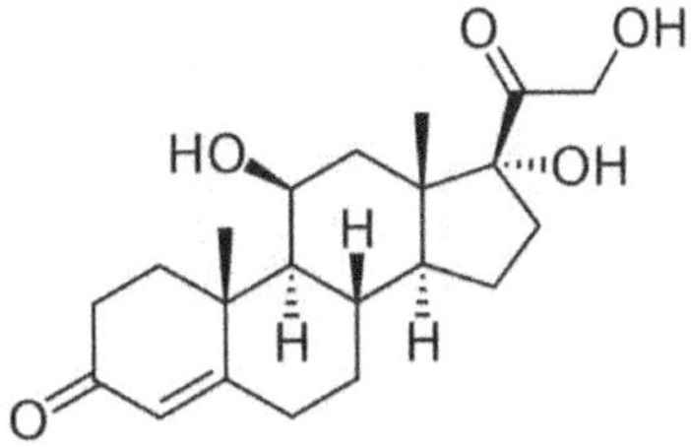

cortisol

Figure 3. Structures of the compounds in licorice that block the cortisol breakdown enzyme as well as the structures of aldosterone and cortisol. Structures from Wikipedia.
http://en.wikipedia.org/wiki/File:Glycyrrhetinic_acid_structure.svg
http://en.wikipedia.org/wiki/File:Glycyrrhizic_Acid.svg
http://en.wikipedia.org/wiki/File:Aldosterone-2D-skeletal.svg
http://en.wikipedia.org/wiki/File:Cortisol2.svg

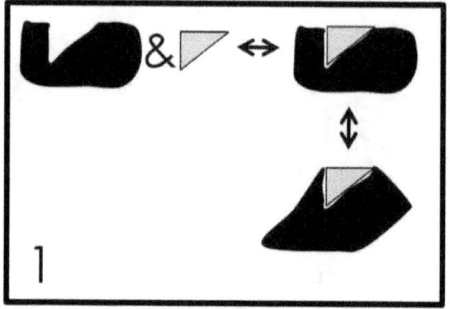
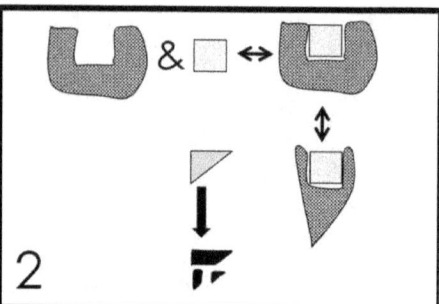
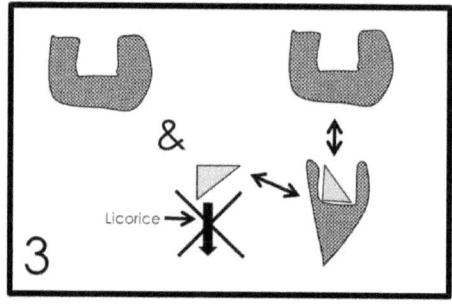

Figure 4. How licorice allows cortisol to activate the aldosterone receptor.
Box 1. Cortisol (triangle) binds to the cortisol receptor (black blob) and then the complex changes shape.
Box 2. Aldosterone (square) binds to aldosterone receptor (checkered blob) and then the complex changes shape. Cortisol is prevented from binding to the aldosterone receptor because an enzyme breaks the cortisol down (fragments).
Box 3. Licorice blocks the enzyme that breaks down cortisol. So even in the absence of aldosterone, the aldosterone receptor will change shape because cortisol can bind to aldosterone receptor and cause the same shape change as aldosterone would.

What might be the cause of Lily's high blood pressure? There are, of course, many causes of high blood pressure and most are still not known. But given the theme of this chapter and the fact that Solomon had low blood pressure because his kidneys put out too much salt and water, you might guess that high blood pressure can sometimes result from the kidneys retaining too much salt and water. That is occurring in Lily's case. Remember that Lily had low blood potassium levels. My guess is that her urine potassium would be in the normal range, 25 to 100 mmoles. But this would in fact indicate a problem. If her blood levels of potassium are low, then she should try to be compensating by decreasing her urine potassium and therefore her urine potassium should be less than 25 mmoles. 25 mmoles of potassium is not normal for her, or anyone else, when their blood levels of potassium are low. One explanation for these symptoms is an overactive adrenal gland. In other words, her glands could have been secreting too much aldosterone. Too much aldosterone would lead to a situation where her kidneys absorbed too much sodium and excreted too much potassium. Too much sodium retention can lead to high blood pressure. However, if her blood aldosterone levels were measured, they would have been normal. So her physicians were initially puzzled. After a few weeks, they were able to determine that she had been given boxes of Dutch licorice for Christmas. Thus she had been consuming a lot of licorice.

How does licorice lead to high blood pressure? Remember, licorice blocks the kidney enzyme that breaks down cortisol. Without licorice, cortisol in the kidney is broken down. With licorice, cortisol in the kidney remains high. Cortisol can bind to the aldosterone receptor. So her kidney cells "think" aldosterone is high and they respond by retaining sodium and urinating potassium. This is one example of pseudohyperaldosteronism. A big word, but if you break it up, it makes perfect sense. The end is aldosterone, so it is a problem with aldosterone. The middle is "hyper", meaning too much, so this is a case of too much aldosterone. But of course it just looks like too much aldosterone, it is actually something else (in our case, licorice, but there are other causes). So "pseudo" for false is in front. It is a false case of too much aldosterone, that is, it looks like too much aldosterone but it is something else.

What happens if one becomes dehydrated?
Initially, there may be a decrease in blood volume, but if blood volume decreased too much, you couldn't get nutrients to your cells. So the body allows water to shift from outside the cells to the blood. So now the space outside the cells has a higher concentration of sodium. Water shifts from the cells to the space outside the cells. This loss of water means the cells shrink, just like if you removed water from a water balloon (and didn't put any air in). This cell shrinkage makes many cells function differently than normal.

Many of us are familiar with a dehydration headache, but there is surprisingly little literature about it. But about half of patients experience a headache when they are dehydrated. What has been studied more is the effect of dehydration on cognitive abilities. It is very clear that dehydration alters cognitive abilities. What remains uncertain is the minimal dehydration required to observe cognitive difficulties. Many of these studies were not done well. The better designed studies get apparently different results, so we don't have a solid database to say exactly how much small amounts of dehydration interfere with cognitive abilities. There are several difficulties in doing these studies. One problem is how to measure cognitive function for example, what kind of test is best? Another problem is how do you measure dehydration, for example, do you measure how much specific cells shrink? The amount of salt in the cells? Outside the cells? The blood volume? Thus, as recently as 2007, a review of dehydration suggested that much more work needs to be done to determine what is the minimal amount of dehydration to cause cognitive problems. There are a number of theories about why brain function would be abnormal as dehydration occurs, including the shrinkage of nerve cells.

Pilots are warned to remain well hydrated because of the concerns that dehydration will impair cognitive functions. An accepted method to estimate hydration without requiring lab equipment is to observe the color of urine. When a person is well hydrated, the urine should be colorless or slightly yellow.

When the urine color is yellow, the person is dehydrated. Why is this?

The amount of yellow color in urine is a reasonable measure of dehydration because the amount of yellow you produce per day is constant under normal conditions. This is because the yellow color of urine is primarily due to bilirubin and its metabolites. Bilirubin is produced when you break down red blood cells. You break down about 1% of your red blood cells every day. So long as your red cell production and destruction are normal, this is a constant rate. How yellow your urine is, is the ratio of bilirubin excretion and water excretion. Since bilirubin production is essentially constant, and in normal kidneys their filtration rate is constant, the primary variable for how yellow is the urine, is how much water you are excreting.

Physiology

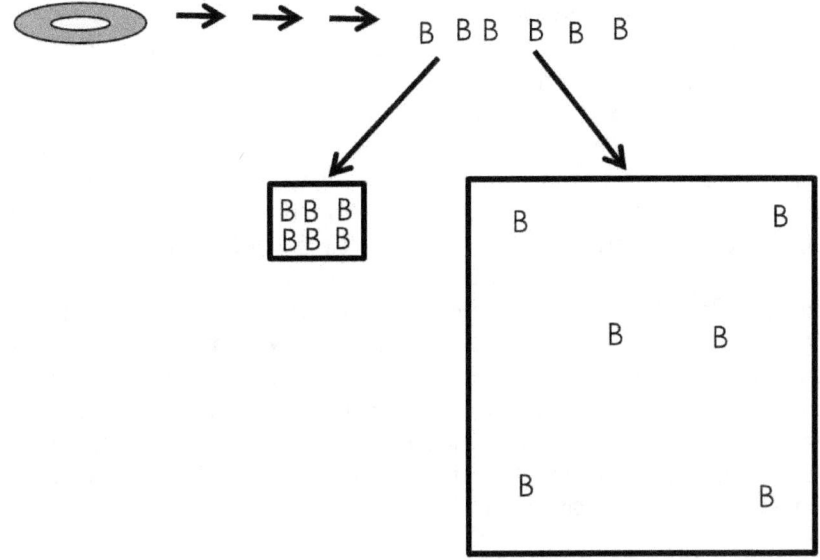

Figure 5. The hemoglobin in red cells is broken down to bilirubin (B). In normal individuals this occurs at a constant rate (about 1% of red cells broken down per day). If the volume or urine fluid is low (left), then the urine has a high concentration of bilirubin and will be very yellow; this is an indication of dehydration. If the volume of urine fluid is large, then the urine is pale or colorless which is an indication of good hydration status. Note that the concentration is different in the two states but that the content (total number of B's) is the same.

Pilots are also told that if they are thirsty, then they are also dehydrated. There are cells in the brain that sense the dehydration and activate the thirst center. This is a key method for the body to regulate water balance. The kidneys can only concentrate urine so much and so there will always be some urine fluid loss per day. In addition, we lose fluid from sweating and exhaling the humid air from our lungs. We can't avoid these losses, so intake is the method to fix the problem. Unfortunately, in modern society, many have learned to ignore or overlook the thirst reflex. However, you know that your dog does not have to check the color of its urine to determine if it should drink water; your dog is in tune with its thirst center.

Can you get too much water?

If you drank blood, so that the salt concentration was the same as in our blood, then your kidneys can urinate out the excess salt and water, but it does take time. However, because the kidneys have a minimum salt amount that they can excrete, all urine is going to contain some salt. Therefore, a person that drinks too much water without salt can have real trouble. Unfortunately, advice was given to marathon runners, endurance athletes, and soldiers in the desert, to drink as much as possible. Some of these people pushed themselves too far and drank too much water without any salt. So even with time, their kidneys could not excrete all the extra volume. These patients can get problems because their ratio of salt to water in the brain gets too low. In fact, some have died from this. Gatorade™ was one of the first companies to realize that it was important to replace both water and salt so they have tried to design their drinks to replace both the water and the salt lost in perspiration.

Fish and aldosterone

In some fish, cortisol has several roles. One role, as in other vertebrates, is to raise blood sugar. Another role is to increase Na retention and K excretion, which is the role that aldosterone plays in mammals, but fish don't have aldosterone. How did the aldosterone regulatory system evolve? The starting point is an ancient animal that makes the cortisol receptor, has an enzyme that makes cortisol and has a way that the cortisol receptor activates gene transcription. In this ancient animal, when cortisol binds to the cortisol receptor, genes regulating pathways that increase glucose levels and genes that increase Na reabsorption are turned on. We move from this ancient animal to modern humans, where there are now 2 receptors, one which when activated turns on genes regulating pathways that increase glucose levels and another which when activated turns on genes regulating pathways that increase Na reabsorption. This part seems to have been figured out and I'll detail the hypothesis in the next paragraph. In addition, we need to have an enzyme that makes aldosterone and an enzyme that is expressed in the kidney that breaks down cortisol. The current evidence is that the new receptor evolved first, a long time before an enzyme evolved to make aldosterone. I'm still trying to find out if anyone has determined when the cortisol breakdown enzyme was first expressed and also when it was localized to the kidney.

Physiology

How did we get 2 receptors with different selectivity? As you might guess, the first step was gene duplication. That is, ancient animal 2 had an error in duplicating its DNA and it ended up with 2 copies for the ancient receptor on a chromosome instead of one. I would have guessed that the ancient receptor bound cortisol and that through a series of mutations, one copy of the duplicate receptor gained the ability to bind aldosterone. But that is not the current theory. The current evidence is that the ancient receptor didn't have a very precise binding pocket so it could bind both aldosterone and cortisol. After duplication, one of the receptors kept this ability and the other receptor had a few mutations (one at a time) that resulted in a receptor that binds cortisol well and aldosterone very poorly which is our present day situation.

The ancient receptor being able to bind aldosterone and cortisol didn't cause any problems, because these ancient animals didn't have any aldosterone! The current theory is that it pays to have receptors and enzymes that aren't too specific, just specific enough to get the job done. So the ancient receptor only provides a selective advantage if it binds cortisol and not a steroid like, say estrogen or testosterone that the animal also makes. For the ancient receptor, there is no advantage to selecting against aldosterone because it would never encounter aldosterone in real life. And by remaining somewhat non-selective, the receptor could respond to aldosterone when an enzyme finally evolved to make it.

This discussion of the evolution of the aldosterone system is meant to illustrate several points: one is that a complete evolutionary theory needs to provide a sequence of single mutation steps that gives rise to a complex new trait. In this case, we now have the single steps that give us the aldosterone receptor.
1. Gene duplication to give us A and B copies of ancient receptor.
2. Mutations that alter binding of B receptor to decrease aldosterone binding.
3. Result: Receptor A binds both aldosterone and cortisol. Receptor B binds cortisol much better than aldosterone.

Later, the enzyme that makes aldosterone evolved. Also later the enzyme to break down cortisol in the kidney evolved. As far as I know, we don't know which order these two steps occurred.

A second point is that there is an advantage to not being too specialized. Since the ancient receptor was not too specialized, in the A version, it was able to accommodate a new hormone, aldosterone, when it "showed up".

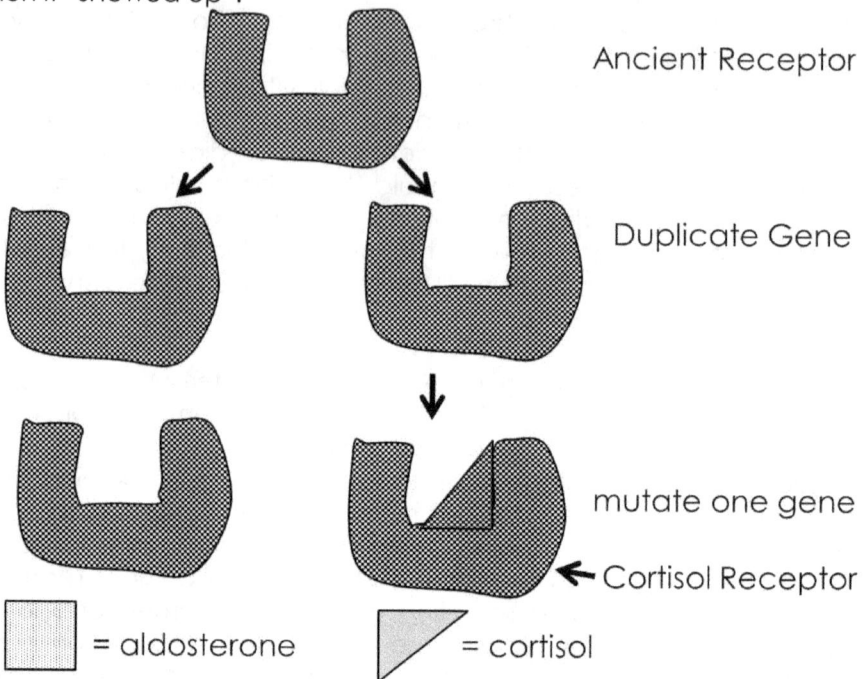

Figure 6. Evolution from a common receptor for cortisol and aldosterone to current status. The ancient receptor could bind both aldosterone (square) and cortisol (triangle). This lack of specificity caused no problems because the ancient animal did not make aldosterone. The ancient receptor duplicated. Then one version of the ancient receptor mutated so that it only bound cortisol and not aldosterone. So today we have a receptor that only binds cortisol and a receptor that binds both cortisol and aldosterone; the latter one has specific actions because in the kidney, there is an enzyme that breaks down cortisol before it could activate the ancient, non-specific receptor.

I think this sort of thought experiment is very interesting. It provides an understanding of why the system is the way it is. And it also provides insights into the fact that there are some pathological conditions where aldosterone is too high and this damages other organs, for example, the heart. In the heart the aldosterone receptor is also expressed, but the heart does not have the enzyme to breakdown cortisol. This means that when a person is stressed out and cortisol rises, it activates what we call the aldosterone receptor and over the long term, this turns on genes that actually hurt the heart.

AQUAPORIN DISCOVERY LED TO NOBEL PRIZE
The discovery of aquaporins is an interesting tale. This is a story of perseverance, luck, hard work, and key insights. As a result Agre was awarded the Nobel Prize. Agre started out trying to purify a different protein from red cell membranes. Red cell membranes are a convenient model system as it is easy to obtain red blood cells. In addition, mammalian red cells don't have any internal organelles, so there is only one type of membrane: the one on the outside. Furthermore, the red cell membrane has many fewer proteins. When red cell membrane proteins were first separated by size on a gel and stained with the usual blue stain, only 7 major bands were seen. Since no one knew which function went with which band at that time, the bands were numbered, Band 1, Band 2, etc.

In the course of trying to purify his protein of interest, Agre wanted to be sure that it was really pure. Therefore, instead of using the standard blue stain, he used a silver stain, which is much more sensitive. He became quite frustrated with his attempts because he was down to just 2 bands, the one he wanted and another one. The other one had not been mentioned by anyone else. He went back and examined the gels from other workers as published in their research papers. Sure enough, when they did a silver stain, the band that bothered Agre was there. However, because it was not seen with the blue stain and they weren't interested in it, no one else mentioned it. Agre realized that red cells had lots of this protein, even though it didn't stain well with the blue stain. He was intrigued by this protein: Why were there so many copies in the red cell? In addition, why was it so odd, in not binding the blue dye, but binding silver very well? So he started dividing his time between his original protein of interest and this annoying protein. He purified the annoying protein and sequenced it. It had a very interesting sequence, so he continued to study it. He eventually identified it as the first water channel protein and called it aquaporin.

Agre went on to determine that aquaporin defects were present in some diseases. Furthermore, he wanted to understand its molecular structure. To do this, he needed to have a large amount of the protein and make a crystal. With a crystal, each molecule is aligned in a regular pattern. Details of the structure can then be determined by bombarding the structure with x-rays. By this time, he had realized that eyes were an even better source of aquaporin than red blood cells. So he sent his students to the local slaughterhouse and they collected a lot of cow eyes. In the lab, they ground up the eyes and purified the aquaporin. It crystallized easily which was unusual for a mammalian membrane protein at that time. When they examined the structure in detail, it didn't quite make sense from what he expected from the amino acid sequence of cow aquaporin. After a lot of head scratching, they realized that they had inadvertently purified aquaporin from the bacteria (E. coli) that contaminated the eyes at the slaughterhouse. This was fortuitous because at that time it was much easier to crystallize bacterial membrane proteins than mammalian membrane proteins. This combination of work, identifying a novel protein and its function, determining its role in pathological conditions, and determining its structure and basic mechanism, resulted in his Nobel Prize.

Let's return to Solomon. Solomon presented because he had excess urination. This could have been because the kidneys were not getting the right message from a hormone. He needs to retain salt and so his aldosterone levels should be increased. His values were almost ten times higher than normal! So the message is there, but the kidneys have a genetic defect that prevents them from reabsorbing sodium and chloride. There are a number of different transport systems for reabsorbing sodium. A mutation that makes any one of them function poorly can lead to excessive salt and water lost in the urine. Diuretics also target these transporters and high doses of diuretics also lead to excessive salt and water loss. Some of these transporters also transport potassium and thus these patients are also at risk for low blood potassium unless they make sure their diet contains adequate potassium. Solomon needs to be sure he eats enough potassium and drinks enough water each day to remain in both water and potassium balance.

Physiology

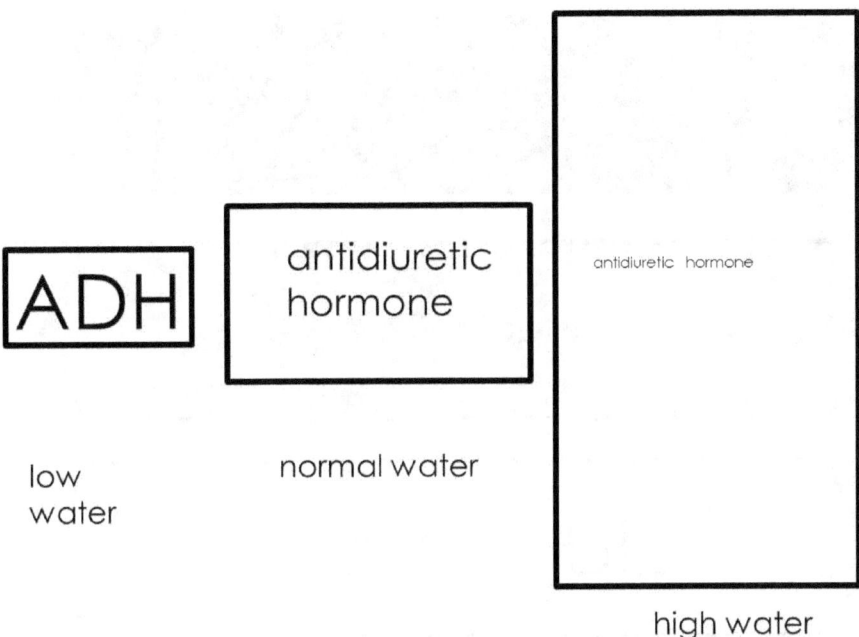

Figure 7. How changes in blood water (volume) change ADH secretion. In the middle is the normal state. When the amount of water decreases (smaller rectangle on left) then antidiuretic hormone (ADH) secretion is increased, which leads to less water loss in the urine.

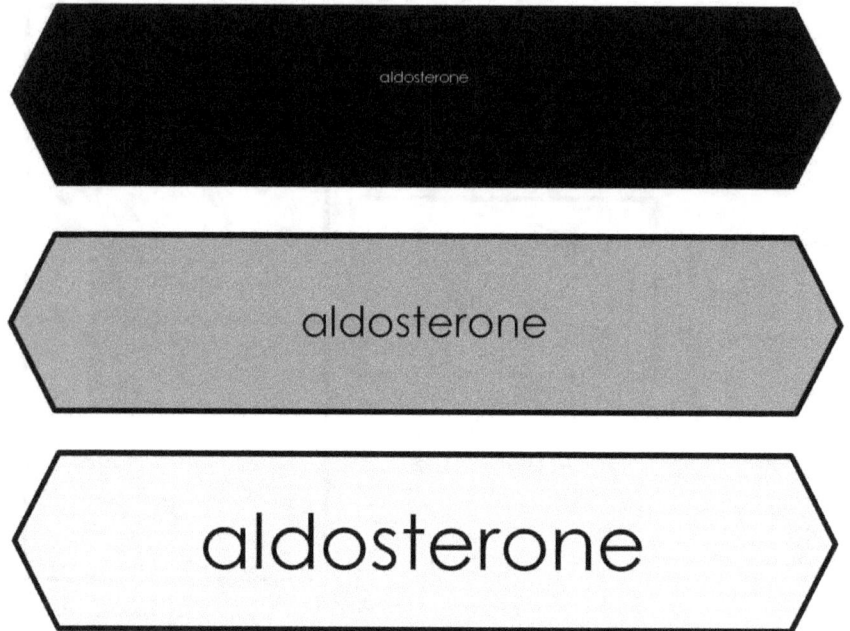

Figure 8. How changes in salt concentration alter aldosterone secretion. In the middle is the normal state. When the concentration of salt decreases (lighter filled box on bottom) then aldosterone secretion is increased, which leads to less sodium loss in the urine. When the amount of salt increases (darker filled box on top, then aldosterone secretion is decreased which leads to less sodium reabsorbed so more sodium is lost in the urine.

Physiology

In summary, we have learned the two major systems the body uses to regulate water and salt balance.
- If there is no intake of salt and water, then the kidney cannot correct the problem, one does need to drink or take in salt. But the change in kidney function can conserve salt and water while the person goes to find some.
- Low water simulates the thirst center
- Low water increases antidiuretic hormone secretion
- Antidiuretic hormone is a water soluble hormone
- ADH secretion increases the pathways for water movement in the last part of the kidney tubular; because of the osmotic gradient, water is reabsorbed when ADH is high and the pathways are present
- Low salt stimulates the salt appetite
- Low salt increases aldosterone secretion
- Aldosterone is a steroid hormone and like all steroids is lipid soluble
- Aldosterone secretion increases the pathways for sodium reabsorption in the kidney
- These system also responds to too much water or salt by turning off the pathways.
- Overconsumption of real licorice can mimic an overactive aldosterone receptor because licorice inhibits the enzyme that breaks down cortisol in the kidney. Cortisol can then bind to the aldosterone receptor and activate it.

From Anemia to Vampires

Literature Case
A young woman, Iris B. Doce, went to her health care provider because she was feeling fatigued often. The health care provider checked her blood and she had anemia. This means that the fraction of her blood that is red blood cells is lower than normal. The health care provider then gave her some medication. Iris returned in two months; while she was still anemic another property of her blood had changed. After a second medication, her blood was back to normal and she was feeling much better.

Anemia remains one of the most common nutritional problems in the world. In this chapter, we will discuss the physiology behind this real case from the medical literature because it was unusual, complicated and had a surprise. Then we move on to consider the mythical stories of vampires and their purported desire for blood. We end with a discussion of why males, on average, have a larger fraction of their blood volume as red cells than females.
The major sections in this chapter are:
- Blood components and hematocrit
- Menstrual blood loss
- Regulation of body iron
- Review of major steps regulating iron absorption
- Iris after 2 months
- Vitamin B12
- Main points so far
- Vampires are in the title, why is that?
- Males vs. females and red blood cells

Blood components
How did Iris' health care provider determine that the fraction of Iris' blood that was red cells was lower than normal[1]? Blood can be considered to be made up of 3 major components: red blood cells, white blood cells, and plasma[2]. Red blood cells are the most dense,

---

[1] Hematocrit is the jargon term for the fraction or percent of blood that is red blood cells.
[2] Plasma is what is left when the red and white blood cells are removed. It contains antibodies, ions, pre-clotting factors, proteins, and intercellular

white blood cells the next most dense and plasma the least dense component of blood. By centrifuging the blood, the red cells end up in the bottom of the tube, the white cells are a narrow layer in the middle and the plasma is the top (and largest) layer.

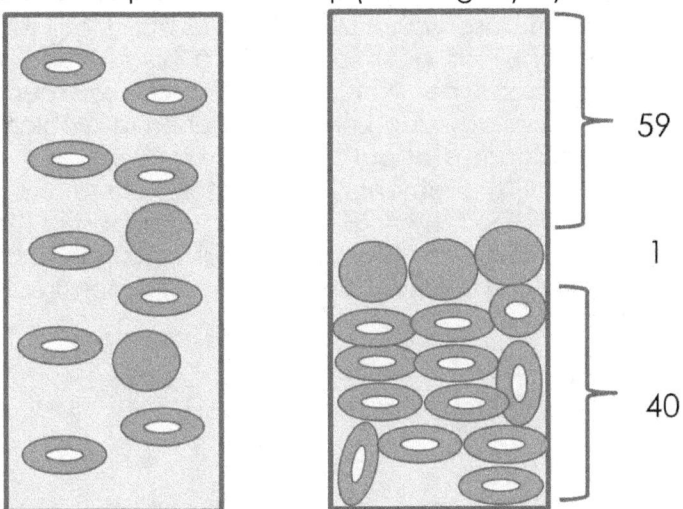

Figure 1 Blood and its components.
Tube A: Tube of blood as taken from body, a "homogenous" mixture of cells and plasma.
Tube B: Tube of blood from normal person after centrifuging. Centrifuging causes the densest objects, the red cells, to go to the bottom and they make up about 40% of the total. The next densest objects are the white blood cells and they make up about 1% of the volume. The top 59% is the non-cell portion, called the plasma. When the clotting factors are removed from this non-cell portion (plasma), the remaining solution is called serum. The ratio of red cells (40) to total volume (40+1+59) is the hematocrit (40% or 0.40).

The normal range for the fraction of blood that is red cells depends upon whether one is male or female. This should immediately raise a question in your mind: is this difference in normal range due to culture/environment/diet or due to biology? We will discuss the details of normal range as well as answer the culture/environment/diet vs. biology question in the last section of the chapter.

signaling molecules. Intercellular, like interstate, means signaling between cells. Intracellular is signaling within one cell.

Menstrual blood loss
Often, many clinicians think of menstrual blood loss as the cause of anemia in menstruating women. When I was teaching medical students in the problem based curriculum, this was often their response when a female patient had anemia. I was a bit surprised since I donated 100 mls of blood per week for two decades when I was researching blood physiology and I never had anemia problems. The standard literature amount is about 75 mls of blood loss per month during menses in healthy women[1]. Furthermore, there are many people who lose more blood every 56 days or so and most of them do not become anemic. Red Cross blood donors give 500 mls (2 ½ cups) up to every 56 days, which is 2.5 to 5 times the normal menstrual blood loss, see Figure 2.

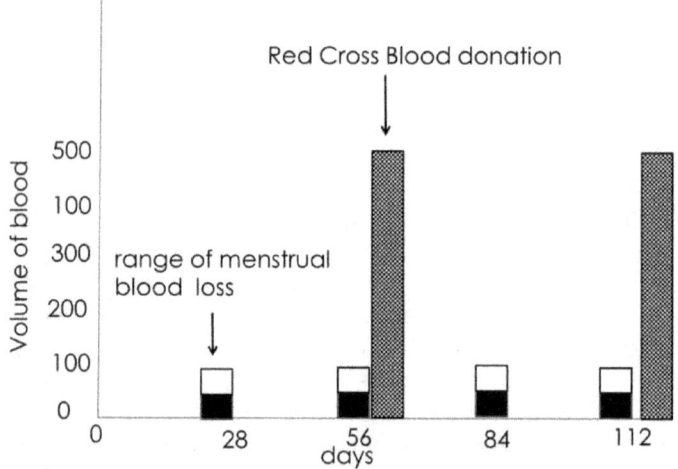

Figure 2 Red Cross donations result in much larger blood loss than menstrual blood loss.
Checked bars: Red Cross blood donations: 500 mls every 56 days.
Black and white bars: estimated range of normal menstrual blood loss: 50-100 mls every ~ 28 days.

Addition of iron in the diet often solves the anemia problem. This means that, before the extra iron, the anemic person was in negative iron balance, excreting more iron than they were absorbing. This

---

[1] There is not a lot of data on the amount of blood loss during menses. 75 mls is about 3/8ths of a cup.

would be analogous to being in debt; one is spending (excreting) more money than one is earning (absorbing). Does this mean the problem is too much spending (excreting)? Well, it depends upon one's point of view, since the relevant parameter is the balance and both contribute. Clearly, we have examples where more blood loss (more spending) does not cause debt, the Red Cross blood donors. Yet also, more iron intake (more earning) does fix the anemia problem in some cases. So whether the problem is cultural (women tend to eat less red meat than men do) or biological (men, except blood donors, do not normally lose blood) depends upon one's perspective.

Regulation of body iron
Iris's doctor did a few tests. The lab tests confirmed that red cells only made up about 0.2 (1/5th) of her blood. The normal range for women is from 0.36 to 0.46. However, it was also noted the size of her red cells was less than normal.

Even though both her diet and menstrual blood loss seemed normal, the measurement of her blood iron and ferritin were below normal. Ferritin is a protein that binds to iron and typically is low when the body iron stores are low. The doctor decided to have Iris increase her iron intake[1].

Body iron amounts, like body calcium amounts, need to be tightly regulated. As discussed in the chapter, Bubbling Beverages Bad for Bones?, parathyroid hormone directly regulates how much calcium is lost in urine and indirectly regulates how much calcium the intestine absorbs. Iron regulation is handled differently and the complete story was worked out in the early 2000's. For a long time, it had been known that urine contains essentially no iron. Iron is not water-soluble in contrast to calcium. Sweat has been checked and it contains essentially no iron. Other studies showed that the amount of iron in blood regulated how well the intestine absorbed iron. However, what

---

[1] Each of the following would give APPROXIMATELY 100% of the recommended daily iron for men and non-menstruating women: 9 ounces of beef or pork, 24 ounces of chicken or tuna or white turkey meat, 2 cups of baked or black beans, 8 slices of bread, 1 cup of fortified oatmeal, 8 cups of brown rice, 1 cup of soybeans, 2 cups of spinach. Menstruating women need roughly twice as much iron.

hormone[1] told the intestine that the body had enough iron has only been recently worked out.

The current model for iron regulation is that the key regulator is hepcidin[2], which is made by the liver. When iron levels are high, the liver releases hepcidin into the blood stream. One of hepcidin's actions is to bind to an iron transporting protein on the membrane of the intestinal epithelial cells.

Let's review just a bit about the epithelial cells here so you don't get lost with where the iron is and where it needs to be transported, see Figure 3. Starting from the inside center of the intestine (called the lumen), there is the luminal membrane of the epithelial cell, then the inside of the cell, then the serosal[3] (also called basolateral) membrane of the epithelial cell, then the interstitial fluid, then the endothelial cells of the capillary, then the blood. Most small molecules in the interstitial fluid can easily diffuse from the epithelial cell to the endothelial cells and get into the blood; thus, we often don't mention this process and just refer to molecules that have just come out of the serosal side of the epithelial cell as going to, or being in, blood.

---

[1] A **hormone** is a molecule released into the blood and signals some cells elsewhere in the body. As the hormone travels throughout the blood stream it passes through capillaries. In capillaries, some of the hormone comes out of the blood and can end up next to cells. If the cell has a "receptor" for the hormone, then those cells can respond and change their behavior.

[2] Hepcidin is a hormone as it is released into the blood by the liver and has effects on intestinal cells. Another way to characterize hepcidin is by its structure; hepcidin is a **peptide** which means it is made up of amino acids. Proteins are also made up of amino acids; proteins just have more amino acids than peptides. It would make life easier if we just dropped the word peptide and used short protein instead!

[3] Serosal has the same root as serum, the fluid of blood.

Physiology

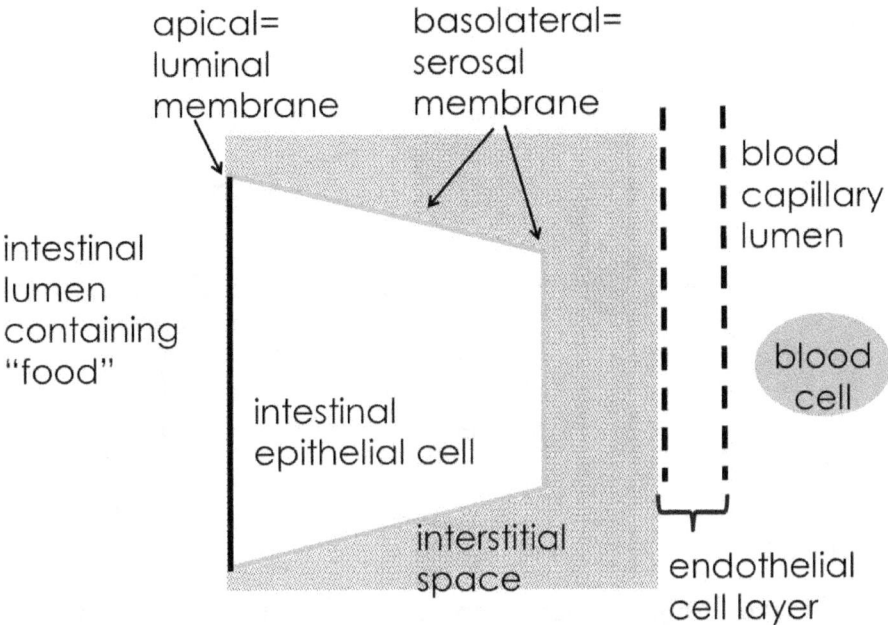

Figure 3 Intestinal cell terminology

Iron needs help from a transporter to get across the luminal membrane from the intestinal lumen into the cell. The iron also needs help from another transporter to get across the serosal membrane from the cell to the "blood". This serosal membrane transporter is called ferroportin. (Get it? ferro- for iron and portin for trans-portin'.) Figure 4 has a diagram of the key steps for iron absorption in the intestine.

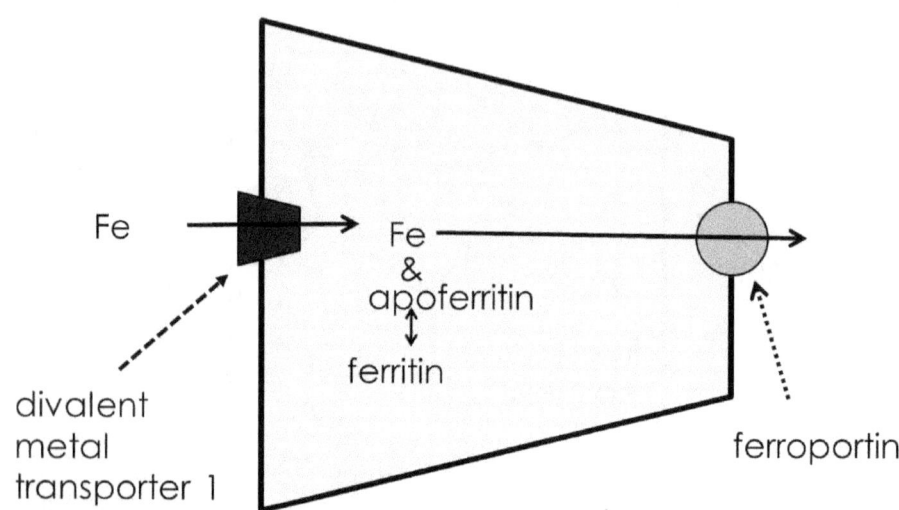

Figure 4. Iron transport across an epithelial cell. After iron enters on the transporter called DMT1, it binds to apoferritin to form ferritin. At the blood side, some iron comes off ferritin leaving apoferritin; the iron is transported out of the cell on a transporter called ferroportin. The protein that transports heme into the cell has not been identified. Once heme gets inside the cell, it is broken down. The iron still exits on ferroportin.

A high level of blood iron causes the liver to release hepcidin. Hepcidin binds to ferroportin, the serosal transporter, to form a complex. This complex, and its surrounding membrane, is taken into the cell. This process is called endocytosis and results in vesicles inside the cell. So what? Well, ferroportin's job is to transport iron across a membrane. When ferroportin is in the outer cell membrane, then iron is transported out of the cell and the iron can then get to the blood. However, if ferroportin is in vesicles, then the iron is just transported from the vesicle to the inside of the cell and the iron can't get to the blood. [1] One of the amazing things about intestinal cells is that we get

---

[1] Remember that the cell has a membrane of lipid molecules that separate its inside from its outside. In this outer membrane are many proteins. The process

rid of them every 3-5 days.¹ Where do the intestinal cells go? Well, the intestinal cells are sitting inside the intestine, so they just fall off and join the rest of our digestive products. Most of the sloughed off intestinal cell contents are digested and the iron just travels through the rest of the gut and goes out in the feces.

To review the major steps regulating iron absorption (see Figure 5):
- high blood iron leads to high liver excretion of hepcidin
- high liver excretion of hepcidin leads to high blood hepcidin
- high blood hepcidin binds to ferroportin on the outer cell membrane
- ferroportin-hepcidin complex moves to vesicles inside the cell
- dietary iron can not get out of the intestinal cell
- no dietary iron moves into the blood (and it didn't need the iron, since blood levels were high)
- the intestinal cells are sloughed off, containing the iron, and the iron passes through and it lost in the feces

---

of pinching off some membrane (including some membrane proteins) is called endocytosis. This pinched off membranes form internal closed structures in the cell, vesicles, often depicted in 2-D as circles. Here is an analogy. The cell is a house. Vesicles are like closets in the house. Ferroportin is like specific windows (ferro-windows) in walls that let iron get through the wall. An epithelial cell layer is approximately like a series of row houses-they are touching each other. If we want to transport iron from the backyard (intestinal lumen) to the street (blood), we need a way to get iron into the house. Once iron is in the house, we need a way for iron to get out of the front wall of the house. We start with the condition that there are ferro-windows in the front wall of the houses, so iron can get out. Then a signal (hepcidin) comes from the recycling center (the liver) that we don't need any more iron in the streets. Hepcidin binds to the ferro-windows and the window and walls move to the inside and become a new closet (vesicle). Now there are no ferro-windows in the front wall of the house and no more iron gets into the street. Iron could go from inside the closet to inside the house, but it can't get to the street.

¹ In contrast to the short lifetime of intestinal cells and their fate of being sloughed off, red blood cells last 120 days, and the body recycles most parts of the red cell.

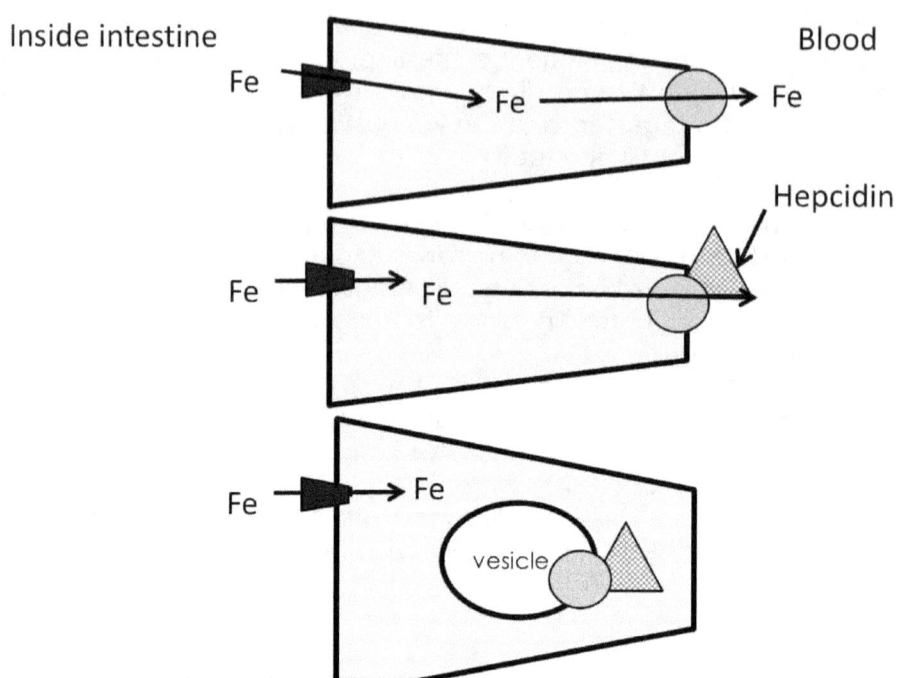

Figure 5. Three epithelial cells illustrating iron transport.
On the left (apical) membrane of the cell are iron (Fe) transporters as black quadrilaterals. On the right membrane of some cells are ferroportins as gray circles. Not shown are the iron-binding-proteins inside the cell.
The top cell is in from a person with low body iron and low hepcidin levels, so the cell can transport dietary iron from the intestine to the blood.
In the middle cell, the body now has plenty of iron and hepcidin levels increase. The hepcidin has bound to the ferroportin. When this happens, the ferroportin will be internalized.
The bottom cell shows the internalized ferroportin and there is no way for the iron to get to the blood; it is stuck in the cell.

This regulation of the number of transporters in the outer membrane illustrates 2 themes that reoccur several times in this book.
1. Specific example: For iron to move from the inside of the cell to the blood, 2 things are needed. One needs a gradient, that is, more iron inside the cell than outside. Secondly, one needs a pathway. Missing either one means no movement.

General statement. Movement requires a gradient (or force or pressure) AND a pathway.[1]

2. Specific example: The amount of iron transport across the serosal membrane is regulated by how many ferroportin transporters are in the serosal membrane; ferroportin transporters can be removed from the outer membrane by endocytosis forming vesicles inside the cell. General statement: Transporters are removed from the outer membrane to vesicles by endocytosis; transporters are inserted into the outer membrane from vesicles by exocytosis.

Figure 7 is a figure showing these concepts; insulin regulation of skeletal muscle sugar transport and antidiuretic hormone regulation of kidney water transport will have figures similar to Figure 7.

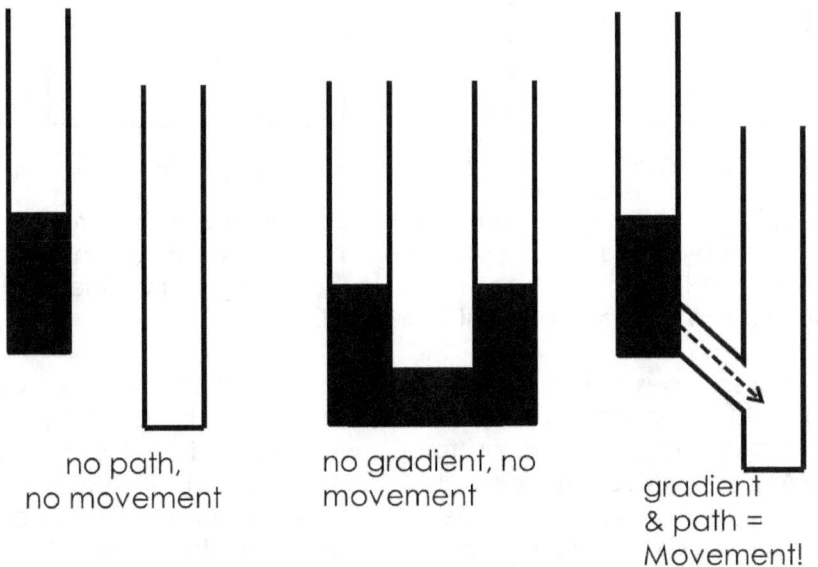

no path, no movement

no gradient, no movement

gradient & path = Movement!

Net movement requires both a gradient and a path

Figure 6. Key General Principle. Movement requires both a path and a gradient. Most remember that a gradient or force is needed to get flow, but often the need for a path is forgotten. If there is a beverage nearby right now, it is presumably off the flow. Why isn't the liquid running down to the floor-because there is no path!

---

[1] See Figure 8 of Orange you Glad? for another example.

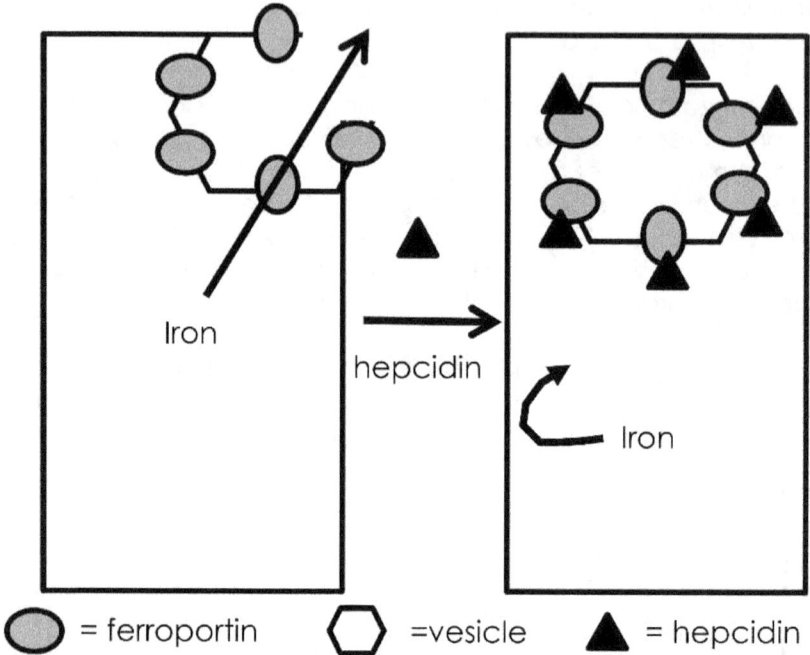

Figure 7. In the square cell on the left, the oval ferroportins are in the surface membrane which provides a path for iron to move. On For the cell on the right, all the ferroportins are in the hexagonal vesicles in the cell and so iron cannot move out of the cell.

Iris, of course, does not have high iron in her blood, she has low iron. The model for iron regulation I just outlined predicts that, if Iris' liver is working properly (and it is as far as we know), it should not be releasing hepcidin, so hepcidin will not be complexing with ferroportin. Therefore, there should be plenty of ferroportin in the serosal membrane ready to take iron out of the cell and into the blood, if Iris were eating enough iron.

Iris after 2 months

After taking iron pills for two month, Iris went back to the doctor not feeling any better. The fraction of her blood that was red blood cells remained at about 1/5th instead of the normal 2/5th, which was frustrating to both Iris and her doctor. However, the doctor looked closely at the lab values of the blood and noted a surprising change. When she first went to the doctor, the red cells were smaller than

Physiology

normal[1]. Small red cells are characteristic of iron deficient anemia. At this follow up visit, the cells were still abnormal-but now they were too large![2] One of the major causes of having each cell being larger than normal but having too small a fraction of blood that is red cells is vitamin B12 deficiency.

Vitamin B12
Vitamin B12 is also called cobalamin which is a contraction of cobalt and vitamin because this vitamin contains the element cobalt. How do you absorb B12? What can go wrong? What are sources of B12?[3]

Even though essentially no nutrients are absorbed in the stomach, the stomach is important for us to get B12. Why is this?

First a reminder: vitamin B12, like all water-soluble vitamins, needs help getting across the intestinal cell membranes. Why? Let me give you an analogy. Water soluble compounds are like fish; they both like being in water. The cell membrane is like a piece of land that separates two bodies of water (the outside of the cell and the inside of the cell). A fish needs help getting across this piece of land just as a water-soluble molecule needs help getting across the cell membrane. A bucket can help carry the fish across the land, just as transporters can help carry the water-soluble molecule across the cell membrane. Alternatively there can be a tunnel that allows the fish to move from one side to the other, see Figure 8.[4]

---

[1] The fancy term is microcytic; micro meaning small and cystic meaning cell.
[2] The fancy word is macrocytic.
[3] To get 100% of one's daily B12 one could each one of the following: 1 serving of 100 % B12 fortified breakfast cereal, 4 to 6 ounces of fish (trout, salmon, tuna), 3 double cheeseburgers, 12 ounces of steak, 5 cups of milk or yogurt, 8 ounces of cheese or 10 eggs.
[4] Sometimes, to check on one's understanding, it helps to think about another case. We just discussed water-soluble compounds, how about fat-soluble compounds? See Orange you glad? for the analogy for fat soluble compounds!

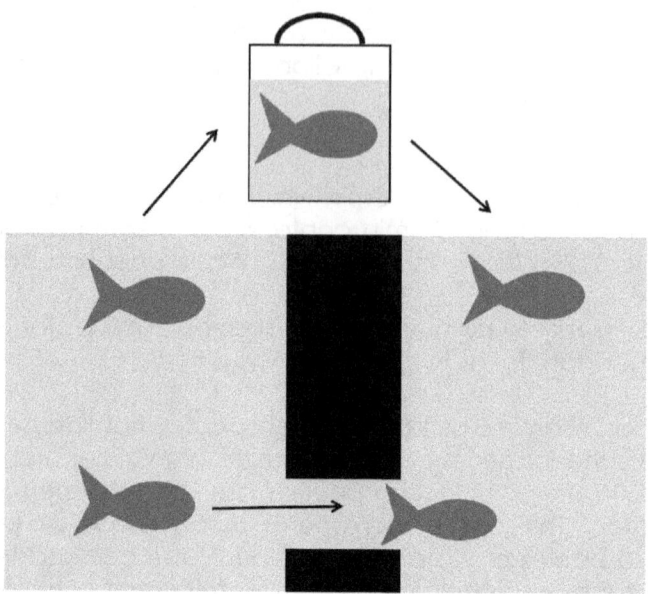

Figure 8. Transport of water soluble compounds across a lipid (oil) membrane.
A bucket or a tunnel can serve as a carrier or transporter that helps water soluble compounds (fish) cross the fat, or lipid, cell membrane (land). These carrier or transport proteins are found in membranes, such as the cell surface membrane and in cell organelles.

Being water soluble, B12 needs a bucket, a membrane transport protein. A general rule is that all water soluble compounds are absorbed in the small intestine and that is true of B12. However, most water soluble nutrients are absorbed in the first 2 segments of the small intestine.[1] B12 and bile salts are unique in only being absorbed in the third, and last, segment of the small intestine.

Even though B12 is absorbed in the small intestine in the portion farthest from the stomach, the stomach is also important for B12 absorption. How can this be? Well, it turns out that the transporter in

---

[1] The first segment is called the duodenum, the second segment, the jejunum, and the third segment, the ileum.

the ileum doesn't recognize naked B12. The transporter is looking for B12 cloaked with a protein called Intrinsic Factor. Moreover, Intrinsic Factor is made in the stomach! In fact, lack of stomach Intrinsic Factor production is one of the leading causes of B12 vitamin deficiency, see Figure 9.

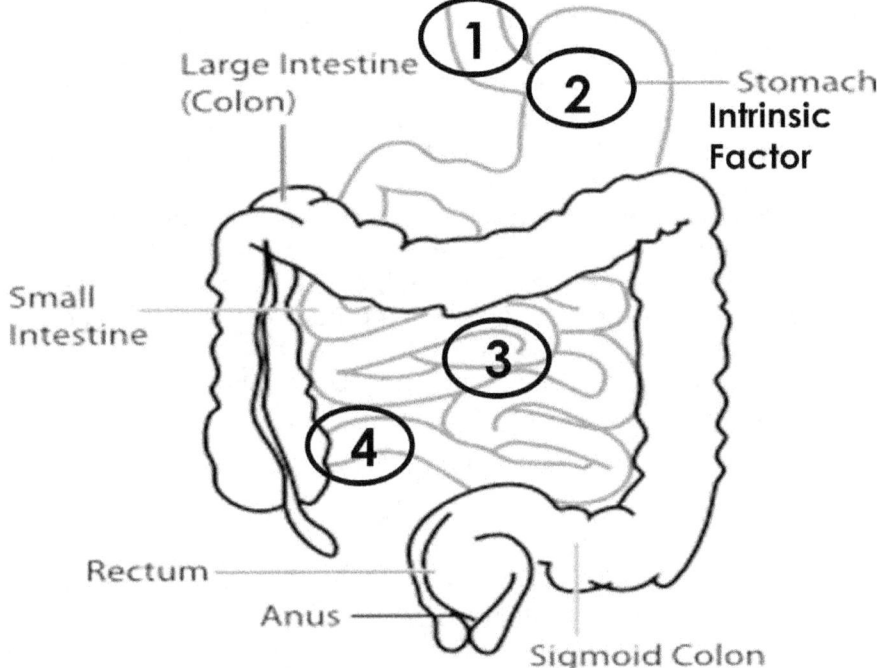

Figure 9. The major steps in vitamin B12 absorption.
1. B12 is eaten in diet.
2. Stomach secretes Intrinsic Factor.
3. Someplace in the small intestine, Intrinsic Factor and B12 get together.
4. At the end of the small intestine just before the colon (ileum), B12 is absorbed. The transporter only recognized the complex of B12 and Intrinsic Factor.

The digestive tract outline is from Wikipedia; I have added B12 info. http://en.wikipedia.org/wiki/File:Digestive_system_diagram_edit.svg

Macrocytic anemia means the red cells are large and the fraction of blood that is red cells is low. The occurrence of macrocytic anemia is thought to be on the decline in the U.S. but this might come at some cost. By the way, Iris lived in Taiwan. If Iris had lived in the U.S. she might not have had the macrocytic anemia even if she ate the same types of food in both places. Why? Because the U.S. has started fortifying all wheat products with folate, B12 deficiency no longer produces macrocytic anemia. Folate enrichment is done to decrease the incidence of spina bifida. Spina bifida is a condition in which the spinal cord doesn't close properly in the developing fetus. This can occur if the mother is folate deficient. However, high folate levels also prevent macrocytic anemia in patients that are vitamin B12 deficient. That sounds like a good thing since you avoid both spina bifida and macrocytic anemia. Those are both good parts. However, B12 is also important for healthy nerve function and folate does not help restore that. Therefore, a person that is B12 deficient and folate rich will have normal red cells but abnormal nerve function. At first, this abnormal nerve function can be hard to detect, as the initial deficits are easily overlooked, for example, occasional memory loss. Unfortunately, the nerve damage is irreversible, so by the time the nerve damage gets severe, there is a possibility that giving back B12 won't restore the function to the original state, see Figure 10.

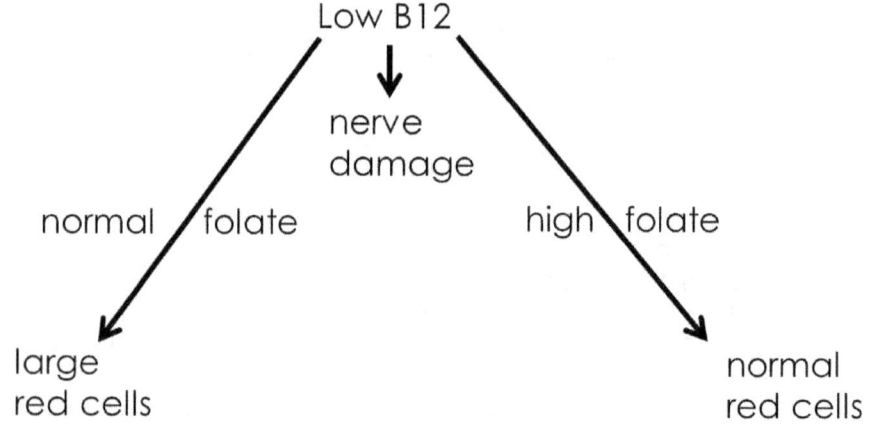

Figure 10. B12 and folate effects.

Physiology

With low B12 and normal folate, one gets both big red cells (easy to detect) and nerve damage (hard to detect).
With low B12 and HIGH folate, one still gets the nerve damage, but the red cells are now normal in size.
With low folate, babies are at higher risk for spina bifida.

However, in Iris' case, it turned out her diet was low in vitamin B12. This is a common problem in people on a pure vegetarian diet because plants don't make B12. You don't need very much B12 and healthy bodies have enough B12 stored to last for years, so one can go a long time without B12 before running into problems. As mentioned above, the neurological problems start out with subtle effects and, unfortunately, are irreversible.

Main points thus far:
- Iron levels in the body are a balance of iron intake and iron excretion.
- Unlike many compounds, no iron is lost in urine. The way iron is lost is through the feces, sloughed off skin, and blood loss.
- Hepcidin is a liver hormone that regulates the amount of iron absorbed from the small intestine into the blood.
- Hepcidin binds to ferroportin. Ferroportin is the membrane transport protein that transports iron out of the intestinal cell and into the blood. When hepcidin binds to ferroportin, the transporter moves into the cell; the iron then remains inside the intestinal cell and is lost in the feces when the intestinal cells die and are sloughed off.
- Iron deficient anemia results in red cells that are too small.
- B12 is also important for making red cells, as well as for proper nerve function.
- B12 absorption requires Intrinsic Factor secretion from the stomach and B12 absorption occurs in the ileum, the last segment of the intestine.
- B12 deficiency results in red cells that are too large and in nerve damage. The large red cells can be avoided, even if B12 deficient, if the diet is high in folate.
- No one has yet described a regulatory pathway for regulating B12 absorption or excretion and one may not be needed.

Vampires are included in the title, why is that? As you probably know, vampires are purported to drink blood. To make biochemistry more interesting, some scientists have speculated that vampires thirst for blood might be the result of having porphyria.[1] There are some interesting similarities between the traits of vampires and the symptoms and treatments of porphyrias, but what we've learned about iron and digestion should reveal one of the major flaws in this line of thinking.

Porphyrias are a set of diseases that involve enzymes in the pathway to make heme. Heme is an iron binding molecule. One protein that has heme groups is hemoglobin.[2] Hemoglobin binds to oxygen and allows the blood to deliver oxygen to the tissues. Heme is also important in other proteins, for example, myoglobin and cytochrome c oxidase.

There are 8 steps needed to make heme; in the different porphyrias one of the eight enzymes is defective. Many of the symptoms are the result of the buildup of the intermediate that would have been modified by the defective enzyme. As an analogy, imagine an assembly line for building a car; there could be 8 steps, for example, the chassis, the seats, adding the doors, adding the trunk, putting in the engine, adding the hood, adding the lights, and painting and say it is done in that order. Then if the person who is to put on the trunk isn't working properly, there will be a buildup of which type of car precursor? The one with a chassis, seats and doors.

It is true that some porphyrias result in the accumulation of molecules that are light sensitive. When light shines on these molecules, they become reactive. This can cause skin problems. Therefore, patients with these types of porphyrias, like vampires, avoid sunlight.

It is also true that in some porphyrias, treatments involve giving heme (like giving the factory finished cars). One might jump to the conclusion that vampires' drinking blood is therefore a "treatment" for their porphyria, but this conclusion is scientifically defective. Let's see where it goes wrong. It is true that blood contains heme; the problem

---

[1] Porphyria is based on the Greek word for purple; in some of these diseases, the patients' urine is sometimes purple.
[2] Heme is the Greek word for blood. Hemoglobin is the major protein in red blood cells; it accounts for their red color.

is that the heme you eat does not end up in your blood. Dietary heme is transported into the intestinal cell. Inside the intestinal cell, the heme is broken down. The iron is removed, and then iron and the other breakdown products are transported from the intestinal cell to the blood. Since patients with porphyria usually have plenty of iron, they don't need more iron; what they need is heme since they have a hard time making heme. But the dietary heme is broken down in the intestinal cells (see Figure 11). Therefore, dietary heme doesn't help, does it?

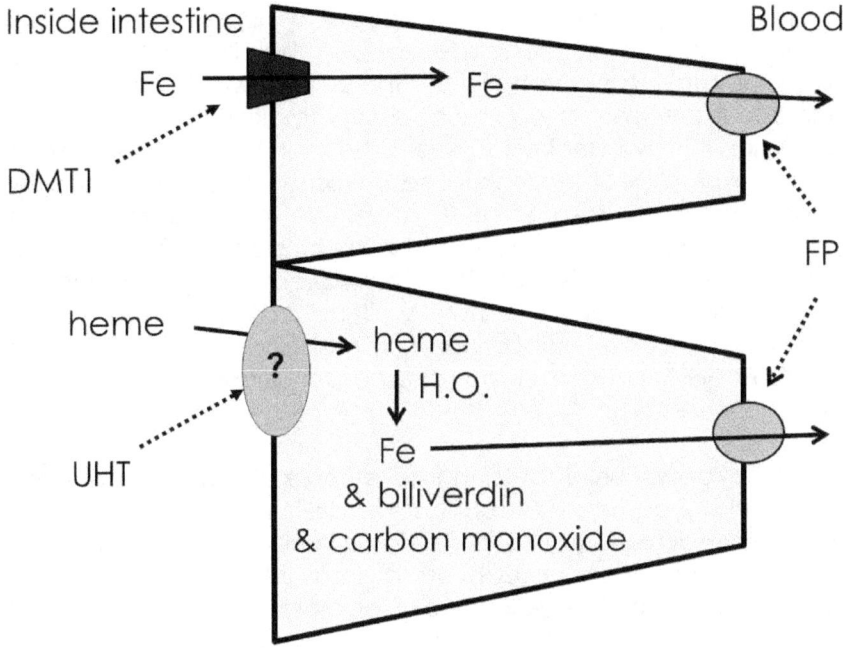

Figure 11. Contrasting iron and heme absorption.
The top cells shows that iron (Fe) is transported across the apical membrane by divalent metal transporter 1 (DMT1) and across the basolateral membrane by ferroportin (FP).
The bottom cell shows that heme is transported across the apical membrane by an as yet unknown heme transporter (UHT). A transporter is required because heme is water soluble. Inside the intestinal cell, heme is broken down to iron (Fe), biliverdin, and carbon monoxide. The iron is transported across the basolateral membrane

by ferroportin (FP) to get to the blood, depending upon the blood hepcidin levels.

You may have been surprised in Fig. 11 to see that one of the breakdown products of heme is carbon monoxide, but that is correct. We do make a little bit of carbon monoxide and there is good, but not conclusive evidence, that sometimes carbon monoxide is used as a signaling molecule between cells. In fact, I missed a chance to perhaps have been one of the discoverers. The hall I worked on also housed the Pulmonary Medicine department. I would often exchange greetings with members of that department when we passed in the hall. One of the physicians told me about a puzzle. One of their patients had high carbon monoxide levels in their blood. Normally, this is a sign of smoking cigarettes. But this patient was in the Intensive Care Unit and there was no way he could be smoking. In fact, he was getting oxygen. They checked to see if the air intake for the hospital was close to where people smoke and it wasn't. They even checked the oxygen tanks to see if there might be a small amount of contaminating carbon monoxide. I was as puzzled as the physician was. But in retrospect, it is likely that the patient was producing the carbon monoxide; in fact it is now well accepted that in certain severe disease states, blood carbon monoxide levels increase because the body's pathways to make carbon monoxide increase in some disease states.

Another breakdown product of heme is biliverdin and if you know Latin or Spanish, you might recognize "verdin" as similar to the word for green. Biliverdin was first identified as a green compound in bile. In the colon, biliverdin in converted into bilirubin and bilirubin accounts for some of the yellow color of plasma and urine. In the colon, bacteria metabolize the bilirubin into compounds that are brown, hence the brown color of feces.

The idea that vampires drink blood to treat their porphyrias, their inability to synthesize adequate heme, doesn't make digestive sense. While there is heme absorption by small intestinal cells, the heme is broken down inside the cell and the iron gets to the blood, but not the heme. Hence some modern porphyria patients get heme intravenously.

Males vs. females and red blood cells

Finally, let's return to the issue of why, on average, males have a larger fraction of their blood volume as red cells than females. First, let me point out that the ranges overlap, just as height does; on average men are taller than women, but there certainly is a large amount of overlap.

Second, let me talk about what "normal" means in medicine and physiology as it is DIFFERENT from the lay use of the term. Normal is the unfortunate term used to describe a particular distribution that often occurs in nature; sometimes the term Gaussian is often used after one of the first "discoverers". For a large population of adults, height often follows a normal distribution. So what is an abnormal height? Many scientists and statisticians have a criteria that if the probability of something happening by change is less than 5%, then it did not happen by chance. This has carried over to medicine and the "normal" range is the range that 95% of the population has. Just because you are outside the 95% does not make you unhealthy or "abnormal" in the lay sense of being weird. I'm 6 foot 4 inches and my wife is 5 foot; we are both outside the normal range for height, but I would not call our height unhealthy or weird. As another example, red hair only occurs in about 1% of the population, which makes it rare and unusual, but not unhealthy or abnormal.

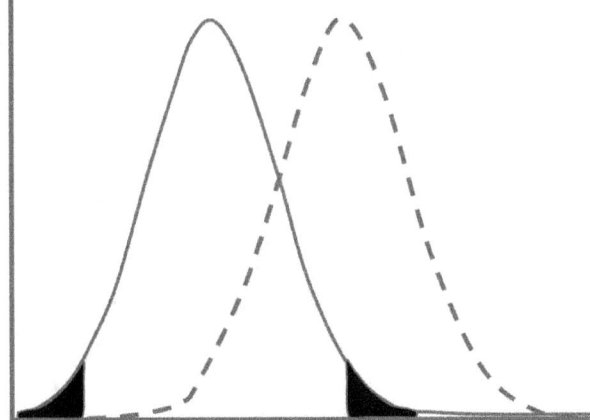

Figure 12. Two normal distributions. While the most common value (the value at the peak) is different, you can see there is substantial overlap between the solid and dashed lines. The filled in portion at the bottom of the solid line is about 5% of the total amount of area under the curve; if someone had values for a parameter in this range,

they would be considered statistically outside the statistical normal range.

One big difference between men and women is their hormones. If you give people (male or female) some extra testosterone, they make more red cells. So at least part of the difference in the amount of red cells is biological; I think it is difficult, in the general population, to rule out some cultural effect as women are more likely to have less iron in their diet.

How does testosterone increase the fraction of blood that is red blood cells? We don't know yet. The most obvious possibility is that testosterone has a direct effect on making red cells. Alternatively, testosterone could regulate other hormones that are already known to regulate making red cells. One of these is erythropoietin. So far, there is no conclusive evidence for this. From this chapter, you might guess another hormone, that is, hepcidin, since one needs iron to make red blood cells. In fact recent data suggest that at least in some men, testosterone decreases hepcidin, which is consistent with additional iron being transported into the blood, which might lead to making more red cells.

Physiology

Chapter Review

In summary, Iris B. Doce was unusual in that her anemia was due to 2 separate causes: iron and B12 deficiencies. In her case, it appears that the deficiencies were due to lack of these compounds in her diet.

- Movement of iron and B12 across cell membranes required both a concentration gradient and a pathway. The pathways for the two compounds are different.
- Iron absorption across the intestine requires transporters in the apical as well as the serosal membranes.
- Hepcidin is a hormone secreted by the liver that regulates the amount of serosal membrane iron transport to the blood.
- When body iron stores are high, hepcidin increases.
- An increase in hepcidin leads to a decrease in serosal membrane iron transport by ferroportin.
- When iron stays inside the intestinal cells, it does not get to the blood but is lost in the feces.
- B12 absorption requires the stomach to secrete Intrinsic Factor; Intrinsic Factor is needed to cloak B12 in the intestine and allow it to be recognized in the terminal part of the small intestine where it is absorbed.
- Heme is broken down inside our intestinal cells therefore we are unable to absorb the heme from our diet into our blood stream.
- Scientists often use the word "normal" is a statistical sense. If a particular parameter has a bell shaped curve, then often the 5% largest and smallest values are "abnormal". This is different than the lay use of the word abnormal which often implies a "defect"

Physiology

Got Cows?

Fictional Cases

1) 5 kids always drink milk at breakfast and met at each other's houses after school to have milk and cookies. Even in high school, they enjoyed glasses of milk when very thirsty. However, getting together after college graduation, many adamantly refused milk. They wouldn't tell their friends why, but it was because when they had milk, the noticed that they had lots of farts, mild diarrhea and often a bloated feeling. Why does milk in a majority of the world's adult population cause these symptoms and why do some adults not have this problem?

2) Adam is having severe neurological problems. His health care provider has determined that the immediate cause of the problems is high blood ammonium levels. The health care provide will need to do many more tests to determine why the blood ammonium is high; in the meantime, and in order to avoid ammonium getting so high as to put Adam into a coma, the health care provider gives Adam some lactulose. The lactulose gives Adam diarrhea. Why would the health care provider be pleased to see the diarrhea?

3) Angie is in the Peace Corps and providing health care for some patients who have cholera. In the US, some cholera patients get IV fluids to prevent dehydration, but sterile saline was too expensive where Angie is. She cooked up rice soup with plenty of salt to drink and fed it to her patients. They all survived. Did her rice soup play a role in their survival? How did the cholera cause the diarrhea?

4) Cynthia Francis presented to her health care provider with night blindness shortly after the fall clock change. The health care provider determined that the problem had probably been going on longer, but because it now was getting dark earlier, Cynthia Francis noticed the problem. Cynthia Francis was eating plenty of yellow and orange vegetables, so her diet did not seem to be the culprit, as least for the night blindness. But in discussing her diet and digestive habits, the health care provider determined that Cynthia Francis had stools that floated in the toilet and that often left a greasy residue, so much so that Cynthia Francis usually used the rest room at school and not at home, for bowel movements to avoid getting teased or scolded at home. What was causing the diarrhea in this case?

In this chapter we will discuss some of the physiological causes of diarrhea and learn more about how the digestive system works.
- What is it in milk that causes the farting and loose stools?
- How do those that can drink lots of milk avoid the problem?
- What is this genetic basis for the difference in milk digestion across populations?
- Do you think lactose persistence is a dominant or recessive trait?
- Why not just use glucose and galactose, why bother to make lactose?
- Do other foods have related problems?
- Do animals have related problems?
- What about the enzyme to make lactose?
- GI Water balance
- How does cholera cause diarrhea?
- What caused Cynthia Francis's diarrhea?

What is it in milk that causes the farting and mild diarrhea when some adults drink too much?
There are two hints about the problem. However, before we get to the hints, let's ask, **what is the value of drinking milk?** For modern, as well as ancient people, milk is a great source of calcium and protein.

So the first hint is to think about the many areas where they don't drink milk as adults, what do they use for a protein (and calcium) source?

In parts of India, yogurt is a popular food. Yet many adult Indians don't drink milk. So what has the yogurt bacteria done to the milk to make it so that Asian Indians can eat yogurt and not have runny stools?

The second hint comes from looking at grocery store shelves. If you look closely at the types of milk available, you might see some lactose-free milk.

Lactose is the type of sugar in milk; you've probably heard of sucrose and maybe even glucose. All three of these are sugars and they all end in –ose, which is a chemical suffix to indicate a carbohydrate. Many things related to milk use the prefix "lact", such as lactation. Thus lactose is just Latin/chemistry for milk sugar. Why would milk sugar be a problem and table sugar (sucrose) or blood sugar (glucose) not be?

Let's think about the symptoms; the people who can't drink milk as adults fart and have loose stools. That sounds like a digestive system[1] problem. And since farting and pooping occur at the end of the process, it would be a good first guess that the problem occurs in the large intestine or colon. And in this case, that is correct, but there are definitely diarrheal problems that occur in the small intestine or are the result of pancreatic, liver or stomach problems. If you eat sucrose, table sugar, you have enzymes in your small intestine that break it down into glucose and fructose. Your intestine has transport proteins that allow you to transport these sugars across your intestinal epithelial cells and into the blood. Children, and those adults that can easily drink milk, have an enzyme, lactase, the breaks down lactose into glucose and galactose[2]. They then absorb glucose and galactose. What about those of us that can't drink milk as adults? We lack the enzyme lactase. (Yogurt bacteria have a similar enzyme to break down lactose, which solves the problem. And lactose-free milk has been treated with the enzyme. In fact, you can buy the enzyme at some stores and make your own lactose free milk.)

If you (like me) lack lactase, then you can't break down lactose. That means the lactose passes from your small intestine to your colon instead of being broken down in the small intestine. Now I have a surprise for you. You might think most of the cells in your body are yours, with your genes, but it turns out your colon probably has more bacterial cells than you have cells in your body. Fortunately, all these bacteria usually get along and are actually quite beneficial to us. However, the bacteria need food and some can use lactose. When the bacteria break down the lactose, they produce a number of products, including protons (H^+), hydrogen gas (H_2), and several short-chain-fatty-acids.

All 3 of these are interesting. Let's start with the hydrogen gas. This can account for some of the farting. It also provides a convenient test,

---

[1] Another term for digestive system is gastrointestinal system. "Gastro" is a prefix referring to stomach. "Intestine" refers to guess what?
[2] "Galact" is the Greek term for milk.

at least in theory, to identify who lacks lactase. Hydrogen gas is easy to measure. None of your cells produce hydrogen gas but some of the bacteria can produce hydrogen gas from lactose. Hydrogen gas would not be easy to collect as it exits the anus, but you can actually collect it in your breath! How can that be? Well the hydrogen gas produced by the bacteria in the colon gets mixed up with all the other colonic contents and some can diffuse across the epithelial layer and get into the blood stream, which delivers it to the lungs, where it again diffuses across another layer of cells and then is exhaled. Hydrogen gas can get across the epithelial layers because it dissolves well in oil (lipids) that make up cell membranes. Sugars do not dissolve well in oil. You can test this by taking a teaspoon of sugar and adding it to a ½ cup of oil. It won't all go into solution, but it will easily dissolve in a ½ cup of water if you stir a bit. Glucose can go across the epithelial layer in the small intestine because those cells have proteins that transport it across the cell membrane lipid layer; glucose won't cross the epithelial layer in the colon because those cells do not have the transport proteins.

Next up is short chain fatty acids. This is the cause of the runny diarrhea. In the colon, the epithelial cells don't let most water-loving molecules across, but they do let water go across. If there are more particles per volume on one side of the colon (say inside) than on the other (say, the blood side), then water will move. Water moves until the amount of water inside the colon is equal on both sides. The force that moves the water is termed osmotic pressure. [1]

---

[1] Osmolarity is the measure of the number of particles in a solution. You might remember molarity from chemistry. Molarity measures the number of **molecules** in a liter of water. Well osmolarity is the number of **particles** in a liter of water. (It would make more sense to me to call this particlarity.) For many solutions, their osmolarity is their molarity. For example, a 300 millimolar sucrose solution is essentially 300 milliOsmoles of sucrose. But 300 millimolar sodium chloride (NaCl) is about to 600 milliOsmoles of salt. Why is this? Well, when you put the NaCl into water, the sodium (Na) and chloride (Cl) go their separate ways and are now two particles.

Physiology

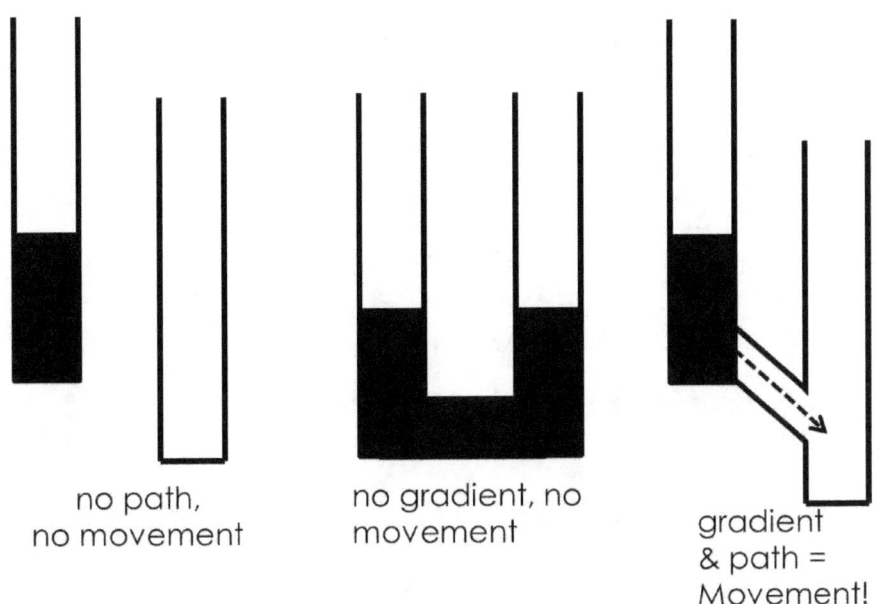

Net movement requires both a gradient and a path

Figure 1. This is a reminder that for molecules to move there needs to be a gradient and a pathway. In intestinal cells and most other cells, water moves through proteins called aquaporins[1].

---

[1] "Aqua" is, of course, refers to water and "porin" is meant to imply a pore, or channel, that takes water. More on aquaporins in the chapter, Soy Sauce and Licorice as Medicine.

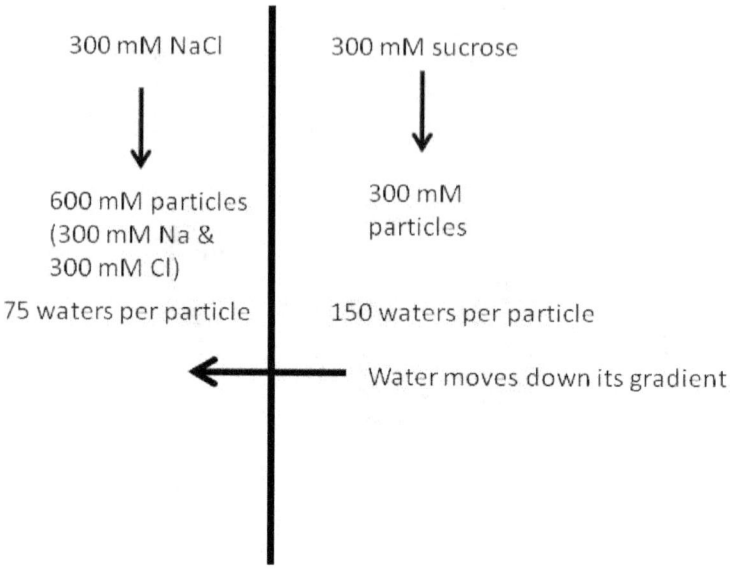

Figure 2. Even though one adds 300 mM to each side of the membrane the left side will have a greater osmolarity (more particles) because the NaCl dissociates into 2 particles, whereas the sucrose remains 1 particle. Because of the different number of particles, there is a gradient for water to move (an osmotic gradient). If there is a path, then water will move down its gradient.

Figure 2 is an example of a laboratory situation where water would move across a dialysis membrane because there is an osmotic gradient.

In the colon, before the bacteria starts to work on the lactose, there are as many particles on the "food" side as on the blood side. Then the bacteria get to work and break down lactose into several different smaller molecules, mostly short chain fatty acids. Now there are more particles on the food side than on the blood side, so water moves from the blood to the food side. And the stools are more watery than before. In fact, one of the jobs of the colon normally is to remove water from the food side. As you might guess, this first involves moving some molecules so that there are fewer particles on the food side and that would create a force to cause water to leave.[1] It is the short chains fatty acids and other particles generated from the breakdown of lactose that account for the mild diarrhea.

Last, the presence of protons as a product of bacterial digestion of sugars doesn't cause the colon any problems. But it can be of use to clinicians. Protons can bind to ammonia (NH_3) to form ammonium (NH_4^+). [1]Like many charged molecules, NH_4^+, will not go across the cell membrane lipid layer. But NH_3 does. As I'll explain in a minute, this results in NH_4^+ accumulating in the compartment with more free protons (which would be the acidic compartment).

We are constantly producing a little bit of NH_3 and we urinate most of it, so the blood concentration stays low. But in some disease states, the blood levels of NH_3 can get high. This is not good; among other things, it can make parts of the brain malfunction. While a clinician is trying to figure out why the NH_3/ NH_4^+ is so high, as in Adam's case, she might give the patient an undigestible sugar. When the sugar gets to the colon, the bacteria break it down, producing protons and other things.

Before the sugar is given, the pH is about the same inside the colon and in the blood. Because the NH_3 is the same on both sides, and the proton concentration (pH) is the same, the NH_4^+ concentration has to be the same. After the sugar is given, there are now more protons on the food side of the colon. This drives the reaction on that side toward NH_4^+, reducing the NH_3 concentration. So now there is a gradient for NH_3 to move from the blood to the colon. The result is that there is more nitrogen (as NH_3 & NH_4^+) in the colon and less in the blood. And of course, the nitrogen (as NH_3 & NH_4^+) is then excreted. You might be thinking, gee, if you can trap NH_3/NH_4^+ with acid, why not trap in the stomach. That's a great thought and works as far as the acid and NH_3 & NH_4^+ go. But think about what happens to stuff in the stomach. It moves on to the small intestine, which is essentially the same pH as

---

[1] We should point out here that there are some epithelial layers that don't let water cross. A key one is the kidney; you need to be able to separately vary how much water and how much salt you urinate see Soy Sauce and Licorice as Medicine.

[1] I remember the names because ammonia is smaller than ammonium and 3 is smaller than 4.

blood. So any NH_3/ NH_4^+ trapped in the stomach is returned when it hits the intestine and the blood value goes back up. Trapping of ions by pH is important in the kidney and also accounts for some drug-drug interactions in the stomach.

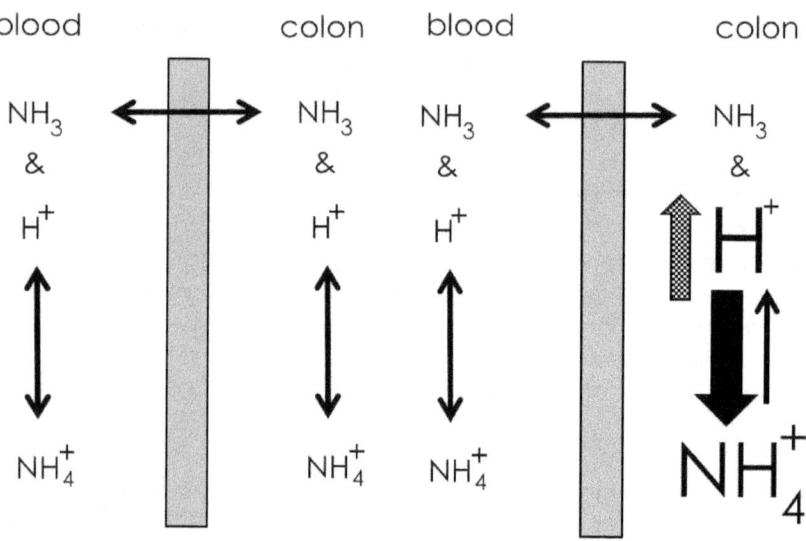

Figure 3. Trapping ammonium by increases H^+ (protons). The key feature is that ammonia (NH_3) can cross the membrane, but ammonium (NH_4^+) cannot. On the left we have the normal state where the H^+ concentration is about the same in blood and colon. The NH_3 moves until it is the same concentration on both sides (there is a pathway, and if there is a gradient, it will move.) Since H^+ is the same on both sides, the reaction between H^+ and NH_3 yields the same amount of product, NH_4^+. In contrast, on the left side, the H^+ concentration has been increased, for example, by bacteria metabolizing lactulose. The increased H^+ then means more NH_3 reacts H^+, forming more NH_4^+; since it cannot cross the membrane, it is trapped and will be lost in the feces.

While we are thinking about too high blood NH_3/NH_4^+, let's mention another key feature/concept of physiological problem solving: when blood levels are too high, you can break the problem down into four categories of possibilities. First, maybe you are eating too much, which is highly unlikely for NH_3/NH_4^+. Second, you are producing too much which is certainly possible with NH_3/NH_4^+. Third, you aren't breaking it down fast enough, and often this involves the liver. Fourth, you aren't getting rid of enough. Which organ does that? The kidney! So now we understand why some people get bloated, fart, and have runny stools after drinking milk: it's because colonic bacteria breakdown the lactose into products that cause gas and an osmotic gradient that leads to diarrhea.

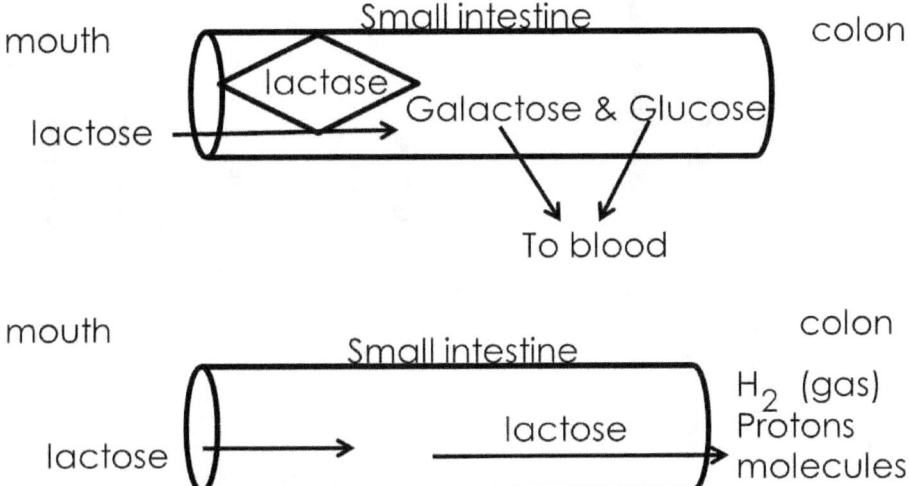

Figure 4. Difference between lactase persistence and no persistence.
Top: If lactase is expressed, lactose is broken down in small intestine. No lactose gets to the colon.
Bottom: If no lactase is present, then lactose gets to colon where bacteria break it down causing farts, bloating, and runny stools.

Is lactose present in all dairy products?
This is basically a chemistry question. Lactose is water-soluble, so dairy products that have little if any water will have little if any lactose. Butter is one example as it is basically all fat. Cheese is lactose-free for 2 reasons: one is that it has little water and two is that the bacteria that make cheese convert the lactose into other products.

What is this genetic basis for the difference in milk digestion across populations? In almost all baby mammals and people, lactase is made; it is a very rare baby of any mammalian species that lacks lactase. In all adult animals and most adult people, lactase is not made. But there are important exceptions; some examples are shown in Figure 3. It used to be that the term "lactose intolerance" was used to describe those adult humans that didn't make lactase, but since that describes the majority of the world, the term lactase persistence to describe adults whose intestines persist in making lactase, is now preferred.

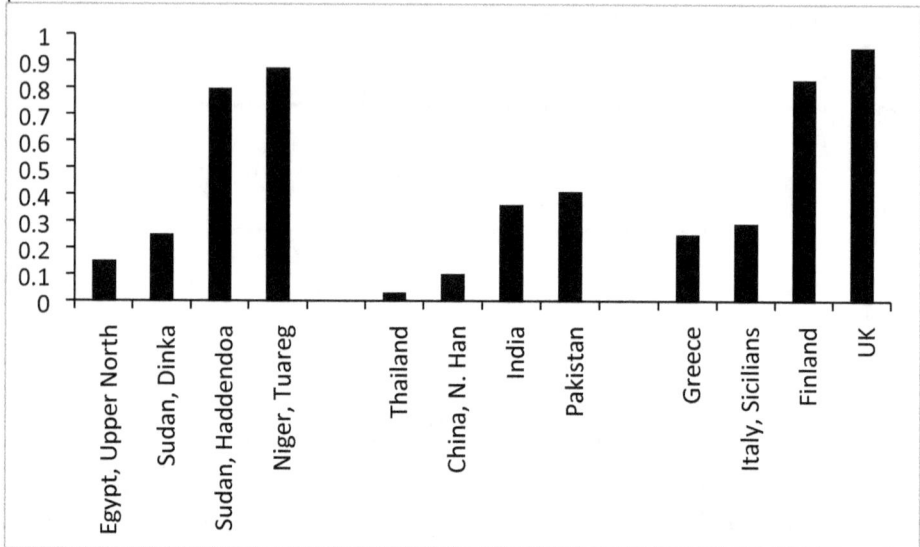

Figure 5. Fraction of population in which substantial lactase activity persists as adults. Shown are two extremes for Africa, Asia, and Europe. Note that because Asia has a much larger population than Europe, the absence of lactase persistence is the most common condition in people.[1]

The fact that it is fairly common suggests that the mutation that resulted in lactase persistence provides a selective advantage for the people that have it. One must remember that the selective advantage, and the selection, occurred a while ago. As an analogy, think about many of the roads connecting old towns in this country.

---

[1] Data selected from Supplementary Table 1, using only studies with n>100. Itan Y, Jones BL, Ingram CJ, Swallow DM, Thomas MG. A worldwide correlation of lactase persistence phenotype and genotypes. BMC Evol Biol. 2010 Feb 9;10:36.
PubMed PMID: 20144208; PubMed Central PMCID: PMC2834688.

Physiology

The original paths in this country were probably Indian trails, which in the east became cow paths, and then horse paths and some are now roads. As roads for cars they seem to meander, but they presumably made sense when the Indian used them again and again to walk; for example, they were easier to walk, avoided some dangerous areas or had food and water available along the way. Thus we should not ask what advantage does lactase persistence provide for someone on today's diet, but rather what was the advantage in ancient times. We would be confused trying to understand why there is a winding road between 2 towns if we only thought of interstates and not about the transportation situation when the path was first "developed".

The following are some plausible ways that lactase persistence might have been selected for. Developing these ideas allowed me to learn some interesting information and apply some physiological knowledge. You might guess that the selective advantage that milk provides might be related to the fact that milk is a good source of calcium. Calcium of course is important for bone health. Many of your grandmothers are probably concerned about their bone health. What ethnic group is particularly prone to problems with bone health? While osteoporosis can affect any ethnic group, and males as well as females, women of Scandinavian descent are at highest risk. What about living in Scandinavia might make it tougher to make good bones? In addition to calcium, bone health depends upon vitamin D, see the chapter Bubbling Beverages Bad for Bones? Light converts a cholesterol-like compound into a precursor of the active form of vitamin D. In Scandinavia, in winter, there isn't much sun, and in ancient times people probably stayed indoors a lot and also wore lots of clothes to stay warm. Also, apparently, thousands of years ago, a gene (or several genes) was mutated in this population that increased the risk for osteoporosis. So any women inheriting this mutation would have a better chance of survival if she also inherited the mutation for lactase persistence. Her ability to drink milk as an adult would increase her calcium intake and counteract some of the risk for osteoporosis from her other mutation(s). This would therefore explain why lactase persistence is so common in people of Scandinavian ancestry. Obviously, this is merely a plausibility argument, but I think it raises some interesting issues. The argument relies on the correlation that women of Scandinavian heritage are at highest risk for osteoporosis and people of Scandinavian heritage are most likely to have lactase persistence (at least for European populations). This explanation also makes the prediction that the gene for increasing the risk for osteoporosis was mutated before the gene that led to lactase persistence. It is possible that future scientists will sequence DNA from ancient Scandinavians and be able to test this hypothesis.

However, this argument, that lactose persistence allows adults to get more calcium and reduce the risk of osteoporosis, doesn't work for the ethnic groups in Sudan and Niger that also have lactase persistence. These groups are not known to be at higher risk for osteoporosis and being close to the equator, they get plenty of sunlight. However, they have been raising cows in this region for thousands of years and having milk as a protein source presumably provided an advantage during times when it was hard to get other protein. Of course, people with lactase non-persistence can get a lot of protein by eating cheese (or yogurt) and avoid the side effects of lactose in the colon, since cheese and yogurt don't have lactose. But cheese and yogurt require more care to make. So this is plausible, but hardly conclusive.

Why don't all ethnic groups have lactase persistence if it is such an advantage? First, to be an advantage, the group needs to have cows. American Indians and indeed all the original inhabitants of the Western Hemisphere didn't have access to cows until the Europeans came. So even if the mutation arose, it would not have provided an advantage. In other areas, for example, southern Europe, there are 2 straightforward explanations. One is that the mutation never occurred in this population. If there is no mutation, then it can't be selected for. If there had been substantial intermarrying between, say Italians and Scandinavians, then you would think that would introduce the mutation into the gene pool. So the second explanation is that it does not confer an advantage; presumably Italians get enough sun, calcium, and protein from other sources and don't have the gene for increased risk for osteoporosis.

Do you think lactose persistence is a dominant or recessive trait?
The best way to answer this question is examine the inheritance pattern, which we'll do shortly. But first, let's see if we can predict the inheritance pattern based upon our understanding. Remember, we have 2 copies of each gene, one from our mother and one from our father. A dominant trait is one in which the person shows the effect with only 1 copy of the gene, whereas for a recessive trait, the person needs both copies of the gene. If we consider lactase persistence as

the effect, then we would say that lactase persistence is dominant if you need only 1 copy of the mutated gene to be able to digest lactose. If you have 1 copy of the mutated gene, then that gene is expressed in adulthood, and therefore its product, lactase, should be active and able to digest lactose. This of reasoning predicts that lactase persistence is dominant. And indeed, that is basically what is observed. Why did I qualify the previous sentence with "basically"? Well, it turns out that a number of factors apparently influence how well an adult can digest lactose and there are different degrees of tolerance/intolerance. So while this particular mutation would give you expression of lactase as an adult, how much lactose it takes for a particular person to experience symptoms shows a graded effect and is not all or none. Being graded seems odd for an either/or situation and suggests that there are many other factors regulating the level of expression of lactase.[1]

---

[1] Recently, a mutation that appears to account for lactase expression in adults has been identified. Surprisingly, it is ~ 14,000 base pairs upstream from the lactase genes. Normally, expression is controlled by base pairs much closer to the gene. In Europeans, there has been identified one mutation, C/T-13910. At the moment, the data suggest that this base pair is important for the transcription factor, Oct-1. Transcription refers to the process of converting DNA to mRNA. mRNA is then translated into a protein. Both DNA and RNA are sequences or polymers of 4 nucleoside bases (A, T, C, and G); proteins are sequences of 20 amino acids. I have 2 analogies for this process and I think the second one is better. Do you see why I would think that? If so, do you agree? Can you think of a better analogy? My first analogy stresses the fact that the building blocks are different for DNA, RNA vs. protein. It would be like saying transcription is converting text from Spanish to English and translation is converting text from English to Braille. The second analogy stresses the fact that it is protein that has the function. In this analogy, transcription is converting building instructions from Spanish to English and translation is converting the English instructions into a Lego structure.
In non-European populations that have lactase expression as adults, other mutations, at T/G_{-13915}, C/G_{-13907} and T/C_{-13913}, seem to account for this difference. The fact that there are different mutations in different ethnic groups suggests that the mutations occurred after people migrated to the different areas.

When writing this chapter, other evolutionary questions occurred to me and thinking about these reveals more interesting science. The only place lactose is made is in the mammary gland and it is of no use to mammals until it is broken down inside the intestine to glucose and galactose.

Why not just use glucose and galactose, why bother to make lactose?
One reason that lactose is an advantage in making milk is that it is not transported out of the cell, unlike glucose and galactose. In fact, lactose is made in an intracellular compartment called the Golgi. Lactose is beneficial because the mother can store the sugar for the baby throughout the day in the Golgi of the mammary gland. I think lactose provides an advantage because it is easier to store than glucose or galactose.

Making lactose requires a special enzyme and as we have discussed, lactase is a special enzyme to breakdown lactose. Both of these enzymes are needed for lactose to be useful in mammals.
Were these enzymes present in animals that don't make milk? If so, what was their function in animals that didn't produce milk and didn't have mammary glands?

What about the enzyme to make lactose?
The enzyme that makes, or synthesizes, lactose from glucose and galactose is called lactose synthase, a name that makes sense. Interestingly, galactose is one of the 3 main simple sugars found in most life forms, another being glucose. It is found not only in animals but also plants. So its name, galactose, is misleading, in the sense that it is not only found in milk, but most other living tissue, too.

Lactose synthase is actually the complex of two proteins. One of these proteins has the main catalytic activity and it is called galactosyl-transferase. This name should help you understand its function: to transfer galactose onto another molecule. For lactose synthase, the galactose is transferred to glucose. Galactosyl-transferase is present in most cells, because galactose is transferred to a lot of proteins. It plays an important role in a variety of cell processes.

The other protein that makes up lactose synthase is lactalbumin. "Lacta" again means milk and albumin refers to any water soluble protein, so lactalbumin is the major water soluble protein in milk. This protein is only expressed in mammary glands, which is why lactose can only be made there. It would be interesting to determine what the evolutionary precursor in non-mammals was that became lactalbumin. Lactalbumin is apparently evolved from lysozyme, an enzyme that is expressed in lysosomes, another intracellular compartment. And it would also be interesting to know the changes that occurred so it would only be expressed in mammary glands. By the way, it is unusual to have the specificity of an enzyme changed by what other proteins bind to it, but this is one example.

Since milk with lactose would be useless to babies that didn't have lactase, it would be interesting to determine which came first: the ability of the intestine to make lactase or the ability of the mammary gland to make lactose or whether both occurred at the same time.

Do other foods have related problems?
First, let's review the principle for why lactose causes digestive problems. First, the compound we eat is not digested and absorbed in the small intestine. Second, the compound is digested by bacteria in the colon and the bacterial digestion products lead to the problem. So these digestive responses can occur for a variety of foods, depending upon exactly which enzyme you have in your intestine and how much is expressed relative to how much of the substrate you eat. A common one, that is the source of lots of jokes, is beans such as kidney beans and pinto beans. These contain complex carbohydrates some of which require enzymes that most (maybe all) people lack. Hence the resulting gas. You can buy various products; one is called Beano that contains the enzyme to break down the offending carbohydrate.

Do animals have related problems?
Certainly, it is expected that adults of most animals will lack lactase. In fact, the problem of lactose induced digestive problems was realized about 100 years ago with dogs. This led to the finding that the lactase enzyme is present in intestines of baby animals but not adults.
Starlings apparently lack sucrase and one would predict that they would therefore have digestive symptoms if they ate much sucrose. I think it would be an interesting experiment to feed starlings seeds

coated in sucrose. Would the starlings associate their somewhat later digestive distress with eating those seeds? If so, would the distress be enough to make them want to avoid eating those types of seeds in the future?

There is a tribe in South Africa that lacks the sucrase enzyme and they have one of the lowest colon cancer rates known. Food for thought.

GI Water balance

First, a quick analysis of water balance in the gastrointestinal tract. As a ballpark figure, each of the major exocrine gland systems in the GI tract secrete about 1 liter of water per day. [1] What are these glands?
Salivary
Gastric (secreting water as well as hydrochloric acid (HCl)
Pancreas (secreting digestive enzymes, bicarbonate to neutralize the acid, as well as water)
Liver (secreting bile acids, bicarbonate to neutralize the acid, as well as water)
In addition, you probably drink about 2 liters of water per day. Only about 1 liter of water a day normally passes from the small intestine into the colon (the junction is called the ileocecal sphincter). The small intestine therefore has a net absorption of water of $(4 \times 1 + 2 + 2) - 1 = 7$ liters of water per day.

If water is moving across the intestine, you know 2 things, right? For movement to take place there must be a gradient and a path. The path for water is aquaporin and they are present in the intestinal cell membranes. The gradient is the osmotic gradient. In the early parts of the intestine, the osmotic gradient is due to the particles generated by digestion. Starches are essentially polymers of glucose or other simple sugars and proteins are essentially polymers of amino acids. A single starch or protein may consist of 100-1000 simple sugars or amino acids, respectively. Where you to eat a single protein or starch molecule, that would be only 1 additional particle and not be much of an osmotic gradient. But as you digest the protein or starch, you will generate 100 to 1000 particles.

---

[1] Often you'll see the number 1 ml/min; a quick calculation will reveal that this is about 1440 mls per day, which is pretty close to 1 liter, especially considering that the gland secretion rate changes throughout the day.

Digestive products would create an osmotic gradient to draw water from the blood to the intestine. That's why you often get thirsty after many meals. But does the intestine create an osmotic gradient to draw water from intestine to blood? The simple sugars and amino acids do not stay in the intestine, they are absorbed across the intestinal membrane, thus moving particles from intestine to blood and reducing the number of particles (and osmolarity) of the intestine. Another way for the intestine to generate an osmotic gradient is to move salt. If salt (NaCl) is moved from intestine to blood, water will follow and this is the driving force for much of the net water absorption in the small intestine.

Importantly, the absorption of glucose and most amino acids requires sodium. There is a sodium gradient, that is, there is a higher concentration of sodium in the intestine than in the cell. This sodium gradient provides a driving force that allows the uphill movement of glucose and amino acids. This means, by the end of the small intestine, there can be very little glucose in the intestine and yet a high concentration inside the intestinal cell and in the blood.

How does cholera cause diarrhea?
The cholera bacterium has a protein that binds to the outside of the intestinal cell. When this happens, it activates the cell, for example, increases cellular cAMP.[1] An increase in cAMP leads to an increase in chloride movement from the blood to the intestine; on the serosal side, this is accomplished by a sodium/chloride cotransporter and on the apical side by a chloride channel, which will be important in the Cynthia Francis case. As one moves the negative chloride from the blood to the intestine lumen, one ends up with more negative charge in the lumen than the blood. This creates a gradient for the positively charged sodium. It turns out that the space between the intestinal cells will allow some sodium to move. Thus there is a gradient (more negative charge in the intestine) and a pathway (a small leak between the intestinal cells) and sodium movement follows the chloride movement. This leads to more particles in the intestinal lumen and water follows.

---

[1] cAMP is cyclic adenosine monophosphate.

When cholera activates the cells, they really get turned on! Patients with cholera often produce 1 liter of diarrhea per HOUR! Imagine twelve 2-liter soda bottles as that is the volume of diarrhea in 1 day. The cholera itself does not usually kill anyone; it is the dehydration and loss of minerals that causes death.

An effective treatment for cholera is to give salt water with sugar. If the solution of salt and sugar is isosmotic, then there is no gradient at the beginning and the blood does not lose any more water. The intestinal cells absorb the glucose (with its helper Na)[1]; this means particles are moved from intestinal lumen to blood and water will follow.

Why not give a solution with very high glucose, wouldn't that give even bigger movements of water? In the long term, this strategy seems to make sense as the more glucose absorbed, the more particles moved and there is an additional gradient to move water. However, think about what happens at the beginning of the intestine. If the solution the patient drinks has a high concentration of glucose, then it has a high osmolarity. That creates a gradient to move water from blood to intestine, exactly the WRONG way to help the patient; they already have too little water in their blood. However, the idea is on the right track. Here's a way to state the problem: you want a solution with lots of sugar but that is isosmotic. Is there a way to package the sugar so that there is a lot of sugar but not much osmolarity? Remember that osmolarity is the number of particles; so why not have a polymer of 1,000 glucoses? A polymer is one particle. However, as it is broken down, one produces 1 glucose, and another, and another, and so on, until 1000 are produced. About as fast as the glucose is produced, it is transported from intestine to blood, thus there is never an osmotic gradient pulling water out of the blood!

What are glucose polymers? Starches are essentially polymers of glucose. Rice is an excellent starch source and is easy to store and

---

[1] Of course if one moves the positively charged sodium from lumen to blood, that creates a net positive charge on the blood side. (Sugar is not charged). The net positive charge creates a gradient for the negative chloride; it can also move through the spaces between the intestinal cells.

prepare, so it is a common ingredient in the solutions given to cholera patients.

What caused Cynthia Francis's diarrhea?
An important piece of information in this case is that Cynthia Frances has night blindness. We know from the chapter Orange you glad?, that vitamin A deficiency can cause night blindness. Cynthia Frances is getting enough carotene and Vitamin A in her diet, but she apparently is not absorbing it. Remember in the chapter, Orange you glad? that vitamin A deficiency can happen if a person is not absorbing the fat in their diet? If fat is not being absorbed in the small intestine, then it arrives in the colon; some bacteria can break down fat and one of the byproducts is probably a gas. Cynthia Frances's floating stools could either be because they are high in fat; since fat is less dense than water, it will float[1]. Another possibility is that the stools have lots of gas/air trapped inside and that could make them light and they would float. In theory, one can tell which the case is by leaving the stool in the toilet for a few hours; the air should leave and the stool sink, but the fat will stay. But it is not necessary to do that in this case; the greasy stool also hints at a fat digestion problem. Do you remember the 3 organs important for fat digestion? The pancreas secretes the lipases to digest the fat, the liver secretes the bile salts to emulsify or dissolve the fats and the small intestine provides the surface area for absorption. In Cynthia Francis' case, the problem is her pancreas; she has cystic fibrosis. In cystic fibrosis, there is a mutation in a specific chloride channel, CFTR.[2]

In cystic fibrosis the CFTR chloride channel does not work properly. This means that in many epithelial cell layers, chloride does not move across the layer properly. If chloride does not move properly, then there is not an osmotic gradient and water does not move. In the

---

[1] Examine an oil/water salad dressing, without shaking, and you'll see the oil layer on top.
[2] CFTR was named after the disease that guided its identification, cystic fibrosis. At the time it was identified its exact function was not known so it was called cystic fibrosis transmembrane regulator. Cystic fibrosis got its name because Dorothy Anderson notices a series of patients with similar symptoms and they all had fibrotic cystic ducts on autopsy. The cystic duct connects the gall bladder to the common bile duct.

pancreatic and cystic ducts, when this happens, there is a high concentration of bile salts or pancreatic enzymes and these damage the duct walls, creating fibrosis. In Cynthia Francis' case, the fibrosis (scar tissue) blocked enough of the duct that her pancreatic enzymes did not get to the intestine and couldn't properly digest the fat.

The most common presenting symptom for cystic fibrosis is respiratory infections. The CFTR chloride channel is also present in the lung and its normal function is important to provide fluid on the airway side of the lung cells. Some lung cells secrete mucous into the lung and this provides important protection, but if there is not enough water, the mucous stays in the lung too long and the dark, moist environment along with the nutrient mucous provides a breeding ground for bacteria, leading to respiratory infections.

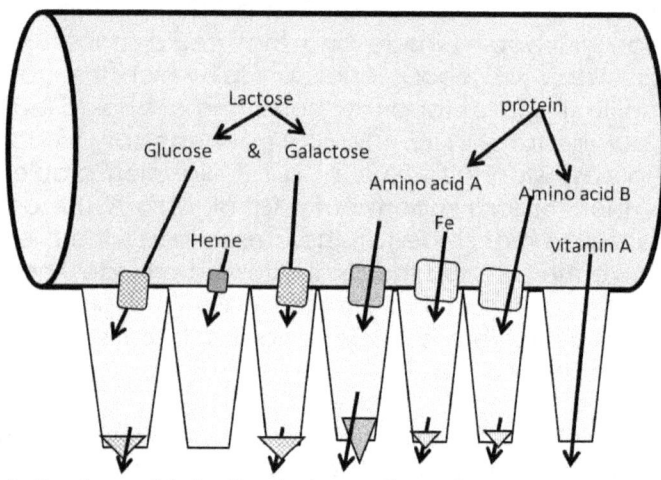

Figure 6. Review of intestinal absorption. Complex sugars and starches are broken down to simple sugars (monosaccharides). For example, the disaccharide, lactose, is broken down to glucose and galactose. These simple sugars are transported into the intestinal epithelial cell and then out of the cell and into the blood. Proteins are broken down into their amino acids and the single amino acids are transported into the intestinal epithelial cell and then out of the cell and into the blood. Iron is transported into the intestinal epithelial cell and then out of the cell and into the blood. It is the step for iron movement from cell to blood that hepcidin regulates. In contrast, heme is transported into intestinal cell where it is broken down and some of the pieces (e.g., iron) are transported into the blood.

Physiology

In summary, here are some key concepts from this chapter.
- Complex molecules are broken down into parts before being absorbed, for example starches to glucose, lactose to galactose and glucose, proteins to amino acids
- Absorption of water soluble compounds normally occurs in the small intestine
- If substantial amounts of nutrients arrive in the large intestine, bacteria digest them, often generating hydrogen gas, protons and short chained fatty acids. The latter create an osmotic gradient.
- In the GI tract, aquaporins provide a pathway that allows water to move in response to osmotic gradients.
- Epithelial cells use sodium to drive uphill transport of glucose and amino acids.
- Epithelial cells move Na and Cl to create osmotic gradients to drive water movement; in the small intestine sometimes cells move Na and Cl and water from blood to lumen and sometimes the other direction.
- Lactase is an enzyme in the intestine that breaks down lactose; it is present in all mammalian children. Some human adults have a mutation that allows lactase production to persist into adulthood and these adults avoid the problems associated with eating and digestion milk.
- Cholera activates the pathways that move Na and Cl from blood to lumen, creating an osmotic gradient to move water; the activation is so strong that there can be tremendous loss of water and minerals.
- Salt/starch/water solutions can be used to help maintain hydration in cholera patients and restore minerals
- Cystic fibrosis is a disease caused by a defective chloride channel, CFTR. This channel is important in lung, GI, and sweat epithelia for the movement of chloride, which drives sodium movement and this osmotic gradient drives water movement. When defective, water movement is inappropriate and allows for bacterial colonization of the lungs and fibrosis if the pancreas and cystic duct.

Physiology

How Sweet It Is

Fictional Cases
Our 3 patients are Diana Inspired, Dixon Mel Won and Diamond Melanie Zwei. All 3 are 15 years old and present complaining that they are always thirsty and therefore drinking water all the time. Of course, that means they are also urinating a great deal. As we will discover, the extra urination came before the excess thirst. Diana Inspired's excess urination and thirst started shortly after she hit her head in a car accident. Dixon Mel Won's problem started when he was quite young and, if he does not take his medications, can get quite severe. Diamond Melanie Zwei's problem has started pretty recently and she has only recently been diagnosed.

In this chapter, we are going to consider
- renal regulation of urine formation (with much more detail in the chapter Soy Sauce and Licorice as Medicine)
- blood sugar regulation
- glucose, glycogen, and fat
- theories about the causes of type II diabetes
- is there a type 1.5 diabetes?
- drugs for treating type 2 diabetes

We will do this looking at these 3 fictional patients and also the hummingbird.

Renal Regulation Of Urine Formation And Antidiuretic Hormone
The kidney, of course, controls urine production. However, a pituitary hormone, antidiuretic hormone, regulates the volume of urine produced by the kidney. Anti-diuretic hormone nicely describes the function we are most interested in at the moment, but this hormone also constricts blood vessels[1]. Antidiuretic hormone tells the kidney to

---

[1] Antidiuretic hormone is also called vasopressin. This is because physiologists realized that the pituitary secreted many hormones and named them for their effects before the biochemists purified the compounds. Thus, if you grind up pituitary glands from one animal and inject the extract into another animal you will see many effects; two are a decrease in urination and an increase in vasoconstriction. Eventually, the biochemists isolated the compounds and they turned out to be identical. In other situations, the anatomists discover a novel

retain water, not to diurese or produce urine. In the case of Diana Inspired, an accident had damaged her pituitary, such that it no longer secreted antidiuretic hormone. Thus, her kidneys don't get the message to retain fluid and the kidneys end up producing more urine than they "should". This decrease in body water leads to increased thirst and she drinks. Her disease is called Diabetes insipidus. Diabetes means "to flow through" and all 3 patients have different forms of diabetes; diabetes insipidus is rarer than the other forms. Diabetes insipidus can be caused by the inability to properly secrete antidiuretic hormone or the inability of the kidney to respond to the hormone[1].

Dixon Mel Won and Diamond Melanie Zwei both have diabetes mellitus; mellitus refers to the sweetness of the urine[2]. Normal urine does not contain sugar, but often the urine of patients with diabetes mellitus is sweet; this was noticed even in the Middle Ages, as flies would congregate around the sweet urine. The excess urinary sugar suggests that both Dixon and Diamond have problems with sugar. One convenient screening test for diabetes mellitus is to measure fasting blood sugar levels.

For the rest of this chapter we'll focus on diabetes mellitus.

Blood Sugar Regulation

Both Dixon Mel Won and Diamond Melanie Zwei have high fasting blood glucose levels[3]. Glucose, because it is a small molecule, is easily

---

structure and later the physiologists and cell biologists discover its function; in that case the name of the structure is often the discoverer's name (Isle of Langerhans) or a Greek letter (beta cells).

[1] More on antidiuretic hormone in Soy Sauce and Licorice as Medicine.

[2] The Latin word "mellitus" means sweet; the Latin word "insipidus" means "bland", that is, not sweet.

[3] Glucose is the name for the primary blood sugar. Most dietary sugar is sucrose which has to be broken down to glucose and fructose; in the chapter, when I say blood sugar, I mean blood glucose. Sugar refers to many different chemicals and we'll concentrate on only 3. Table sugar (sucrose) is a chemical contraction of blood sugar (glucose) and fruit sugar (fructose). When table sugar arrives in your intestine, you break it down into its 2 parts. Then each one is absorbed

filtered by the kidney. Thus, glucose enters the tubular fluid. The kidneys normally reabsorb this filtered glucose, just as the kidneys reabsorb calcium (see the chapter Bubbling Beverages Bad for Bones?) and most other nutrients. However, Dixon and Diamond often have such high blood glucose levels that the amount filtered overwhelms the kidney's capacity to reabsorb the glucose. This glucose therefore remains in the tubular fluid, draws out water, and contributes to the increased amount of urine.

Why don't non-diabetic individuals have high blood glucose? Actually, many non-diabetic people do have high blood sugar levels after a meal, particularly if the meal or drink is high in simple sugars. However, as the concentration of glucose increases in the blood, the pancreas secretes insulin. In non-diabetic individuals, the secretion of insulin quickly leads to blood sugar removal from the blood. In the presence of insulin, blood sugar goes into skeletal muscles and liver. Once inside these cells, blood sugar is used to form glycogen, a starch-like molecule that provides a burst of fuel when needed. The rest of the blood sugar is converted into fat, see Figure 1.

---

into the blood stream; so my use of the term "blood sugar" for glucose is slightly misleading in that fructose is also a sugar found in blood. (A bit like using the term lineman to refer to players on the line in football, but the split end, even though he lines up on the line, is not lineman.) But most of the time, the main sugar in blood is glucose. Most of the scientific literature concentrates on glucose, so we will too.)

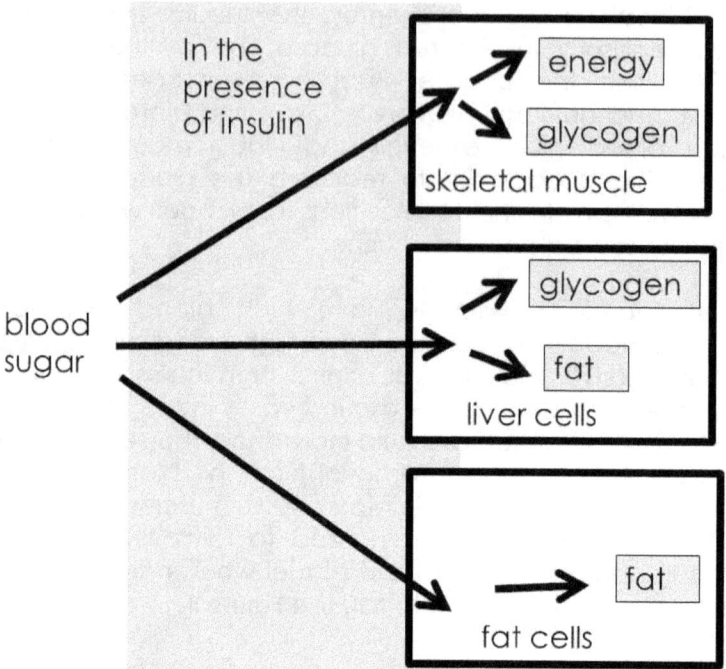

Figure 1 Where does blood sugar go in the presence of high insulin?
Glucose goes into skeletal muscles where it provides fuel and excess is converted to glycogen.
Glucose goes into liver where it is converted to glycogen and excess converted into fat.
Glucose goes into fat cells where it is converted to fat.

One screening test to determine if a patient has diabetes is to measure the fasting blood glucose levels. High fasting blood glucose levels is one way to define diabetes mellitus. Interestingly, hummingbirds also have high fasting blood glucose levels, but they do not seem to have any of the pathological symptoms of patients who poorly control their blood sugar. As we delve into the consequences of high blood glucose in these patients, the hummingbird will provide some interesting insights and also a puzzle.

After determining the fasting blood glucose levels, sometimes the next step is to do a glucose tolerance test, to determine how well blood glucose levels are regulated. For this test, the patient is given a known amount of sugar water. After drinking it, his or her blood sugar is monitored for several hours. Figure 2 shows the blood sugar levels in a normal patient, Diamond and Dixon. As you can see, the blood glucose levels go much higher in the two patients with diabetes than in the normal patient.

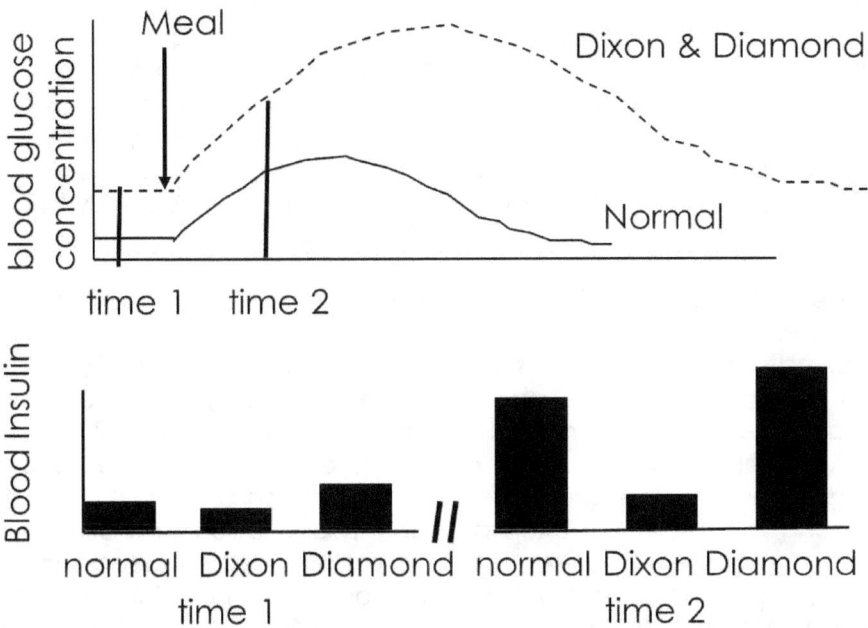

Figure 2. Top. Changes in blood sugar after a sugar meal. In all patients, blood sugar went up after the meal. In patients with diabetes, the blood sugar went higher than normal. Note that the fasting blood glucose levels were also high for both patients with diabetes. At the time 1 and time 2, blood insulin was measured.
Bottom. The insulin concentration before and after a meal.
The normal patient had a substantial increase in insulin, which helped to cause a reduction in blood glucose.
Dixon's initial blood insulin was low and it did not increase much even when glucose was high.
Diamond's initial blood insulin was high and it went even higher after the meal; even though insulin was so high, glucose was elevated.

As soon as the blood glucose levels start to go up, blood insulin levels also go up in the normal patient. Insulin increases the movement of glucose into cells and so helps to keep blood glucose levels lower. You would be correct to suspect that insulin is part of the problem related to high blood sugar for Dixon and Diamond.

When blood glucose levels go up, Dixon's blood insulin didn't change very much. Why might that be? The pancreas secretes the insulin thus there are two possibilities. Either the pancreas is not responding to the blood glucose levels properly or the pancreas does not have insulin to secrete. In Dixon's case, the problem is that his body is destroying the pancreatic cells that secrete insulin and so his pancreas does not have insulin to secrete. Dixon Mel Won has diabetes type I, which is an autoimmune disease where the body's immune system has mistaken his pancreatic beta cells as foreign invaders and is trying to destroy them.

Since the function of insulin is to lower blood sugar, Dixon Mel Won can be treated with insulin shots. If he manages the shots correctly, and eats appropriately, his blood glucose will usually remain about normal and he has the potential to have a normal lifespan and a healthy life. This is a complicated procedure that the normal pancreas usually handles without any conscious effort. In the non-diabetic person, insulin is being secreted almost as soon as blood sugar goes up and the insulin secretion probably is also a function of how fast blood sugar rises. So Dixon has to guess how fast his blood sugar will go up. He can probably figure that out when he drinks a soda by doing some measurements. But unfortunately, those measurements will not accurately predict the blood sugar rise if he has the soda and French fries, even if we ignore the digestion of the French fry starch to sugar. This is because the French fries will alter the time the soda spends in the stomach and the rate of sugar breakdown in the intestine for absorption to the blood. Furthermore, the stress hormones also regulate blood sugar levels; when you are stressed out, you increase your blood sugar levels so you are ready to fight or fly. Dixon has to anticipate how much and how fast his blood sugar will rise, which is determined not only by what he eats, but how stressed out he is.

This stress response might suggest to you another reason for why patients might have high blood glucose, in addition to our first reason which was that too little insulin was secreted. The stress catecholamines (epinephrine and norepinephrine) tend to increase blood sugar. Cortisol, another stress hormone, will also raise blood sugar. A rare cause of too high blood sugar is when the adrenal cells are overactive and release too much adrenaline or cortisol. However the most common reason for high blood sugar is that insulin is not as effective as in a normal patient and this is called type II diabetes. A third system is also important for increasing glucose, the hormone glucagon. Glucagon is produced by the pancreatic alpha cells and is the first line of defense against low blood sugar levels.

If a non-diabetic's sugar starts to drop too low, glucagon is secreted and glucose is produced, for example, by breaking down glycogen in the liver. Unfortunately for Dixon, as the beta cells in his pancreas are destroyed so are the alpha cells. The alpha cells produce glucagon. So if he makes a mistake and injects too much insulin, not only does the injected insulin make the blood sugar fall (as it would in a non-diabetic) but Dixon won't have glucagon to counteract the insulin, whereas a non-diabetic person would.

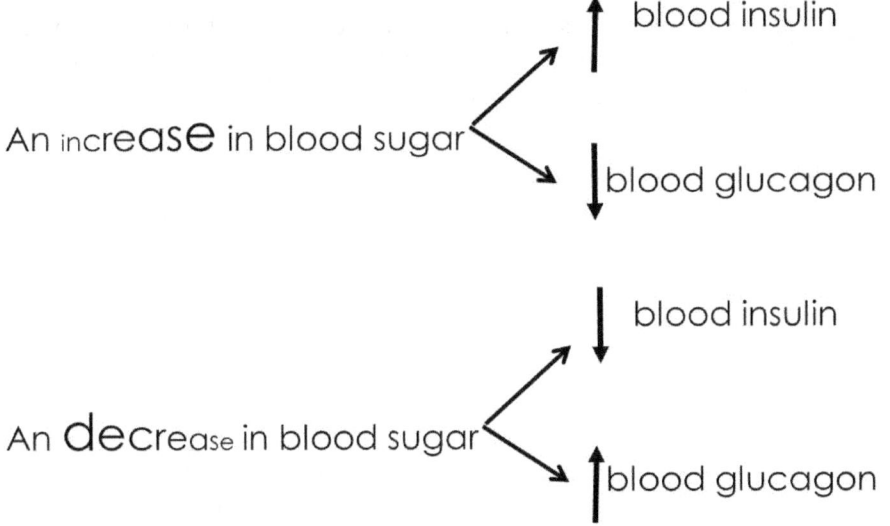

Figure 3. This figure illustrates how blood sugar influences insulin and glucagon secretion.
High blood sugar leads to an increase in insulin and a decrease in glucagon.
Low blood sugar leads to a decrease in insulin and an increase in glucagon.

You would be correct to realize that it is not actually the insulin levels that control blood glucose, but the ratio of insulin to glucagon; there are rare cases where it is important to measure both to figure out what is going on. But in most cases, the insulin levels are sufficient for the clinician to figure out the problem. However, if a patient has high blood glucose, normal insulin, and does not have type II Diabetes, checking glucagon levels is a good next step to figuring out the problem.

Diamond Melanie Zwei also had an abnormal glucose tolerance test but she had higher than normal blood insulin levels[1]. What could be her problem? Insulin is supposed to lower glucose levels. As shown in the Figure 2, insulin is there, but glucose levels are not going down quickly enough. This is termed insulin insensitivity, or insulin resistance, many cells are not responding to insulin like they should and taking glucose out of the blood into the cells.

How do cells respond to insulin? As mentioned, one of the responses is to increase glucose influx into skeletal muscle and liver cells and to store the excess glucose as glycogen.

---

[1] She also had normal blood glucagon, cortisol, and catecholamines.

Physiology

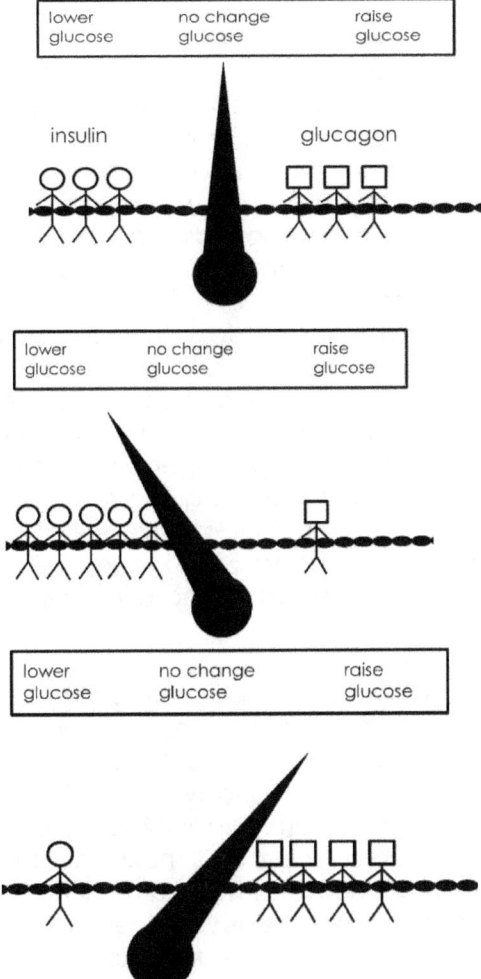

Figure 4. How the ratio of glucagon to insulin affects blood sugar, glucose. Glucagon is on the right side (square heads) of the tug-of-war and insulin on the left side (circle heads). The arrow points to the response. **Top:** Glucagon and insulin are in balance and there is no change in glucose. **Middle:** Insulin is having a much larger effect than glucagon and therefore blood glucose will be lowered. **Bottom:** Insulin is having a much smaller effect than glucagon and therefore blood glucose will be increased.

Glucose, Glycogen and Fat
Before we talk more about glucose storage, we should review why glucose is so important. Glucose is the main energy source for most cells. Once inside a cell, the cell converts glucose to ATP and ATP fuels most processes in the cell, including contraction in muscles. But some cells are hybrids and can use more than one fuel. The nerve cells in the brain can use compounds called ketones when glucose gets very low. Acetone is one ketone and sometimes you can smell it as a fruity smell on someone's breath when they are producing ketones. Ketones are made from fat. Surprisingly you don't convert fat directly back to glucose. Heart and skeletal muscle cells actually use free fatty acids for fuel as well as glucose. Skeletal muscles have an additional fuel storage, which is very helpful to athletes; some of the excess glucose is stored as glycogen in muscle and this can be used to fuel the muscle when blood glucose drops. The other major place for glycogen is the liver. In fact, glycogen is the first choice for storage of excess glucose, but after the liver and skeletal muscle have their quota of glycogen, the rest of the excess is stored as fat in fat cells[1].

There are two very big physiological differences between glycogen and fats. One is that the body does convert glycogen back to glucose when more fuel is needed, but the body does not convert fats back to glucose, but rather to fatty acids. The second difference is that the body has a maximum capacity for glycogen; once you've filled up that storage area, you don't make more glycogen. As too many of us experience, you don't have a maximum capacity for fats; if your calorie consumption exceeds your fuel usage too often and too long, the body will continue to store the excess fuel as fat and more fat...

---

[1] The fancy word for fat cells is adipocytes; adipo is the same prefix as found in adipose, a fancy word for fat tissue

Physiology

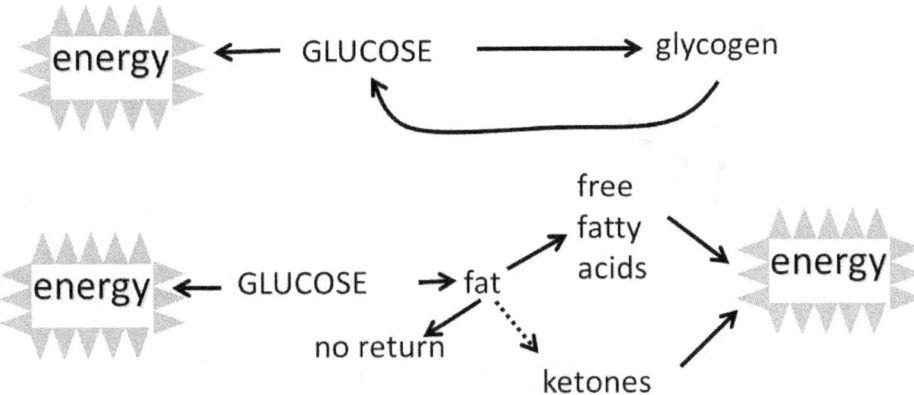

Figure 5. Glucose for energy and fuel storage.
TOP: Glucose can be stored as glycogen. Glycogen can be broken back down to form glucose.
BOTTOM: While glucose can be stored as fat, the fat cannot be broken back down to glucose. Fat can be converted to free fatty acids or ketones to provide energy. Skeletal and cardiac muscle can burn free fatty acids. The brain can use ketones. Ketones are only formed under low insulin conditions.

All cells in the body need to have energy. However, some cells are more important than others and a regulation system is in place that usually allows the most important functions to get glucose first; some of your money expenditures are more important than others and hopefully you've worked out paying the important bills first. In the body, insulin and other hormones regulate which cells get glucose and when to store the excess. To understand how insulin signals to cells and what goes wrong, we need a bit more background information on receptors.

Receptors
In all insulin responsive cells, insulin binds to an extracellular receptor. One primary division for receptors is whether they are intracellular or extracellular. Insulin, parathyroid hormone, norepinephrine, and acetylcholine all have extracellular receptors. Steroids and vitamin A have intracellular receptors.

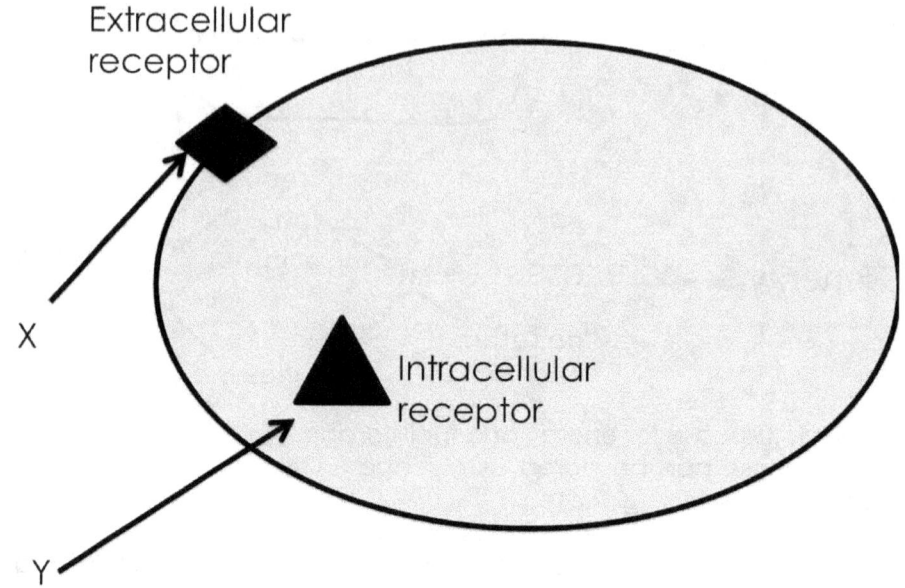

Figure 6. An extracellular receptor for X and an intracellular receptor for Y. Examples of X are insulin, parathyroid hormone, and antidiuretic hormone. Examples of Y are vitamin A, vitamin D, and steroid hormones.

Receptors are proteins that bind signaling molecules and this is an important concept. When binding happens, the receptor protein undergoes a change in shape[1] and this leads to other changes in the cell. Extracellular receptors sit in the cell membrane and bind to signaling molecules outside the cell, and this leads to a key shape change on the inside of the cell membrane surface, which triggers a host of events inside the cell. I don't have a single analogy that catches all aspects of this, so I'll use several analogies, each of which catches part of the idea.

You've probably heard of the lock and key analogy for enzymes and this also works for the recognition of molecules by receptors: the molecule has to fit into the receptor just right to allow it to change shape (allow the lock to turn and unlock the door). A few receptors

---

[1] When a protein changes shape chemists call it a conformational change.

are actually like doors, and after the key has turned and the door is open, it allows specific molecules to get into the house. One example is the nicotinic acetylcholine receptor on muscles which allows sodium to go into muscle cells. But the insulin receptor isn't quite like that, it's more like the key in a car that turns on the engine; the insulin binds to the outside of the engine compartment (the cell) but has an effect on the engine (the inside of the cell).

A third analogy that examines more the difference between extracellular and intracellular receptors uses sport. Extracellular receptors, including the insulin receptor, are like coaches: The signal comes from the booth (blood), the coach hears it, and orders a change on the field; the coach (extracellular receptor) doesn't actually go onto the field and make the play. The coach (extracellular receptor) just relays it. There are intracellular receptors, particularly for steroid hormones and lipid soluble hormones such as vitamins A and D. In this analogy, a steroid would be like a running back that runs onto the field and talks to the QB (who would be the intracellular receptor) and the QB would then change the directions he gives to the players. The intracellular receptor would then change what genes are expressed by the cell.

Insulin Receptor

The insulin receptor is a tyrosine kinase, so when insulin binds on the outside, on the inside the tyrosine kinase activity is increased[1]. In skeletal muscles this sets up a cascade of events that result in the skeletal cell putting more glucose transporters into the outer membrane. So now more blood glucose can enter the cell. The insulin/tyrosine kinase cascade also increases the activity of some of the enzymes in the glycogen synthesis pathway. This combination leads to glucose movement out of the blood and increased production of glycogen in the skeletal muscles.

---

[1] Kinase means an enzyme that transfers phosphate. So a tyrosine kinase transfers phosphate to tyrosine amino acids on the acceptor protein.

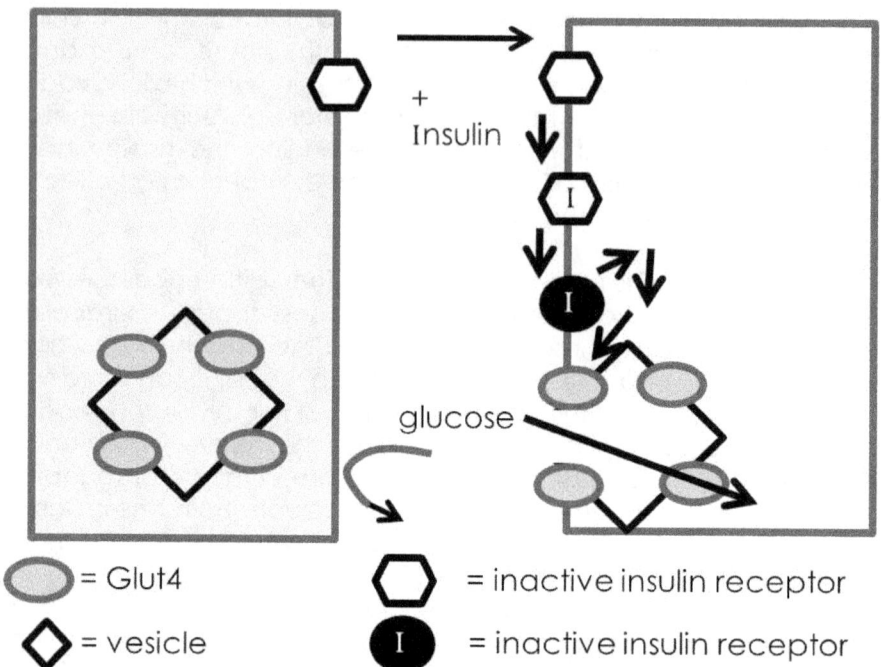

○ = Glut4 ⬡ = inactive insulin receptor

◇ = vesicle ⬢Ⅰ = inactive insulin receptor

Figure 7. Moving transporters between vesicles and surface cell membrane.
On the left is a gray resting square cell. In the resting state, most glucose transporters (ovals) in skeletal muscles are found in the membranes of vesicles (diamond). Vesicles are small membrane enclosed compartments inside the cell. Since there are no glucose transporters on the cell surface, glucose cannot get into the cell. The insulin receptor (hexagon) is inactive. In the presence of insulin we move to the right white square cell. Insulin binds to its receptor (hexagon) which then changes shape (filled circle). The active insulin receptor starts a signaling cascade (arrows) that lead to insertion of the vesicles into the cell membrane and thus this leads to a great increase in transport of glucose into the cell.

Here is one place where humming birds are different; their skeletal muscles do not appear to have insulin regulation of glucose transport. Their skeletal muscles probably always have plenty of glucose transporters in the membrane and they are always doing glucose transport. Also, I don't know if hummingbirds store much glycogen.

Physiology

For an equal amount of energy, glycogen storage weighs more than fat storage. Part of this you might have noticed from nutrition labels; 1 gram of carbohydrate is 4 calories but 1 gram of fat is 9 calories. The other difference is that when glycogen is stored, it has waters around it, whereas fat does not. Since the more the hummingbird weighs, the more energy it needs to fly, it makes sense that any way that efficiently decreases its weight would provide an advantage.

In the liver, while insulin also increases glycogen storage, it's a bit different mechanism. Insulin still binds to its receptor and activates its tyrosine kinase activity. But in the liver, the glucose transporter is a different protein than in skeletal muscle; the glucose transporter in liver is not insulin sensitive.

Why not have insulin sensitive glucose transport in the liver? Would it make the liver more efficient? I don't know the answer to that. But if an insulin sensitive glucose transporter in the liver were more efficient, then one might think it would provide a selective advantage. In order for a gene to be selected, two things must occur. In addition to providing a selective advantage, there needs to be the mutation that provides the advantage. Even if there is a selective advantage, if there is no appropriate mutation for an insulin sensitive glucose transporter in the liver then the liver stuck. It would be interesting to determine when the primordial glucose transporter split into these two forms, Glut1 in liver and Glut4 in skeletal muscle. When did the expression get localized to different cells?

Activating the insulin receptor also leads to an increase in fat storage.

Let's summarize some key points.
- On an organ regulation level, an increase in blood sugar leads to an increase in pancreatic insulin release.
- High blood insulin causes an increase in uptake of glucose into skeletal muscles and any glucose not used for the energy of muscle contraction is stored as glycogen.
- Also glycogen synthesis is increased in the liver; this would lower the cellular glucose concentration and Glut1 would allow more glucose into the cell as the glucose moves down its concentration gradient[1].

---

[1] This is a general principle, if something moves there has to be a path (Glut1)

- Furthermore, in the presence of insulin, some of the glucose is converted to fat and transported to, and stored in, the fat cells.

All of these responses will lower blood glucose levels.

On a cellular level, insulin binds to its receptor, activates its tyrosine kinase activity, which phosphorylates tyrosine on some target proteins which leads to a cascade that
- increases Glut4 in the membrane of skeletal muscle
- increases cAMP in liver cells which leads to activation of glycogen synthesis enzyme
- increases the enzymes to convert glucose to fat.

Since Diamond is insulin resistant, what has gone wrong with this process? The list would include:
- insulin doesn't bind to receptor
- insulin binds but the receptor doesn't activate tyrosine kinase
- downstream cascades are messed up. One possibility is that a phosphatase, which removes phosphates, is too active.
- final process does not work, for example, Glut4 does not work or the enzyme for glycogen synthesis doesn't work.

In different people, there are probably different reasons for the insulin resistance, so lumping everyone into one category, Type II, might not be the best strategy. Figure 3 illustrates two different insulin sensitivity responses.

---

and a gradient (blood glucose > liver glucose).

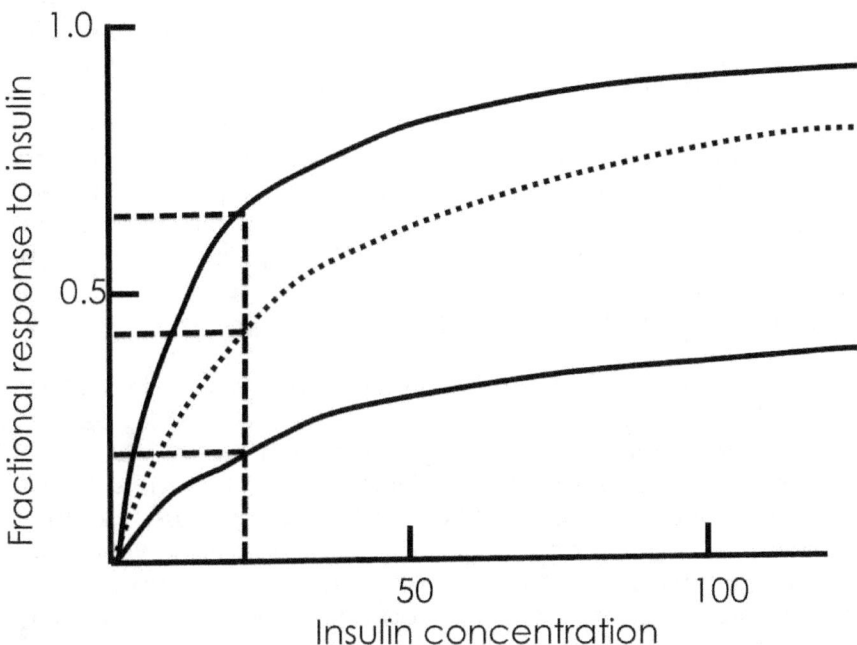

Figure 8. The different between insulin sensitivity and insensitivity. The top curve shows the normal response to insulin. As the insulin concentration increases, the response to insulin increases and then plateaus. At 20 nM insulin, the response is about 0.6. In an insulin resistant person (Diamond) at the same amount of insulin (20 nM), the response to insulin is lower, only to 0.2 or 0.4, versus 0.6 in the normal. There are two different types of insulin resistance: in the middle (dotted) case, at high enough insulin, the response is nearly normal. In the lower case, no matter how much insulin is added, the response is never as great as normal.

What causes Diabetes Type II, the disease that Diamond Mel Zwei has?
We don't know yet, but there are many theories. Over time, two main things seem to go wrong. One is that the skeletal muscle, liver and fat cells need higher and higher insulin concentrations in order to respond. This is termed insulin insensitivity[1]. The other thing that goes

---

[1] As an analogy, as a person begins to lose her hearing, others around them have to speak louder to the person who is now sound insensitive.

wrong is that the pancreatic beta cells are not able to secrete enough insulin. It remains a debate about which one happens first. At the moment it is easier for me to mechanistically explain what happens if insulin insensitivity happens first, just as I can explain why hearing loss comes before people shouting. That makes that theory more appealing. But if someone could show for sure that beta cells became defective before insulin resistance developed, then clearly my mechanism is not accurate. In general terms the mechanism is that if insulin resistance happened first, the body would have higher blood glucose longer than normal. The high glucose causes reactions that damage the pancreatic beta cell.

There appears to be a correlation between sugar intake and diabetes. One theory was developed that the pancreatic beta cell wore out when sugar intake was high. The high sugar intake would require the pancreas to secrete a lot of insulin and all that extra work wore out the pancreas. I did not like this theory, because I could not understand how a gland can wear out. I can understand why a knee or hip joint might wear out because there is mechanical rubbing. But I don't know of any other gland that wears out. Does the saliva gland wear out when someone chews gum too much?[1]

In recent years, there have been studies to try to figure out a mechanism for why the beta cell function deteriorates.

How can high blood sugar damage the pancreatic beta cell? To answer this type of question, scientists try to think of the possible explanations and then design experiments to test them. Glucose could directly affect the beta cell or it could alter something else that then alters the beta cell.

So let's take the first possibility, that glucose directly affects the beta cell. We can divide this into 2 easy groups. Prolonged high blood glucose could either injure the beta cell by interacting with the

---

[1] Of course the salivary gland is an exocrine gland; it secretes a fluid that goes out of the body (since the inside of the digestive tract is really outside the body). So how about an endocrine gland? The parathyroid is an endocrine gland and no one says a low calcium diet wears out the parathyroid gland (see Bubbling Beverages Bad for Bones? for details on parathyroid hormone).

outside of the cell or it could injure by damaging something inside. It is this latter one that seems more likely from the data right now. So why not just block glucose from getting into the pancreatic beta cell? If glucose can't get in, by this model, it can't hurt the cell, and voila, no damage from prolonged high glucose and no diabetes! The problem is that the way that the pancreatic beta cell "knows" that there is high blood glucose is because it comes into the cell and raises the glucose levels. The intracellular signaling details are interesting and help us understand one of the drug therapies for treating some kinds of type 2 diabetes.

To summarize the causes of type I and type II diabetes, in type I there is an autoimmune reaction and the immune system attacks and destroys the beta cells (Dixon Mel Won). When insulin secretion gets too low, these patients need insulin shots. Type II diabetes probably includes many problems (Diamond Melanie Zwei). While we don't understand all the causes and all the initial steps, by the time the patient exhibits symptoms, they usually have insulin resistance (which cells? skeletal muscle, liver, fat) and their beta cells, while they still work, can't produce enough insulin to keep their blood sugar normal. In the early stages, patients with Type II diabetes don't usually need insulin shots, but often the disease progresses and they do need insulin shots.

Type 1 ½?
In order to understand why people are sick, scientists try to identify common characteristics. But often scientific divisions don't encompass the full complexity of nature. We have the same issues elsewhere in society. It is very hard to have a single definition that cleanly divides people into child and adult. Is it the age where they can drive? Vote? Live independently? Drink alcohol? The division of diabetes into type I and type II has helped to make sense of much of the disease, but it is no doubt an oversimplification. Some people have features that are common to both type I and type II. They are sometimes termed Type 1.5 or latent autoimmune diabetes in adults.

As mentioned above, future work is likely to subdivide type II diabetes into different groups. Another common type of diabetes is gestational diabetes, that is, diabetes that develops during pregnancy. In many of these pregnant women, the diabetes stops after childbirth, but in some it seems to progress to type II diabetes. I say "seems to" not to mean that they don't have type II, but that it is hard to determine for any individual if the gestational diabetes progressed to type II or if they would have gotten type II even if they didn't have gestational diabetes. There are a number of other types of diabetes that are not as common, these are detailed at http://diabetes.niddk.nih.gov/dm/pubs/overview/. Note that good definitions are very helpful; can you imagine the confusion if we hadn't understood that diabetes insipidus and diabetes mellitus are different diseases?

Some say there is a correlation between obesity and diabetes. I think a more accurate statement is that obesity increases the risk of developing diabetes. At the moment, about 85% of patients with diabetes are obese. But only 20% of people who are obese have diabetes. Why do only some people with obesity get diabetes? First, there is a correlation between where the fat is and the odds of getting diabetes. If the person with obesity is pear shaped, so more of the extra weight is in their hips, they are less likely to have diabetes than the obese person who is apple shaped, with the extra weight in the abdomen. This suggests that fat cells are different; there is something about abdominal fat cells that makes it more likely to get diabetes. For the apple shaped person with obesity, there are presumably also genetic components that alter the risk as well as other environmental components. One environmental component is exercise. People with obesity who exercise have a lower risk for diabetes than sedentary obese people.

In fact, for any weight, steady exercise reduces the risk of diabetes. Again, the exact mechanism is not clear. But it is interesting that exercise, like insulin, leads to an increase in skeletal muscle glucose transport.

In our society, having too much fat is considered unhealthy. Here is another area whether the hummingbird challenges us. Some hummingbirds fly from Mexico to Texas non-stop. They have to bring their fuel with them. They do this by doubling their body fat from 20% of their weight to 40% of their weight. This is the most efficient way for them to store the fuel they need. Does this increase in fat also increase their risk of having diabetic symptoms? Does their extreme exercise help reduce the risk? It may be that when we find out the answers to these questions about hummingbirds we will gain more insights into the causes and potential treatments for humans with diabetes.

Patients with diabetes monitor their blood sugar and take steps to keep it in the normal range. Depending upon the type and severity of the disease, a patient might do a finger prick blood glucose test several times a day. For monitoring longer term ability to keep glucose in a reasonable range, a different measurement is used. This longer term measurement relies on a few key principles. Glucose can react non-enzymatically with other molecules; a common reaction is between glucose and the amino acid lysine in proteins. This is called glycation[1] and it is not only important for monitoring but also for understanding one theory of what goes wrong in patients who don't control their diabetes.

We are able to measure the reaction product between glucose and lysine. Which protein should we monitor to estimate how often glucose was too high? We can easily take a blood sample, so it may as well be a blood protein. Since we want a long term average, we look at hemoglobin, the protein inside red blood cells. Red cells have a high concentration of hemoglobin. Red cells lifespan is 4 months, so it will give us an average of the glucose over several months. The glycated hemoglobin that is measured is called HbA1c. Hemoglobin has many types, hemoglobin type A1 is one type. I don't know why the "c" symbol is chosen for glycation, but it might have been the third band on the gel. Of course for this measurement to work, the glucose concentration inside the red cell has to closely follow that in the blood. Fortunately it does as the human red cell rapidly equilibrates glucose across its membrane. The rate of glucose transport across rat and mice red cells is much slower. Also the red cell lifetimes are shorter in rats and mice. So I don't think measuring HbA1C is as helpful in rats and mice. Some diabetic researchers study

---

[1] Both enzymatic and non-enzymatic reactions are often linear in concentration and time at low concentrations and short times. But at high concentrations, enzymatic reactions reach a plateau. (The response is very similar to the insulin response shown in Figure 6 of How Sweet It Is) The non-enzymatic glycation reaction is linear in the concentration of glucose and the length of time the protein is exposed to the glucose and so it is a better measurement of the blood glucose concentrations.

pigs and they need to be aware that pig red cells do not transport glucose. How do they get their energy? Pig red blood cells use inosine as their fuel source.

Hummingbirds have lower levels of glycated hemoglobin. This is probably the result of the shorter lifetime of their red cells[1]. Whether they have a faster turnover of other proteins, so that they have less glycated proteins in general is not known. But if so, it would be consistent with the higher protein turnover providing some protection against glycated induced pathology.

One of the things that can go wrong in patients who don't control their diabetes is severe foot damage. In fact the most common cause of foot amputation in the US is diabetes. Another diabetic complication is blindness. How does diabetes cause foot and eye damage? It is thought that high glucose is the initial problem. One theory is that high glucose leads to glycation of proteins. Then either some of these proteins don't function properly, or the glycated products activate receptors that lead to improper functions. There are other theories about what can go wrong with high glucose. In any case, prolonged high glucose leads to damage of the small blood vessels and capillaries. Then some of the blood contents leak out and damage the surrounding tissue. In the eye, this eventually can lead to blindness. Physicians regularly check a patient's eyes to monitor for this possibility.

Patients with diabetes also suffer from nerve problems. We don't have definitive experiments to determine if the nerve damage is secondary to the damage to the blood vessels or whether high glucose can directly lead to nerve damage. But in either case, the nerves don't work. This is most serious for sensory nerves. A patient with diabetes can damage their foot (even a simple blister), but because of the nerve damage not be aware of it. So the blister gets worse and because of the capillary damage it is also harder to repair than for a healthier person. And perhaps because of the high glucose, the blister/wound is at greater risk for infection. In the worse cases, the foot gets so bad it needs to be amputated. There is

---

[1] It is also possible the hummingbird red cells have low glucose transport; under at least some conditions, chicken red blood cells have low glucose transport.

substantial data suggesting that if blood glucose levels are held in check, all this damage can be avoided.

In the early stages of Type II Diabetes, it seems likely that proper diet and exercise is sufficient to control blood sugar. But it is hard for most patients to comply. The goal of a good diet is to keep blood sugar from being very high very often.   The American Diabetic Association recommends a diet that is low in simple sugars. If you eat glucose, or even sucrose, blood sugar rises quickly. The glucose is absorbed directly. Sucrose only needs 1 step to be broken down to glucose and fructose and absorbed.   The ADA suggests eating complex carbohydrates. These complex carbohydrates are eventually broken down to glucose and other simple sugars and absorbed, but the enzyme can only release a single glucose each cycle. Therefore the rate of production of glucose in the intestinal lumen is slow. One measure of how slowly carbohydrates are broken down is the **glycemic index**, which is a measure of the blood sugar after eating a standard amount of carbohydrate.

There are 3 difficulties with the glycemic index. First, most of the data I've seen is based on studies done with fewer than 10 people. Second, each food is eaten in isolation. This makes for a cleaner scientific measurement; all the sugar must be from carrots if one only ate carrots. But it's hard to apply this to real life, because our meals are hopefully a mixture of foods. Third, the studies are usually done with 400 calories of each carbohydrate. Again, this makes scientific sense, but I don't think most people eat an equal number of calories of pasta and broccoli.

My family doctor told me that he had 2 patients with Type II Diabetes who wanted to try the Atkins low carbohydrate diet. On this diet, in the initial stages, a person eats less than 100 calories of carbohydrate a day. That's less than 1 or 2 pieces of bread. Green vegetables have very low carbs, so those can be eaten a lot. But Atkins dieters can't eat any pasta, bread, or potatoes. Thus these patients eat mostly protein (meat) and fat. The patients insisted on trying the diet even though my family doctor thought the diet sounded unhealthy. Nevertheless, he monitored their blood sugar and fat levels. My doctor was astounded to see that they controlled their blood sugar very well. He reviewed his biochemistry and physiology. If the patient only eats 100 calories of carbohydrate each day that is about 25 grams of total sugar. Even if they ate it all at once and it all went directly into their blood, it would only double their blood sugar. Fat and protein are not broken down to sugar in the intestine, so their digestion products will not produce any glucose.

How does a person on the Atkins diet get enough glucose? Actually, a better question is where do the cells get their energy? First it is important to realize that the body does not convert fats to glucose. If the body wants to get energy from fats, it converts them to free fatty acids. Skeletal muscle and cardiac muscle can burn free fatty acids. These patients were eating plenty of fat (and also had plenty of fat stores as they were obese). Some cells can only use glucose, the red cell for example. The body can convert the amino acids in protein to glucose, so this can fuel cells that need glucose. The brain needs either glucose or ketone bodies. If the patients are on a very low carbohydrate diet, their insulin is low and their glucagon is high. They will actually break down some fat to produce ketone bodies. These can fuel the brain. But aren't ketone bodies bad? It is true that some people with diabetes who aren't controlling their blood sugar levels well can get ketone body acidosis (usually called ketoacidosis). Acidosis, producing excess acid, is bad if the kidney doesn't get rid of the acid. Patients with diabetes can also have kidney damage and therefore have a harder time excreting the acid. (In fact, diabetes is a major reason for kidney failure.) But as far as I can tell, ketones themselves are not harmful.

These patients on the Atkins diet controlled their blood sugar so well that they were able to stop taking their medications. They also lost weight, probably because of 2 reasons. With the high fat meals, they felt fuller longer. In addition, most binge foods are a combination of fat and carbohydrate (think ice cream or chips and dip). Thus the patients probably ate less. In addition, their triglycerides decreased significantly. Triglycerides are one form of transporting fat in the blood and high triglycerides are a risk factor for cardiovascular disease. The triglyceride levels dropped even though the patients ate a high fat diet because the liver only makes triglycerides if insulin is increased. With normal blood sugar, insulin is low and the liver does not make triglycerides.

My doctor put several of his patients on this diet and so did many of his friends. Many patients lost a substantial amount of weight. He then attended an obesity conference and one of the key speakers said that the Atkins diet was just a fad and didn't work. But most of the family physicians in the audience asked about the data and the speaker didn't have any. There is still a debate about the long term effects of a very low carbohydrate diet, but there are some studies that confirm the family physician's anecdotal evidence that the low carbohydrate diet does lead to weight loss. It certainly also makes sense that it would help control blood sugar levels, since the only source for high blood sugar levels is dietary sugar and carbohydrates.

I told the Atkins story to make 3 points. One, the body does not break down fat to make glucose, see Figure 5. Two, if glucose does not increase, then insulin does not increase, and therefore triglyceride production is low. Three, even scientists can be biased by their expectation. A diet of eggs and bacon for breakfast, ½ pound of cheese, celery and cucumbers for lunch, , and steak with butter sauce, zucchini with cheese sauce, and a lettuce salad with blue cheese dressing, with a fancy cheese and coffee and cream for dinner doesn't sound healthy. But it can lead to weight reduction. And it certainly keeps blood sugar levels normal. So by these 2 short term measures, it does seem healthy. Long term studies need to be done. There are also other measures of what is healthy to be considered. But it is important to actually try to do experiments to test ideas. And if one says something is harmful, it shouldn't be just because you think it is, it would be nice to have a mechanism for what part of the body or cell is actually damaged.

Drugs for treating type II diabetes
Because changes in diet and exercise are hard for patients to follow for a lifetime and because some patients have progressed too far, drugs are also used for treatment. I want to briefly tell you about 3 classes of drugs.

Incretin mimics are the first class of drugs. Incretins are hormones that are released by the intestine in response to a meal. They affect pancreatic beta cells. They don't directly cause insulin release, but for a given blood sugar concentration, there is more insulin release when incretins are present than when they are absent. This is consistent with the clinical observation that, for the same increase in blood glucose, whether the glucose was given by mouth or directly into the veins, the insulin response was greater when the glucose was given by mouth.

One of the drugs in the incretin class is based on a peptide isolated from the Gila monster.

Sulfonylureas are the second class of drugs. They have 2 effects. They tend to increase insulin secretion by the pancreatic beta cell. They also tend to make cells with insulin receptors more insulin sensitive.

Thiazolidinediones are the third class of drugs. They bind to specific transcription factors[1] called PPARs that regulate many processes in fat metabolism. The normal substrate for some of these transcription factors is fatty acids. PPARs bind to another transcription factor and it is as a pair that they are active. Interestingly, one of the pairs is often the transcription factor that binds vitamin A. These drugs increase the sensitivity to insulin and lead to lower blood glucose levels. One of these drugs, Avandia, was in the news in 2010. While it improved blood sugar levels, it has also been linked to more cardiovascular problems. It has been removed from the market in Europe. In the U.S. it is now more tightly regulated. Related drugs seem to be ok. It is not known whether Avandia does something slightly different when it binds to the PPARs or whether Avandia happens to bind to another protein and that complex has bad effects.

---

[1] Transcription factors regulation transcription (duh!). Remember that transcription is the step of forming RNA from DNA. The RNA is then translated to form proteins and it is proteins that do most of the "work" in the cell. We mentioned that steroids regulate some transcription factors in the chapter, Steroids Make the World Go Round and that vitamin D regulate some transcription factors in Bubbling Beverages Bad for Bones? factors.

Physiology

Chapter Summary

- several hormones regulate blood sugar levels including insulin, glucagon, cortisol, epinephrine and norepinephrine
- an increase in blood sugar leads to an increase in insulin
- an increase in insulin causes a drop in blood sugar by increasing sugar uptake into skeletal muscle and liver, for storage as glycogen, and also by regulating fat synthesis in liver and fat cells
- glucagon, cortisol, and epinephrine all tend to increase blood sugar by promoting glycogen breakdown and/or conversion of amino acids to glucose
- fats are not really converted back to glucose but are metabolized to fatty acids, which can be used by skeletal muscle and the heart
- under low insulin and low glucose conditions, fats are also converted to ketones which can fuel the brain
- high blood glucose levels lead to cardiovascular problems, often related to the resulting high blood fat levels
- high blood glucose levels also lead to capillary, nerve, eye, and kidney damage.
- type I diabetes is caused by an autoimmune response that damages pancreatic beta cells; eventually pancreatic alpha cells are also destroyed.
- type II diabetes is caused by insulin resistance and can eventually also involve beta cell damage
- control of blood glucose levels appears to be a key to maintaining good health
- regular exercise and diets that minimize simple sugars are thought to reduce the risk for type II diabetes as well as improve the outcome for patients.

Both Diana Inspired and Dixon Mel Won will need to replace the hormone that their body can't make, antidiuretic hormone and insulin, respectively. As we learn more about autoimmune diseases, we hope to be able to identify patients with type I diabetes earlier and stop their immune system from destroying their own pancreases. If Diamond Melanie Zwei does not control her blood sugar well, she has the potential to further destroy her pancreatic beta cells and then she, too, will need insulin. The current theories suggest that type II diabetes is as nearly as preventable as most lung cancers. If no one smoked tobacco cigarettes, the amount of lung cancer would plummet. If everyone ate an appropriate diet and had regular exercise, the current theories predict that the amount of type II diabetes would also plummet. But we don't know how to get most people to comply with a healthy diet and to get regular exercise. In the meantime, there are drug therapies that can help improve blood sugar control by either increasing insulin secretion or increasing insulin sensitivity or both. These should at least slow down the progression of the disease.

Physiology

Bubbling Beverages Bad for Bones?

Parent: I just read that teens who drink soda have weak bones, so you can't have any soda anymore.
Teen: No way. I don't believe that study. I get plenty of calcium and soda is not a problem.
Parent: Look at this study. The teens that drank soda had weaker bones.
Teen: Mom, that study, at best, only shows a correlation, not cause and effect.
Parent: What do you mean?
Teen: It might be for the people they studied, that the soda drinkers had weaker bones, but that does not mean it was soda drinking that *caused* the weak bones.
Parent: Well, what else could it have been?
Teen: Well, maybe those teens drank soda and no milk, and therefore didn't get enough calcium. I drink plenty of milk and also eat some foods high in calcium, e.g., salmon, tofu, and broccoli, so it's ok if I drink soda.
Parent: How do you know that?

This interchange is also occurring in the medical literature. In this chapter, I will present both sides of the argument. Both sides usually agree on the data, but the two sides do have different perspectives. One side feels there is no compelling evidence that soda directly affects bone health and therefore they are not ready to indict[1] soda. The other side feels there is no compelling evidence that soda is safe with respect to bone health and the often found correlation between soda use and weak bones is enough for them to indict soda. This difference is similar to whether you presume someone is innocent until proven guilty or guilty until proven innocent, as well as whether you want enough proof to convince a policeperson or a defense lawyer. Another different perspective is whether a correlation between soda use and weak bones is enough to indict soda or whether one needs to provide a clear mechanism by which it happens. Just like for a criminal case, do you merely need to show the defendant was

---

[1] I'm using "indict" in the legal sense; to me it implies that there is a strong case, but that a jury has not yet decided whether the evidence is beyond a reasonable doubt.

present at the scene of the crime or do you need to show how the crime was actually committed?

This chapter has the following sections:
- Population studies
- Calcium threshold and how much calcium needed
- Parathyroid hormone
- Vitamin D
- Calcium regulation
- What is wrong with high blood calcium
- Soda, phosphate and calcium
- Acid and calcium
- Caffeine and calcium
- Summary

For this discussion, we will use a debate between two friends, Connie, who is convinced that drinking soda causes reduced bone density, and Dorothy who is doubtful that soda, itself, is the culprit in bone density issues.

Population Studies
Connie: Many epidemiological studies have found evidence that people that drink soda have reduced bone density.

Dorothy: Remind me, what are epidemiological studies?

Connie: Epidemiological refers to population studies; in many studies, you look at 2 groups of people, say soda drinkers and non-drinkers, and then you look at bone density.

Dorothy: Oh, yes, now I remember. I think a problem with your studies is that the soda drinkers just substituted soda for milk and they therefore didn't get enough calcium. It's not soda's fault, they just need to get more calcium in their diet.

Connie: Well, in some studies they matched the calcium intakes of the volunteers in each group.

Dorothy: How did they do that?

Physiology

Connie: Simple, they just asked people about their diet and calculated their calcium.

Dorothy: People aren't very good about accurately recalling what they ate. I bet you can't even tell me everything you ate yesterday.

Connie: Sure I can. Let's see..... I had cereal yesterday for breakfast. Or maybe it was toast. And I had orange juice.

Dorothy: Was the orange juice calcium enriched?

Connie: I thought so.

Dorothy: Ah, but you don't *know*, do you?

Connie: All right, I agree people aren't always accurate. However, I think both the soda drinkers and non-drinkers were equally inaccurate.

Dorothy: Sorry, I just don't think those are very good studies.

Connie: You think epidemiological studies are all worthless??!!

Dorothy: No, I didn't say that. I do think studies that rely on people's self-report of eating and related everyday activities, such as amount of exercise, or time spent cooking or studying, are inaccurate.

Connie: So what would you suggest?

Dorothy: Ideally, you'd have everyone on a strict diet and they could only eat the food supplied to them; then you'd know exactly what they'd eaten.

Connie: Good luck getting volunteers for that.

Dorothy: Exactly. It would be a good study, but you'd never get enough volunteers. Though I have heard of some studies where they recruit only dieticians, since they are trained to keep track of what food they eat and they would obviously be very aware of the importance of following the prescribed diet. However, it's hard to do a large study with just dieticians.

Connie: So you think these studies are worthless.

Dorothy: No, I agree that epidemiological studies are useful, but they have limitations.

Connie: What's wrong with them?

Dorothy: Basically, they evaluate correlations, not cause and effect. I also want to point out that there are several studies of soda and bone health where they find no correlation, what do you think of that?

Calcium threshold and how much calcium needed
Connie: That might be because in some studies, they are studying people who have a lot of calcium in their diet. If someone has a diet high in calcium, soda may not be such a problem. However, I think if they restricted that data analysis to only the people who were just getting the minimum daily requirement (MDR) for calcium, they would find that those who drank soda would have bone problems because soda makes them urinate more than they should.

Dorothy: What do you mean?

Connie: The official term is threshold effect. If you consume more calcium than you need, you just urinate it out. Threshold is a bit similar to preparing a pan for baking. Often you'll grease the pan first by spreading butter on it and then you'll sprinkle flour on and shake off the excess. The calcium you urinate out is like the excess flour. Once you've sprinkled enough flour onto the pan (had enough calcium for the body) anything more is above threshold and so the excess is shaken off (excess calcium is excreted.) If you only study bakers who use way too much flour, of course all their pans will be coated in flour. But if you study bakers who are using barely enough flour, then you are more likely to see pans with incomplete flour coverage.

Dorothy: How much calcium does one need per day?

Connie: It obviously depends upon how old you are. An average sized adult needs about 1 g per day (1000 mg). Young children need more calcium per weight, but obviously, they are smaller in weight. Toddlers need about ½ g (500 mg) a day. In early elementary school, this increases to 0.8 grams (800 mg) per day. As children start their growth spurt, (roughly tweens and teens), they actually need more than an adult-1.3 g or 1,300 mg per day. This information is simplified from the International Osteoporosis Foundation, http://www.iofbonehealth.org/patients-public/about-osteoporosis/prevention/nutrition/calcium.html#calcium_table

Parathyroid hormone
Dorothy: So what in soda would make you urinate out too much calcium? Do you think soda interferes with the parathyroid system?

Connie: Don't you mean thyroid?

Dorothy: No, parathyroid glands are next to the thyroid (See Fig. 1). Para means beside in Greek. These glands secrete a hormone with the useful name of parathyroid hormone. This hormone is important in regulating blood calcium levels.

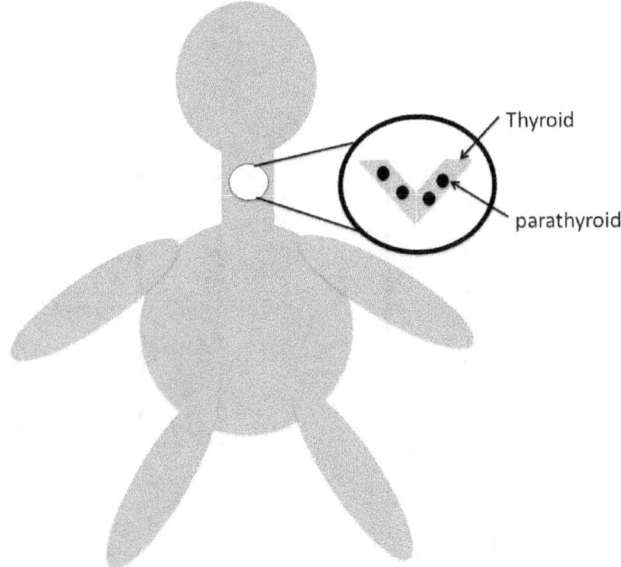

Figure 1. Location of thyroid and parathyroid. Not to scale. Duh!

Connie: I thought vitamin D was important for calcium.

Dorothy: Vitamin D is important. Parathyroid hormone regulates the active form of vitamin D.

Connie: You seem to know a lot about this system, tell me about it.

Dorothy: Let me start by explaining some basic principles. When you are done growing, hopefully your body is in a steady state.

Connie: Steady state, this sounds like chemistry or physics: yuck!

Dorothy: It's just a little bit of physics but it plays a key role in understanding biology and physiology because it makes understanding the problem a whole lot easier. If a system is in a steady state, then input must equal output. For example, if your weight is in steady state, then you are burning as many calories as you are eating.

Connie: Oh, that's easy.

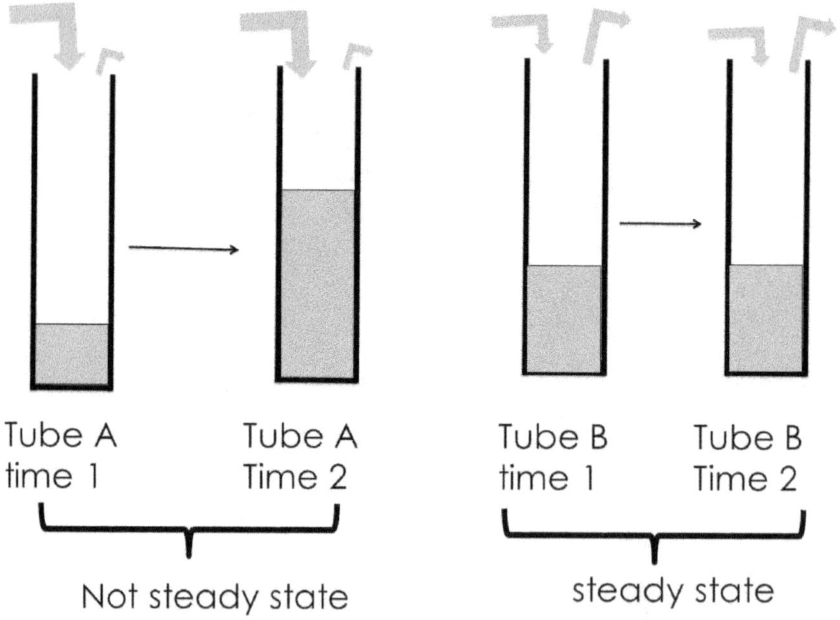

| Tube A time 1 | Tube A Time 2 | Tube B time 1 | Tube B Time 2 |

Not steady state   steady state

Figure 2. In the non-steady state (left), more is going in than going out so the level increases. In the steady state (right), the amount going in matches the amount going out and the level stays the same.

Physiology

Dorothy: Well, now apply it to calcium.

Connie: If I stop growing, the amount of calcium I eat must equal the amount of calcium I get rid of?

Dorothy: Right.

Connie: Why don't I just stop eating calcium? I know when I'm growing, I need calcium so my bones can grow, but once I'm done growing, I don't need to make any more bones.

Dorothy: Nice idea, but the body doesn't work that way. Your body is constantly remodeling your bones by removing some calcium and then putting some more back in.

Connie: Oh, I get it; it would be sort of like constantly repairing any damage to a highway. They go out, locate a crack, remove some concrete, and put in some fresh concrete, so the road stays smooth.

Dorothy: That helps.

Connie: Ok, so now I see that if we are constantly remodeling bone, it makes sense to have some calcium in the diet. I'm not fond of dairy products. What are some other good sources of calcium?

Dorothy: The following table lists how many ounces of each food would, by itself, provide 100% of the recommended amount of calcium for an adult, 1300 mg. [1]

8	American cheese
16	ricotta cheese
17	Almonds
23	plain yogurt
25	tofu with added calcium
30	orange juice with added calcium
30	fruit yogurt
35	Soybeans
47	Spinach
66	bok choy
104	mac and cheese
111	Broccoli

Connie: Could I just take Tums™?

Dorothy: You can, but some people take too many Tums™ and get ill, so don't overdo it. By the way, remember that Tums is an antacid and so will alter stomach acid; this can effect the absorption of medical drugs and herbal medications, so one has to be careful using Tums™ as a source of calcium.

Connie: So how does the body regulate calcium?

Dorothy: Well, as I was saying, when you are in a steady state, the amount you take in must equal the amount you get rid of.

Connie: Yes, I got that.

Dorothy: So how can you get rid of calcium?

Connie: I guess you could urinate the calcium.

---

[1] Adapted from http://www.nichd.nih.gov/milk/prob/other_foods.cfm

Physiology

Dorothy: That's right, for most water-soluble compounds you urinate them out. But the digestive system is also involved because you have plenty of calcium, your intestines stop absorbing calcium and you'll lose it in your feces.

Connie: So what about parathyroid hormone?

Dorothy: When blood calcium levels get too low, the parathyroid cells secrete parathyroid hormone. This hormone has 2 effects on the kidney. To understand the effects, you need to remember the funny way kidneys clean the blood. Blood goes through the **glomerulus** which is a tuft of capillaries in the kidney. As the blood goes through, all the small molecules are filtered out into the tubular fluid. Then as the fluid passes through the tubules of the different parts of the kidney, the stuff you want to keep is reabsorbed.

Connie: Yeah, I've heard about that. One of my teachers compared it to a way of cleaning my room: I take out all the small stuff in my room. I leave behind the furniture, which is similar to the blood keeping the cells and the large proteins. Then I go through the stack of small stuff and pick out the things I want to keep and bring it back into the room. The stuff remaining in the pile (chemicals remaining in the tubular fluid) goes out to the garbage (goes out to the bladder and is in the urine).

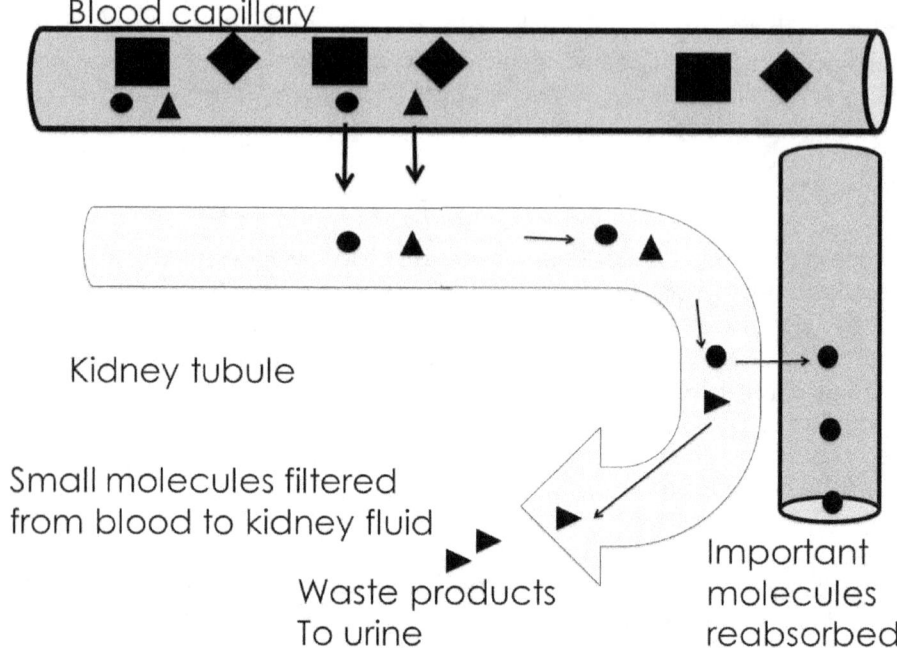

Figure 3. In the glomerulus, small molecules (circles, triangles, for example ammonium and glucose) are filtered out of the blood into the kidney tubules, whereas larger molecules (squares, diamonds, for example, proteins) are retained in the blood. Later in the kidneys, the important small molecules, for example, simple sugars (circles), are reabsorbed. Waste products (triangles, ammonium) stay in tubule and exit in urine.

Dorothy: Exactly. So one effect of the parathyroid on the kidney is to increase the reabsorption of calcium.

Connie: Reabsorption?

Dorothy: Yeah, that's like when you pick the things out of the pile to put back into your room.

Connie: Why don't they just call it absorption?

Dorothy: Absorption is when you absorb the calcium in your intestine; like when you bought the stuff initially. Reabsorption is when you decide to take back something from your cleaning pile.

Connie: Ok, got it. So now the kidney is urinating out less calcium, so that means more stays in the body; I get it. Does parathyroid hormone also lead to more calcium absorption from the intestine?

Vitamin D
Dorothy: Yes, but not directly, that's the job of one-twenty-five-dihydroxy-vitamin-D. It's a bit of a complicated story, so relax for a minute. A compound related to vitamin D, one, twenty-five dihydroxy vitamin D, $(1, 25 (OH)_2 D)$, increases calcium absorption in the intestine. It does this by increasing the amount of the proteins involved in absorbing calcium. Epithelial cells have a side that faces outside the body (for example, the inside of the intestine) and another side that faces the blood stream. They form a tight barrier between inside and outside. Calcium needs to get into the cell on the intestinal luminal side, that's the side facing the intestine lumen or inside the intestine. Then calcium needs to get out the other side, the serosal side[1].

When you eat a meal high in calcium and your body needs calcium, you want to absorb the calcium across the intestinal cells from intestinal lumen to blood. Let's say the intestinal calcium is 10 mM and the blood calcium is 2 mM. The absorption problem is harder than it appears and I'll first tell you what seems to be the "easiest" way and then why it is a problem. The easiest way would be for calcium to move from intestinal lumen to cell, down its concentration gradient, through a protein calcium pathway and then to have the calcium move from cell to the blood side down its concentration gradient, through a protein calcium pathway. If calcium in the cell were 5 mM (or indeed anything between 2.1 mM and 9.9 mM) we would satisfy the need for a gradient.

One key feature of calcium is that it is relatively high outside cells and extraordinarily low inside cells. The blood calcium is about 2 mM, but the free calcium inside most cells is 0.2 uM, about 10,000 times lower! If

---

[1] Serosal comes from the same root that gives us serum for the non-cellular part of blood, so the serosal side is the blood side.

calcium gets above about 2 uM (that is, 1,000 times less than outside) most cells will die in short order. A key piece for understanding the solution is the fact that I used with word "free" for the limit of calcium inside the cell. Unfree calcium is calcium bound to another molecule; in this case the calcium is bound to a protein. Thus even if the free calcium inside the cell is 0.2 uM, the total calcium can be much higher, for example, 10 uM, if 9.8 uM is bound to a protein, conveniently called, calbindin.

As an analogy for free calcium consider free gum as gum not in its wrapper and bound gum as gum in the wrapper. You want most of the gum to be bound and only a bit of free gum which you are about to put into your mouth.

So far, we have calcium in the intestine at 10 mM and free calcium in the cell at 0.2 uM and total calcium inside the cell at 10 uM. The calcium gradient refers to the gradient for free calcium so clearly there is a large gradient on the apical (luminal) side and if there is a calcium pathway, calcium can enter. The next problem is that we need to get calcium from the cell to the blood side. If we merely put in a pathway for calcium on this side, notice the problem. The free calcium on the blood side is about 2 mM, 1,000 times larger than the 0.2 uM inside the cell. So we have a gradient and a pathway, BUT the gradient is in the direction that drives calcium from the blood to the cell. We wanted to move calcium the other way. So rather than have calcium move through a pathway, we have calcium move uphill through a protein call a calcium pump[1].

Here's an analogy for the process. Imagine you want to transport squirrels (calcium) from the road (intestine) through a corn field (cell) and to the forest (blood). Also imagine that the corn field (cell) has a wall (surface or plasma membrane) that keeps squirrels out. If you

---

[1] Calcium pumps are also Ca ATPases since the fuel for the pump is ATP. There are two common groups of Ca pumps that differ on their location. The PM Ca pumps (PMCA) are found in the plasma membrane (outer membrane) and the SER Ca pumps (SERCA) are found in the SER membranes. The ER is the endoplasmic reticulum, an intracellular compartment of cells; in muscle cells it is specialized and called the sarcoplasmic reticulum; "sarco" the Greek word for flesh.

have more squirrels on the road than in the forest, you could just put in a few squirrel paths through the wall. But the problem is that you don't want too many squirrels in the cornfield. So you put the squirrels (calcium) in cages (calbindin) when they are in the cornfield. At the corn field/forest wall, you need to lift the squirrels over the wall to move them out of the field, just as the calcium pump lifts calcium uphill moving out of the cell.

Connie: So calbindin is a protein that binds calcium; I like it when the names make sense: I wonder why they didn't call it calpumpin instead of calcium pump?

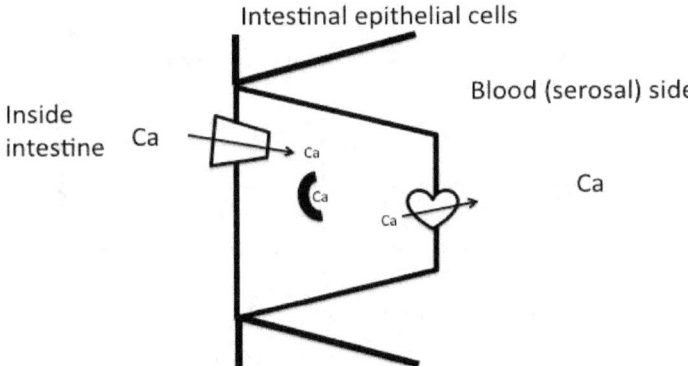

Figure 4. The trapezoid in the middle is one intestinal epithelial cell. Calcium from food goes across the inside membrane. Most of the calcium then binds to calbindin inside the cell, so the free (single) calcium is low. On the blood side, a protein (the heart shape) pumps the calcium toward the blood.

Now I see how one-twenty-five-dihydroxy-vitamin-D works-it increases the expression of the calcium channel, calbindin, and the calcium pump and so increases the absorption of dietary calcium. What's the link with parathyroid hormone?

Dorothy: As you can tell from its name, one, twenty-five **di**hydroxy vitamin D is vitamin D modified with 2 hydroxy groups. (Di stands for 2 just like tri stands for 3.) The numbers just tell you where they are on the molecule. An analogy might be that you decide to put shades on your house, the shades being analogous to hydroxyl groups. To know which windows they are on, you give each window a number starting from the front door and going left. So one, twenty-five means that the first and 25th windows have shades. Twenty-five clearly indicates that it is a big house. For organic molecules, we aren't numbering windows but carbon atoms so we know that vitamin D has at least 25 carbons atoms. The liver has the enzyme that puts the first hydroxyl group on the 25th carbon, to form 25 hydroxy vitamin D. The kidney has the enzyme that puts the second hydroxyl group on the 1st carbon, to form one, twenty-five dihydroxy vitamin D. The amount of the kidney enzyme is regulated by parathyroid hormone.

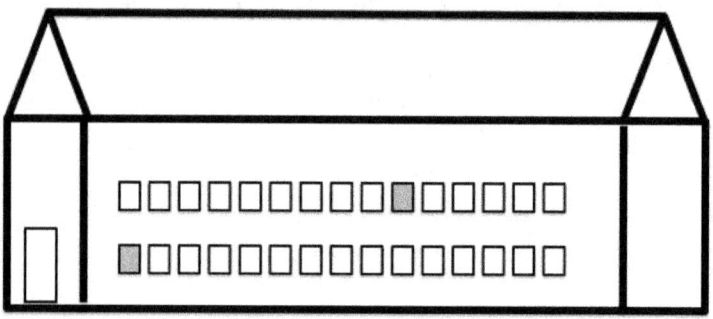

1, 25 dihydroxy vitamin D
1, 25 di-shade house

Figure 5. In this house, windows 1 and 25 have shades. The numbering starts on the first floor, the window closest to the front door. In vitamin D, the carbons are all numbered according to a chemical convention.

Connie: We should review the sequence of events, now that we have all the names and players.

Dorothy: Good idea, but maybe we should briefly touch on how the parathyroid gland knows whether to secrete parathyroid hormone or not.

Connie: I'd guess you could use a calcium receptor to sense the blood calcium, just like the kidney had a parathyroid receptor to sense blood parathyroid hormone levels.

Dorothy: You are right, there is a calcium receptor; but the calcium receptor is a bit unusual. Most receptors bind organic molecules, that is, molecules that contain carbon, and these tend to have specific and complicated structures, so that the receptor recognizes only that specific molecule. For example, the parathyroid hormone receptor recognizes parathyroid hormone which is a complicated protein with 84 amino acids. In contrast, calcium is an element and is simple and spherical. It clearly doesn't have any carbons. It is unusual to have an extracellular receptor that binds an ion. The cells in the parathyroid gland have a receptor that responds to the level of calcium in the blood. When calcium levels get too high, more calcium binds to the receptors. When the calcium receptor is activated, it tells the parathyroid cell to stop secreting parathyroid hormone.

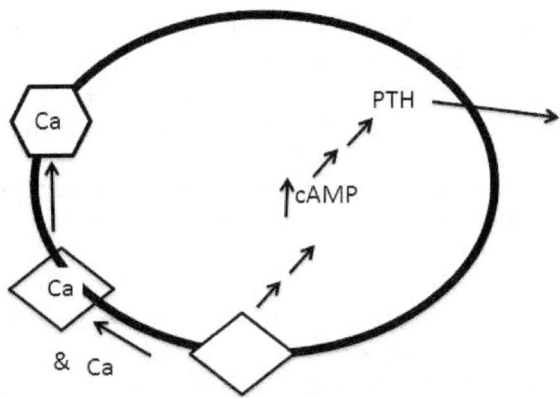

Parathyroid cell

Figure 6. The large circle is a parathyroid cell. When calcium is low, the calcium receptor is empty and diamond shaped. With this shape, the cell increases cAMP. The increase in cAMP eventually leads to the release of PTH out of the cell where it can reach the blood stream. When calcium is high, the calcium binds to the receptor. The receptor changes shape from diamond to stop size. In the stop sign shape, the cell does not increase cAMP and therefore the cell stops (or reduces) PTH release. You might find it interesting to compare this figure with Figure 2 in the chapter, Black and White (and Banning Tanning?)

Calcium regulation
So let's follow calcium and its regulation from when you eat calcium.
- eat food with calcium
- sometimes stomach acid helps dissolve the calcium

- in the intestine, calcium enters the intestinal cells through calcium pathways on the luminal surface, binds to calbindin, and the calcium goes out the serosal side on the calcium pump
- three of these steps involve proteins, the calcium entry pathway, calbindin, and the Ca pump. Probably all 3 are increased by vitamin D.
- calcium in blood is bound to protein, anions, or free (unbound).
- calcium in blood binds to calcium receptor on the parathyroid gland
- calcium receptor with calcium bound changes shape and this causes a decrease in parathyroid hormone release
- so now, when blood calcium is filtered by kidney, since parathyroid hormone is low, more calcium stays within the tubular fluid and is urinated out.
- you are now urinating out more calcium and if you don't match your input with this output, your blood calcium will go too low
- When this happens, less blood calcium binds to the calcium receptor, so these receptors no longer change shape and no longer inhibit parathyroid hormone release. Not that "no longer inhibit" is a double negative, so one gets parathyroid hormone release.
- In the parathyroid cell, when the calcium receptors are turned off, the cells secrete parathyroid.

So now blood parathyroid hormone increases. Remember an increase in parathyroid hormone has at least 2 effects: 1) activation of the kidney enzyme that converts vitamin D to its active form, so intestinal calcium absorption increases and 2) activation of kidney transport proteins so that more calcium is reabsorbed from the tubular fluid and less is urinated out.

Connie: That sounds like a negative feedback loop.

Physiology

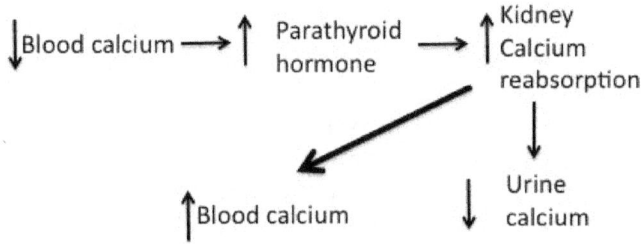

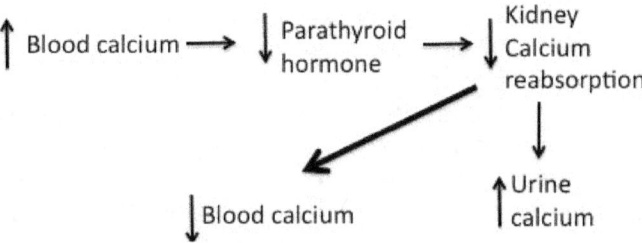

Figure 7 The negative feedback for blood calcium regulation. Top is for when blood calcium is low; bottom for when blood calcium is high.

Dorothy: Precisely! The loop is: blood calcium is sensed by the parathyroid gland which secretes parathyroid hormone which alters kidney and intestinal function, which alters blood calcium.

Connie: Yep, an active calcium receptor turns off parathyroid hormone secretion.
Well that's very interesting, but what about soda and calcium regulation?

Dorothy: I don't think soda has any affect on the parathyroid hormone system.

Connie: Yeah, I've never heard of ingredients in soda altering parathyroid gland or hormone function.

What is wrong with high blood calcium?
Connie: Tell me why can't blood calcium go up by 2 or 3 fold; it is low anyway, and certainly having blood sugar increase 2 or 3 fold temporarily isn't a problem.

Dorothy: There are 2 problems with allowing blood calcium to increase very much. The first problem is that the complex of calcium and phosphate is relatively insoluble in water. It turns out that the normal calcium and phosphate concentrations in blood are slightly below the concentrations where they would precipitate and come out of solutions. An increase of 2 or 3 fold for calcium may well lead to calcium phosphate precipitation.

Connie: What's the second problem?

Dorothy: Well, in many cells there is a small leak to calcium; if you increase the calcium outside the cell, then a bit more leaks in. The inside calcium concentration is about 5000 times lower than outside. If the amount of blood calcium increases, then more calcium will leak in and this will raise the calcium concentration inside the cell. The calcium concentration inside the cell is so low that it doesn't take much of an increased calcium influx to raise cell calcium. Some cells are very sensitive to increases in inside calcium.

Connie: Could you give me an analogy?

Dorothy: Sure: Imagine the cell is the basement in your house and your basement and the swimming pool share a common wall. There is a lot of water outside the basement, just like there is a lot of calcium outside the cell. Your basement has very little water (just as the cell has very little calcium) and you want to keep it that way. However, like most basement walls there are leaks (just like a cell membrane has leaks). In your basement (cell) there are some very water sensitive documents (calcium sensitive proteins/enzymes) and it wouldn't take too much of an increase in the water level in the pool to cause more water to come into the basement and ruin those documents.

Soda, phosphate and calcium
Connie: Ok, so you don't want your blood calcium to increase too much, I get that now. But you mentioned phosphate and that reminds me that I think soda contains phosphate, so I think it is the phosphate in soda that causes the calcium to be lost in the urine. In fact, I read somewhere that the parathyroid hormone controls both calcium and phosphate levels in the blood. So there, it's the phosphate in soda that causes the problem.

Physiology

Dorothy: I agree that, in theory, excess phosphate in soda could cause excess calcium excretion by the kidneys.

Connie: Oh stop using such big words; just say phosphate makes you pee out calcium.

Dorothy: Ok, sure, you think too much dietary phosphate would make you pee out too much calcium, because the dietary phosphate would alter blood phosphate which would alter parathyroid hormone levels and would decrease kidney reabsorption of calcium and therefore you'd have more in your urine.

Connie: You got it! So what do you say to that?

Dorothy: First I agree that you have the pathways correct. But I have an objection to that model. First, you get lots more phosphate elsewhere in the diet, for example, orange juice and milk have more phosphate than soda. And no one claims that drinking too much OJ leads to weak bones.

Connie: Well, maybe no one's looked at OJ and bone density.

Dorothy: You are probably right. But scientists have looked, particularly at animals, and for reasonable changes in dietary phosphate, over the long term, there is no change in bone health.

Acid and calcium
Connie: Well, I still think phosphate is a possible reason for sodas to lead to poor bone health. And I also think acid plays a role. I think acidic diets lead to chronic acidosis and so the kidneys are urinating out calcium and it's lost from bone.

Dorothy: When you say acidic diet, I gather you mean a diet that produces too many protons and therefore reduces the pH[1]. Chronic acidosis then refers to a process that is producing protons for a long

---

[1] Remember that protons are often estimated using a log scale, called pH. Unfortunately the definition $pH = -\log(H^+)$ has a negative sign so that as protons go up, pH goes down and vice versa.

time. I agree that in the short term, if you suddenly increase the amount of acid produced in the body, including by changes in diet, the body tries to fix the acid load (the pH shift) in the short term by putting protons into bone and taking out calcium to buffer the pH in the bloodstream.

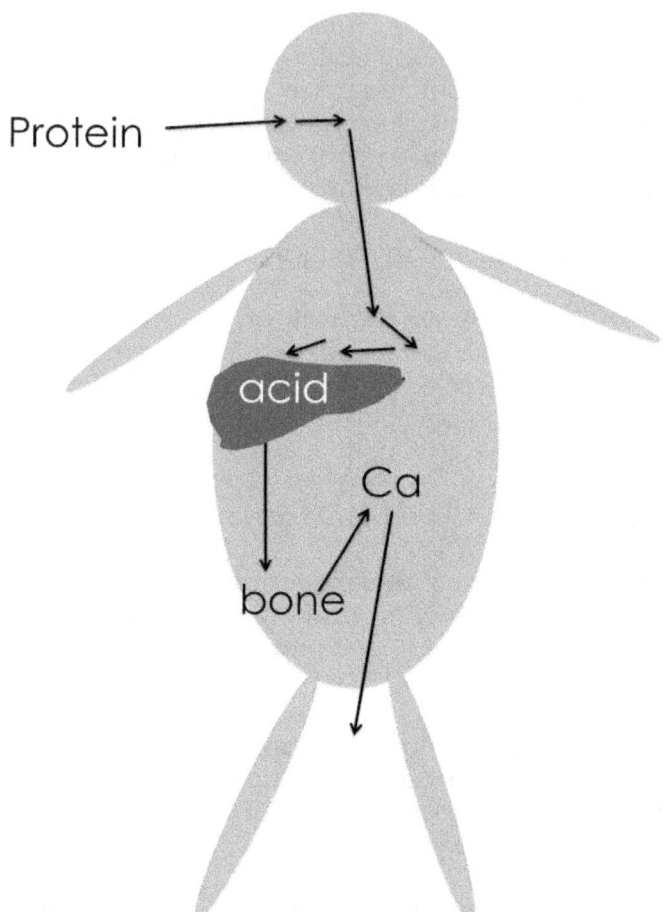

Figure 8. Calcium is lost in urine when there is a short-term increase in dietary acid producing protein. You eat the protein; it is digested and goes into the blood. In the cells, the protein is converted to acid. In the short term, the blood acid goes to your bones and the bones release some calcium which raises blood calcium and you urinate out the "extra" calcium.

Physiology

Connie: Yep, I remember putting an egg in a glass and covering it with vinegar. In about 3 days, the egg shell was very soft.

Dorothy: Did you use a hard or soft boiled egg?

Connie: Either one works. But when I tested the egg at the end, I first put it into a plastic bag so that if it made a mess it was easy to clean up. The point is that the acid (protons, H^+) in the vinegar exchanged with the calcium in the eggshell and as the calcium leached out, the shell "dissolved". Just like in a diet that creates acid, I think the acid will leach out some calcium from our bones.

Dorothy: Well, I disagree. It may be that during the first day or two when the diet has more acid, and the blood calcium is slightly up, the kidneys urinate out the extra calcium. However, I think that within a few days, the kidneys will gear up and be able to excrete the extra acid and you won't need to use bone calcium to buffer the acid load.

Connie: But over the long term, some scientists say our kidneys can't quite adjust. They claim that on a "modern" diet we are slightly acidic in our blood and always losing a little calcium. They say that early humans had a diet that was relatively low in acid (well, actually pretty high in base, due to all the fruits and vegetables people ate). Thus, they say we are designed to eat a diet that has low net acid. They say our modern diet of animal protein and particularly soda adds an acid load which our kidneys can't handle. Thus we buffer the extra blood acid by removing some bone calcium. This makes slightly calcium deficient and we end up with weak bones. So there, that's why too much soda is bad for you.

Dorothy: Your argument that our kidneys can't adjust to a chronic acid load is a very powerful argument. I can't disprove it, but I also don't quite buy it. I think we might both agree that we'd each like more data and information to support our side, but right now, it is not conclusive.

Here's my counterargument, which I should point out, has not been made, so far as I can tell, in the medical literature. Your argument has been published, but I would point out, almost always by the same group, so it's a bit hard to evaluate whether most scientists agree with it or not.

If you look closely at their data you will see that when people are on a chronic high acid diet, their blood acid values change minimally; the values are clearly within the range we call normal. Personally, I don't think there is any way the bones would sense, directly or indirectly, such a small change in blood acid and therefore they wouldn't give up their calcium.

Connie: What about the argument that we haven't evolved/adapted to manage a high acid diet.

Dorothy: Well, it's a very interesting argument, but we don't have any menus from 10,000 or 100,000 years ago. So the estimate of what the diet actually contained is pretty difficult. Even with menus and even knowing how much food we produce today, it requires some assumptions to guess what people actually eat nowadays, so I'm not convinced we know that the ancient diet had less net acid.

Connie: So if you had to guess, what would you guess the ancient diet was?

Dorothy: I'm not saying I know, and indeed, I would say that I would guess there is a 70% chance the scientists you quote are right, but 70% is not a sure thing. My other concern is that many mammals, particularly carnivores, do have a high net acid producing diet and so my bias is that most mammals have the capacity to deal with a high net acid producing diet. Of course, humans could have lost that capacity, just as we lost the capacity to make vitamin C, which all other mammals (except guinea pigs) apparently have retained.

Connie: So I guess we are at a standoff about acid and calcium loss.

Dorothy: I guess so.

Physiology

Connie short & long term; Dorothy short term

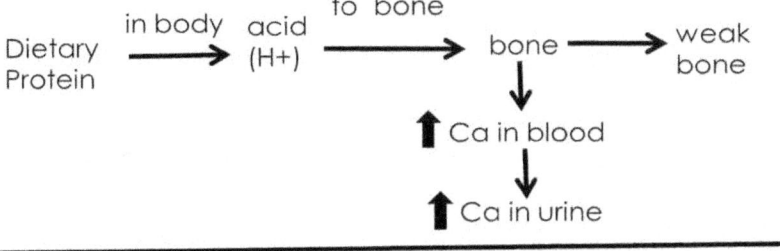

Dorothy long term

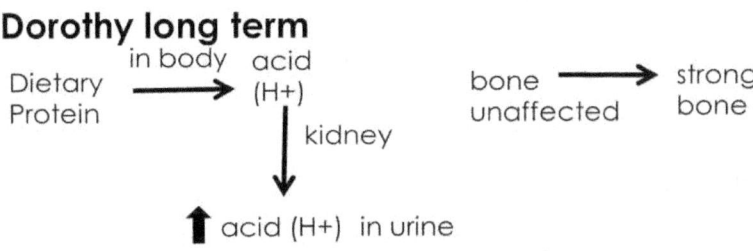

Figure 9. Connie and Dorothy disagree on how the body handles dietary acid loads in the long term. The top diagram shows what happens when there is sudden change in diet that produces an acid load and both Connie and Dorothy agree with this. Connie thinks this also happens over the long term, but Dorothy thinks the bottom pathway kicks in. The difference is that, on top, the acid load is handled by the bones. The bones release their calcium and it is urinated out and the bones become weak. In the bottom scheme the kidneys handle the acid load and pee out the acid and the bones are not affected so they stay strong.

Caffeine and calcium

Connie: Could the caffeine in soda cause bone weakness? Have you heard that caffeine is a diuretic?

Dorothy: Yes, doesn't diuretic mean causing increased urination?

Connie: Yes, diuresis comes from the Greek word for urination, ouresis. Do you want to guess what natriuresis means?

Dorothy: Uresis, that's urine again.

Connie: Yep.

Dorothy: Natri, does that refer to natural?

Connie: Nope, think about your chemical symbols. Do you remember what element Na stands for?

Dorothy: Na, let's see, that's sodium. I always wondered how they got Na from sodium.

Connie: Yeah, natrium is Latin for salt.

Dorothy: So natriuresis must mean more sodium in urine.

Dorothy: Give me one more:

Connie: Ok, kaliuresis?

Dorothy: So kalium, maybe K, so potassium.

Connie: Bingo.
So let's talk about caffeine and diuresis first and then move on to caffeine and hypercalciuria.

Dorothy: Don't you mean calciuresis?

Connie: No, In contrast to the previous terms, for high calcium in urine the word that is used is hypercalciuria, I'm not sure why. "Hyper" is probably familiar to you as it means extra or high, for example, hypertension just means high blood pressure. "Calci" is obviously means calcium and "uria means in urine; so hypercalciuria means high calcium in the urine. [1]

Dorothy: Ok. First, how would you determine if caffeine really were a diuretic, that is, increases the volume of urine?

Connie: Well, I'd drink a caffeinated drink and see if I urinate more.

---

[1] Hypercalcemia would mean high calcium in the blood; the suffix "emia" referring to blood values.

Physiology

Dorothy: How do you know it's the caffeine and not just the drink?

Connie: Ok, I'll drink a caffeinated drink one day and the same volume of decaffeinated coffee the next.

Dorothy: Decaffeinated coffee still has some caffeine; maybe water would be a better control.

Connie: But maybe something else in the coffee besides caffeine is a diuretic.

Dorothy: Ok, decaffeinated coffee is ok. Will you eat and exercise the same each day?

Connie: No way, that would be too boring.

Dorothy: Of course, you'd need to do the study on more than just yourself. So you could take a set of people on a specific strict diet and specific exercise for one day and give half caffeinated coffee and half decaffeinated coffee.

Connie: Yeah, but everybody urinates a bit differently, I have an aunt who urinates...

Dorothy: I don't want to hear about your aunt's urinating habits. These seem like pretty hard studies to find volunteers for. But of these studies that have been done, the evidence is that acute caffeine does cause diuresis, but usually, for chronic users, there is no clear effect.

Connie: Remind me what chronic and acute mean.

Dorothy: Acute means short term, so in this case, it is someone that does not normally drink caffeine and they are given it during the experiment. Chronic means long term or time and again

Connie: I get it because chronic has the same root as chronometer, fancy watches. So I guess chronic is time and time again.

Dorothy: Ha-ha. Anyway, a chronic user is someone who has caffeine every day and the body adapts to this.

Connie: So how about in animals since you can control their diet and drinking.

Dorothy: That's right; in animals acute caffeine does appear to be a diuretic.
In fact, mice have been used to test one theory of how caffeine has its effects. Caffeine, unlike some other drugs, doesn't target only 1 protein. Caffeine probably has 2 important targets in the kidney: it goes to adenosine receptors and dilates blood vessels, so more blood flows to the kidney, so therefore more is filtered. Second, it also inhibits the enzyme that breaks down cAMP, cyclic adenosine monophosphate. So in the mice experiment, they knocked out the adenosine receptor.

Connie: What does knockout mean?

Dorothy: Scientists take some special mouse cells and selectively add a piece of DNA that inserts into the DNA for the adenosine receptor gene. For example, they can put in a stop codon very early in the gene so only the first 10 amino acids are made and not the other 1000. Then this cell is returned to a very early stage of mouse development and the new mouse has some cells that make adenosine receptors and some have the mutation so they can't make adenosine receptors. After appropriate breeding, it is possible to have mice that don't have any adenosine receptors.

Connie: Ok.

Dorothy: So anyway, when the control mice were fed caffeine they increased their urine output but the knockout mice didn't.

Connie: I don't even want to think about how they collected and measured the urine.

Dorothy: One problem with knockout experiments is that many other genes influence how a particular gene works. A sports analogy would be what happens when you remove player X from a team. He might be the best player here on that team and therefore the team is crippled without him. On a different team, player X's removal might not be so important.

Physiology

Connie: So the results of the experiment mean it is the adenosine receptor and not the cAMP that are important for the diuresis effect of caffeine, at least in those type of mice?

Dorothy: I don't think this experiment rules out cAMP, but it means at least adenosine receptor is important. We'll have to wait for experiments that address whether alterations in cAMP metabolism are also involved in the increased urinary excretion due to acute caffeine intake.
My bias is that most of the data suggest that if calcium intake is adequate, caffeine doesn't cause any bone problems.

Back to soda, I don't think there is any convincing evidence that soda directly causes a problem with bones or leads to excessive calcium loss, as long as there is enough calcium in diet.

Connie: We both agree there needs to be enough calcium in the diet. You are not convinced by the data that soda leads to calcium loss. I'm not convinced that there is data showing that soda does NOT lead to calcium loss. I think we are both using a standard of "beyond a reasonable doubt", but we are starting with different premises: you start by assuming soda **does not** have an effect on calcium loss and want data that disproves, beyond a reasonable doubt, that position, that is, you don't see any data that convincingly show that soda does lead to calcium loss. I'm starting with the assumption that soda **does** have an effect on calcium loss and I don't see any data that disproves, beyond a reasonable doubt, to disprove my assumption that is, I don't see any data that convincingly show that soda has no effect on calcium loss.

I do think we both agree that no one claims soda is a health food.

Dorothy: I think you have correctly summarized our positions. How about summarizing the basic physiology of calcium regulation?

Connie: Sure.
- once we stop growing, calcium intake needs to match calcium excretion
- if there is more excretion than intake, we are losing calcium, probably from bone

- the blood levels of calcium are sensed by the parathyroid gland, which releases parathyroid hormone
- the effects of parathyroid hormone that we discussed are to increase kidney calcium reabsorption and to produce more of the active form of vitamin D.[1]
- the active form of vitamin D leads to an increase in the proteins required to absorb more calcium from our intestines
the net effect is that when blood calcium is low, the intestine absorbs more dietary calcium and the kidney excretes less calcium, thus blood calcium hopefully increases.

- [1] There are additional effects of both parathyroid hormone and vitamin D, but they are beyond the needs of this chapter.

Physiology

Steroids Make the World go Round

Literature cases

1. A pet owner, Cathy, brought her 12-year-old cat, Andy, to the vet. The cat had started spraying urine around the house. The owner said that the face also seems larger and rounder and the vet agreed. These observations are consistent with the cat having high testosterone, but the cat had been neutered when it was only a few years old[1]. The vet took a blood sample and the values for androstenedione[2] and testosterone were increased more than 20 fold above the high end of the normal range. Further imaging studies suggested that the cat had an adrenal tumor and it was surgically removed. After about 8 weeks, the cat's owner returned, and indicated that the cat had stopped spraying urine and had become more affectionate again. Its face had also returned toward normal.

2. A 14-year old female tennis star, Tessa, had noticed for about 2 years that she had acne, not uncommon for a teen. She also noticed an increase in facial hair and a deepening of her voice. This is expected for teen males, but not expected for a teen female. Her blood levels of testosterone were also high. What would you expect? Do you think she might have been taking testosterone supplements? How would you determine if she was? She denied taking supplements- is that adequate evidence? What is the tipping point for you to decide that you think she is lying and work harder to find out about your supplement use vs. that you accept her statement and do more (expensive) testing to find out a non-supplement reason for her symptoms? If you preliminarily decide that she is not taking supplements, what could be other causes of her symptoms?

3) Tori went to her health care provider complaining of facial hair growth and deepening voice. With insightful questioning the health

---

[1] In males, testosterone is made by the testes; neutering males involves removing the testes.

[2] Androstenedione is a particular steroid compound and is a precursor to testosterone. The "andro" part is from the Greek for man, since it is present in higher concentrations in men than in women. The rest of the word refers to its chemical structure. Testosterone is named for where it was first isolated from, the testes and the "sterone" suffix refers to its structure as a steroid.

care provider determined that the most likely cause was testosterone transferred from her male partner's towel. Would that really be enough testosterone?

4) Paul, 5 years old, and his mom went to their health care provider because Paul was having breast development. With insightful questioning the health care provider determined that the most likely cause was a hair care product. Do you know of any hair care products made from parts of female animals?

In this chapter we will discuss
- Steroid name confusion
- General principles of steroids and their receptors
- Cathy's cat Andy
- Tessa, androgen effects and "normal" range
- Testosterone, Acne and Hair Growth
- Steroid binding proteins, carriers, and "free"
- Tori and testosterone creams
- Paul as well as phytoestrogens
- Caveat on case studies
- Testosterone and muscle strength and aggression
- Do high concentrations of androgens cause cardiac side effects?
- What about androgens and psychological and behavior problems?
- Bumps on the Penis
- Chapter summary

Steroid name confusion
The word steroid is often used in the news when discussing athletes. In this case, the reference is to anabolic[1] steroids, that is, steroids that build up tissue, particularly muscle. So far all anabolic steroids also have virilizing effects, that is, they produce male secondary characteristics.

You or one of your friends may have asthma and take inhaled steroids to treat it. Or you or one of your friends may have had a severe

---
[1] Anabolic is from the Greek, "to throw" & "up" hence the use for something that builds up muscle making complex molecules from simple ones.

reaction to poison ivy or another severe inflammatory reaction and taken steroids as treatment. In these cases, the steroids were glucocorticoid steroids. Cortisol is an example of a glucocorticoid[1]; one of the effects of cortisol is to raise blood sugar (glucose) (see more in the chapter, How Sweet it Is). Cortisol has nothing directly to do with sexual characteristics. It is often called the stress hormone because it is increased during times of stress. Glucocorticoids also turn down long term defenses, such as the immune system, which is responsible for inflammation. Scientific shorthand sometimes creates some confusion as both glucocorticoids and testosterone are steroids. Glucocorticoids and testosterone bind to different receptors and have different effects. So when someone says they are taking steroids, out of context, it is not clear if they are taking androgenic compounds or glucocorticoid compounds. Aldosterone is also a steroid, it is a mineralocorticoid as it regulates sodium and potassium (mineral) levels.

Many of the hormones in this chapter got their names from the major effect or the effect that was first noticed. We now know they have many effects and it is likely that we will discover even more. Androgens[2] include testosterone; testosterone levels are much higher in males than females (a point that will be important in a later case in this chapter), but females do make testosterone. Similarly, while I'm sure you are well aware that estrogen[3] has many effects in females, it is also made in males. The production of these steroids is controlled by peptide hormones, in particular Luteinizing[4] Hormone, LH, and

---

[1] Glucocorticoid refers to steroids that have actions similar to cortisol; the name emphasis the reaction of raising blood sugar (glucose) and the fact that it was isolated from the cortex of the adrenal gland. The adrenal gland also has a medullar part; I used to get confused about whether the cortex or the medulla was the outer layer; cortex is from the Greek for "bark" and bark is obviously on the outside of the tree.
[2] Androgen is derived from the Greek term for man, "andro", and these hormones, at high concentrations, produce male sexual characteristics.
[3] Estrogen comes from the words estrus and generate; hormones in this class are important for regulation the ovulatory cycle (estrus) and for many female sexual characteristics.
[4] Follicular Stimulating Hormone, as its name suggests, stimulates follicle growth in the ovary. Luteinizing Hormone name may not be as helpful. LH

Follicular Stimulation Hormone, FSH. These clearly got their names from the initial work detailing their effects in females. And while males don't luteinize or have follicles, they do secrete these hormones and they play an important role in regulating testosterone levels.

General principles of steroids and their receptors

Steroid refers to a chemical classification; a steroid has the basic structure shown in Figure 1.

Steroid Backbone

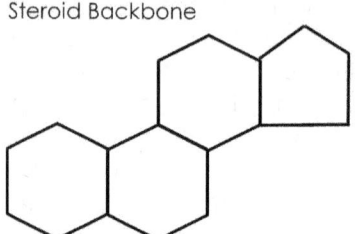

Figure 1 Steroid backbone. Each vertex is a carbon atom. There are 4 "rings"; 3 with 6 carbons and one with 5. Can you find the ring with only 5 vertices? You should be able to recognize this backbone shape in many of the structures in this chapter.

Cholesterol is the precursor for making steroid hormones and has the structure shown in Figure 2

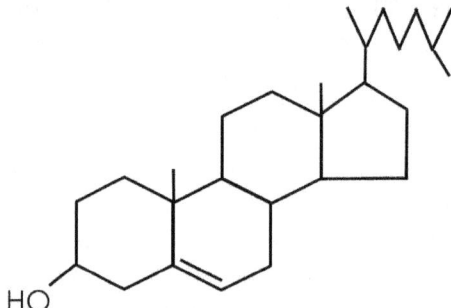

Figure 2. Cholesterol structure which has a steroid backbone within it.

---

regulates the formation of the corpus luteum. The corpus luteum is a set of cells that grows after ovulation. Luteum is the Latin word for egg yolk, thus corpus luteum essentially means, "body, yellow". The color is caused by a carotene like molecule, lutein, which is readily available in such vegetables as kale, spinach, and zucchini.

Physiology

All steroid compounds are hydrophobic and lipophilic[1], in contrast to many other hormones. Insulin and other hydrophilic (water soluble) hormones[2] are made in advance and stored in vesicles[3]. When the hormone secreting cell gets the appropriate signal, the hormone is released. Can steroids be stored in vesicles? You can easily answer this question if you remember a key concept in physiology: if something is to move, there needs to be a gradient and a pathway. If we are to store steroids in vesicles, then the steroids should not move out. Clearly there is a gradient, because store implies more in the vesicles than in the cytoplasm. What about a pathway? Because steroids are lipophilic they can just diffuse out of the vesicle across the bilayer so there is a "pathway,"[4] in the sense that a screen door provides a pathway for air to move into the house. Steroids are made on demand, in contrast to the making and storing of hydrophilic hormones in advance of their need. This difference in how the secretion is regulated is interesting. Hormones like insulin can be released quickly and can have very quick effects. Because it takes time to make the enzymes that then make the steroids, the blood concentrations of steroids increase more slowly, on the order of many minutes, compared to seconds for hormones stored in vesicles. Would there be an advantage to having steroids increase in concentration faster? Probably not, because steroids primarily regulate gene transcription and this is a relatively slow process.

Steroids, being lipid soluble, are able to cross the cell membrane without a transporter. In the cell cytoplasm, the steroid binds to its protein receptor, for example androgen receptor, or estrogen receptor or progestin receptor. When the receptor has bound its

---

[1] Hydrophobic means water-fearing; lipophilic means fat-loving. So this was just a fancy way of saying steroids are not water soluble but are fat soluble.
[2] Other hydrophilic hormones include insulin, antidiuretic hormone, glucagon, adrenaline, and acetylcholine.
[3] Remember that a vesicle is an internal closed structure in the cell with a lipid membrane that separates the vesicle interior from the vesicle exterior; the vesicle exterior is the solution of the cell interior.
[4] I suppose the vesicle could store steroids if there were a high concentration of a steroid binding protein. That is not the case however. However, I don't know that it would be a disadvantage.

hormone, the complex then moves to the nucleus. In the nucleus, the complex can bind to specific regions on the DNA and regulate the expression (transcription) of particular genes. Often, cells have specific other proteins that can interact with the complex and further modulate the expression.

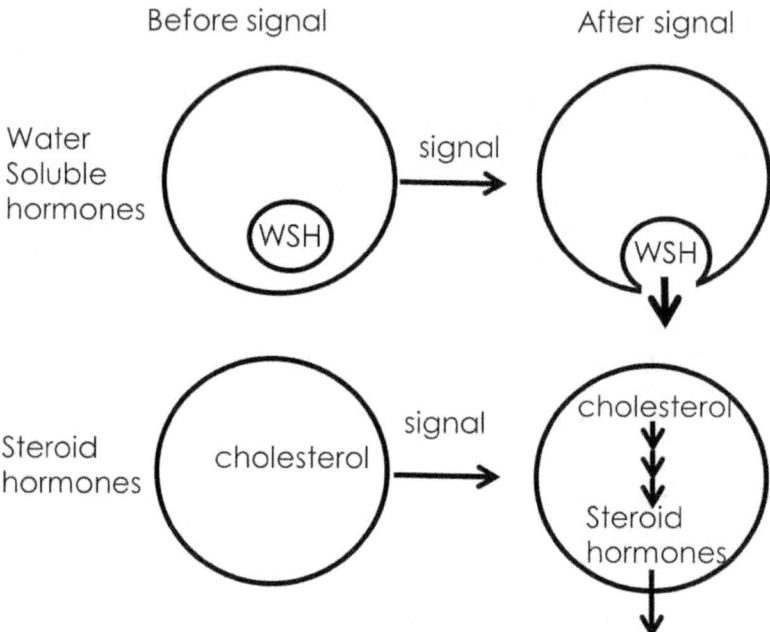

Figure 3. The contrast between water soluble hormones and steroid hormones. Water soluble hormones (WSH) are made in advance and stored in vesicles. When the signal comes, the vesicles fuse with the membrane and the hormones are released quickly. Steroid hormones are not made until the signal comes; the signal increases the rate of conversion of cholesterol to the active steroids which then diffuse out of the cell and into the blood. Water soluble hormones and neurotransmitters include insulin, parathyroid, norepinephrine, and acetylcholine. Water phobic hormones (lipid soluble) include not only steroids, but thyroid hormone and vitamins A and D.

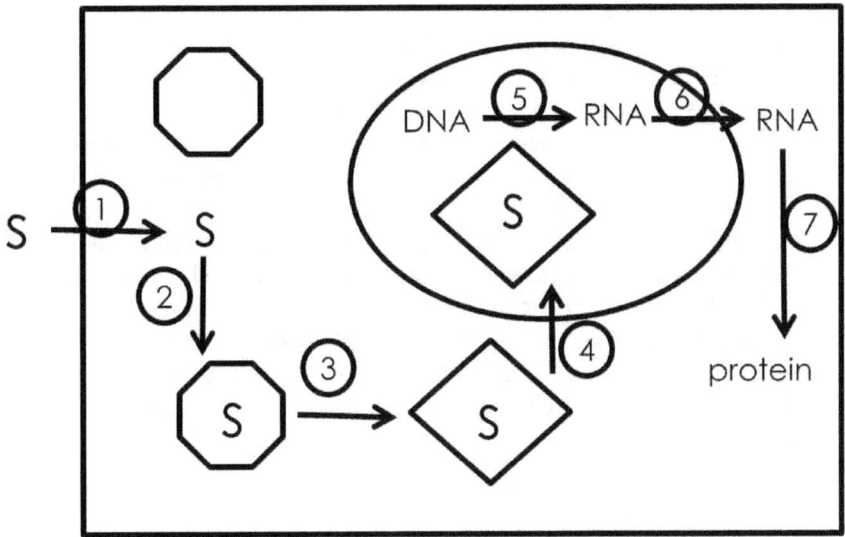

Figure 4. Steroid hormones (S) alter gene expression.
The large square is a single cell
The large oval is the nucleus of that cell.
The stop sign and diamond represent two different shapes of the steroid receptor which are inactive and active forms.
1. the steroid hormone(S) diffuses into the cell.
2. The steroid hormone binds to its inactive receptor (stop sign shape), e.g., androgen receptor, estrogen receptor, progestin receptor, aldosterone receptor, or cortisol receptor.
3. The hormone-receptor complex changes shape from stop sign to diamond and becomes active.
4. The active hormone-receptor complex moves into the nucleus.
5. After binding to DNA, the transcription from DNA to RNA increases for specific genes.
6. The RNA moves to the cell cytoplasm
7. The RNA is translated to protein.
Note that not shown are accessory factors specific to different cell types that can also alter which genes are expressed or turned off.

Cathy's cat Andy
Why would an adrenal tumor be secreting androstenedione and testosterone? The adrenal gland has enzymes that can modify cholesterol to produces a number of steroids, such as aldosterone and

cortisol. However the adrenal gland does not normally produce testosterone[1]. Interestingly, testosterone can be made from compounds in the adrenal gland with only one additional enzymatic step. That is, the synthesis of testosterone only requires that the adrenal tumor make one enzyme that is not normal for adrenal cells. However, in order to be a tumor, the cell is making a number of proteins that are not normal for an adrenal cell. Since the surgery seems to have corrected the problem, the simplest explanation is that the tumor cells did express these extra enzymes and did account for the high blood levels of androstenedione and testosterone. I think that is a plausible explanation, but I suspect a good defense lawyer could find some ways to raise some doubts. What addition evidence could the prosecution get that would make the case even more convincing? [2]

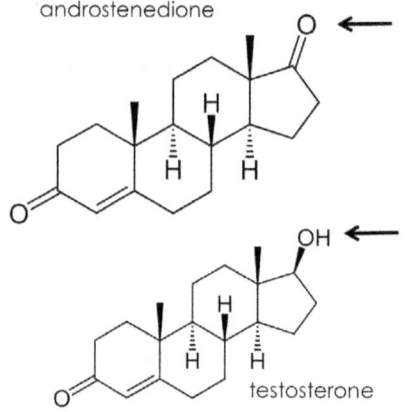

Figure 5. It takes only 1 enzymatic reaction to convert androstenedione to testosterone; the ketone group (= O) is reduced to

---

[1] The statement is a slight oversimplification. The normal adrenal does produce a little bit of testosterone. In intact males, the testes produce a lot more. In females, testosterone is produced in the ovaries and a bit by the adrenal glands. In both cases, it is secreted into the blood. In some other female cells testosterone is produced and has local effects, but is not secreted into the blood

[2] When the surgeons removed the tumor, if they had checked the tumor for the presence of the extra enzymes, that would have strengthened the case.

the alcohol. Androstenedione is normally produced by the adrenal gland.

Tessa, androgen effects and "normal" range
Further testing revealed that Tessa had an adrenal tumor that was secreting testosterone. The health care providers were interested in determining if the tumor might have improved her strength. Before the tumor was removed, the health care providers tested her strength extensively. After they removed the tumor, the hair growth went away, though it did take several months. The clinicians decided it would be best to compare her muscle one year later, so that she would be in the same training regimen as the initial testing. She showed no decrease in muscle strength and did as well in the national tennis tournaments after the surgery as before. This seems a little surprising as one would expect high levels of testosterone to increase muscle mass in females; of course it is possible that the testosterone did give her more strength at 14 and that her strength at 15 reflects a year's growth which happened to roughly match the possible testosterone effect.

A recent study examined the testosterone levels in elite male and female athletes[1] and a summary of the results is shown in Figure 6. Note that some male athletes have testosterone levels below the normal male range and indeed as low as many females.

---

[1] Healy ML, Gibney J, Pentecost C, Wheeler MJ, Sonksen PH. Endocrine profiles in 693 elite athletes in the postcompetition setting. Clin Endocrinol (Oxf). 2014 Aug;81(2):294-305. doi: 10.1111/cen.12445. Epub 2014 Apr 2. PubMed PMID: 24593684.

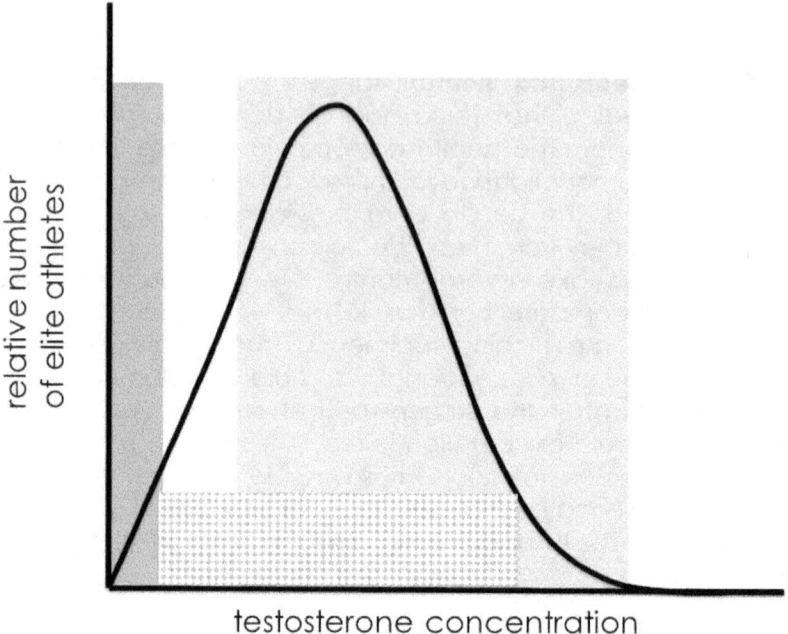

Figure 6. On the horizontal axis is plotted the testosterone concentration and on the y axis the relative number of individuals. The line is a curve fit to the data for elite athletes. The dark gray area is the "normal" testosterone range in females with the lighter gray area the "normal" testosterone range for all males. The bottom checkered rectangle is approximately the 95% interval for these elite athletic males.

Let me talk about what "normal" means in medicine and physiology as it is DIFFERENT from the lay use of the term. Normal is the unfortunate term used to describe a particular distribution that often occurs in nature; sometimes the terms Gaussian or normal are often used. For a large population of adults, height often follows a "normal", Guassian, or bell shaped distribution. So what is an abnormal height? Many scientists and statisticians have criteria that if the probability of something happening by change is less than 5%, then it did not happen by chance. This has carried over to medicine and the "normal" range is the range that 95% of the population has. Just because you are outside the 95% does not make you unhealthy or "abnormal" in the lay sense of being weird. I'm 6 foot 4 inches and my wife is 5 foot; we are both outside the normal range for height, but I would not call our height unhealthy or weird. As another example, red hair only occurs in about 1% of the population, which makes it rare and unusual, but not unhealthy.

In Figure 6, a number of elite male athletes have testosterone levels below the normal range, that is the range that 95% of all males have. Obviously, even with the low testosterone, these men can compete at the highest level. It is interesting to note that even within the elite athletes, one expects 5% of the athletes to have abnormal testosterone values based on the statistical definition of abnormal (seen in less than 5% of the population), not abnormal in terms of weird or unhealthy.

Testosterone, Acne and Hair Growth
We will discuss the details of how testosterone might increase muscle strength later in this chapter, but it is complicated, so we will start by discussing in detail the effects of androgen on the skin glands that secrete sebum and on hair growth as these will illustrate the major principles.

As you know, one of the major annoyances of puberty is the development of acne. In males, one of the contributing factors is the increased levels of testosterone. Testosterone binds to its receptor in the sebum gland cells and leads to an increased production of sebum, by increasing the transcription (DNA to RNA) of particular genes which are then translated into proteins that make the sebum. Sebum is the an oily substance responsible for one's skin or hair feeling greasy. This material is a good food source for some bacteria and some bacteria grow even better when the gland duct is blocked.

Another occurrence during puberty is the growth of hair. Before puberty, the hair on the body is of the vellus type and usually almost unnoticed. An increase in testosterone leads to hair that is thicker, longer and more pigmented and hopefully in the appropriate places. In males, this includes facial hair. In the hair follicle cells, testosterone is converted to dihydro-testosterone. While both these compounds can bind to the androgen receptor and activate it, DHT binds to the receptor better. As shown in Figure 7, there is a concentration range where, even though T and DHT are the same concentration, DHT will activate more receptors than T. The activated receptor goes to the nucleus and turns on the genes responsible for making the hair thicker, longer, and more pigmented.

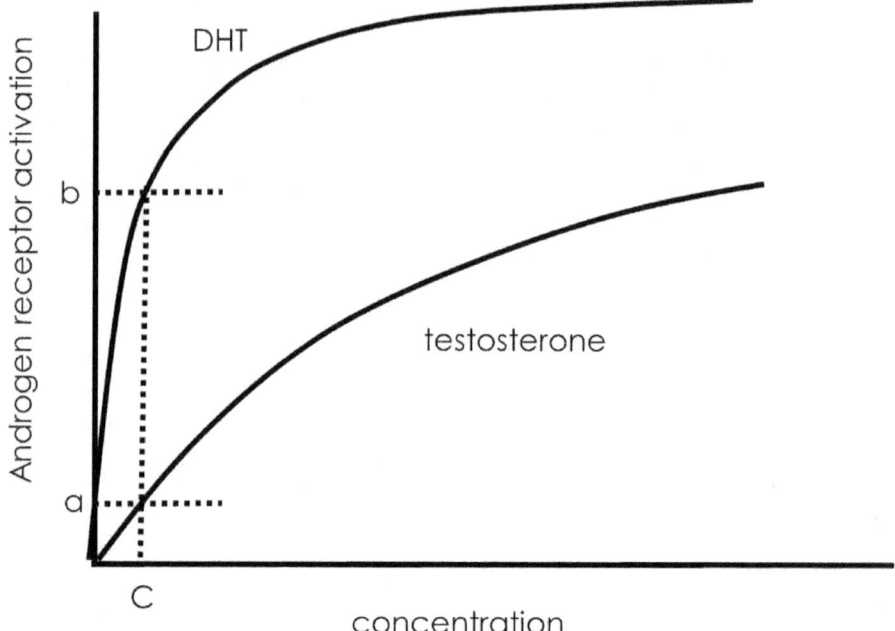

Figure 7. This graph is a way that scientists illustrate that at low doses DHT is more effective than testosterone; this is termed a higher affinity for DHT for the receptor than testosterone. On the x (horizontal axis) is the concentration of testosterone or DHT. On the way axis is the fraction of how many receptors are activated. At concentration C, testosterone would only activate the receptor to a; whereas as the same concentration, DHT would give activation b, about 5 times greater.

What about the facial hair on Tessa? There are a number of possibilities and I think it is interesting to consider them all.

1. High blood testosterone would lead to high testosterone in the cell, which is converted to DHT, which binds the receptor and activates it. The tennis star certainly had high blood testosterone so this could be her problem. In normal women, the face hair follicles express the

androgen receptor but the testosterone and DHT are too low to activate it.

2. Tessa also had high blood levels of androstenedione. Androstenedione can also enter the facial hair follicle cells; these cells have the enzyme to convert androstenedione to testosterone and you know the rest of the story.

3. What about women that have normal blood testosterone and androstenedione levels?

a. It is not the total testosterone and androstenedione levels that are important, it is the free concentration. Most (98%) of the testosterone is bound to plasma proteins, usually albumin and sex globulin binding protein and only 2% is free. What happens if the concentration of the sex globulin binding protein is decreased and if the total testosterone or androstenedione concentration stays the same? The free concentration will increase since less is now bound. This can sometimes cause the unusual hair growth, so these women may have normal total testosterone levels but higher than usual free testosterone levels. More on binding proteins and free steroids is presented in the next section.

b. Even if the free blood testosterone and androstenedione concentrations are the same, there can be excess hair growth. Another possibility is that the enzyme that converts testosterone to DTH is overactive. Or the enzyme that converts androstenedione to testosterone can be overactive.

c. The hair follicle cells may also have different concentrations of the other co-factors.

4. It gets even a bit more complicated, but this complication helps us understand why oral contraceptives are often used to treat excess facial hair growth in females (with the benefit of reduced acne).

a. estrogen can inhibit the enzyme that converts testosterone to DHT. Before menopause, a women's circulating estrogen levels are usually high enough to prevent facial hair growth because this enzyme is inhibited and DHT stays low. After menopause, there may be no change in the amount of testosterone or androstenedione in the blood or in the hair follicle, but because estrogen is lower, the conversion enzyme works better and DHT goes up enough that, in some women, the androgen receptor is now activated, see Figure 8.

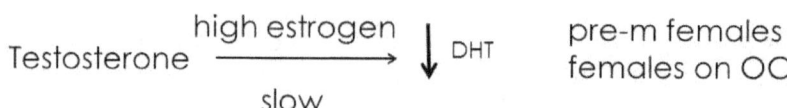

Figure 8. Top. In males and post-menopausal females, estrogen is low so the conversion from testosterone to DHT is fast and they have high DHT levels. **Bottom.** In pre-menopausal females and females taking oral contraceptives, estrogen is high so the conversion from testosterone to DHT is slow and they have low DHT levels.

b. Testosterone can be converted to estradiol by an enzyme called aromatase. The balance of aromatase and the testosterone-to-DHT-conversion-enzyme may be different in different women.

Some oral contraceptives contain two synthetic hormones, for example, ethinylestradiol and progestogen. This pair can suppress ovarian (and some adrenal) testosterone production as well as block testosterone to DHT conversion. In additional oral contraceptives can increase steroid hormone binding globulin. If the total amount of testosterone doesn't change, this leads to a decrease in the free testosterone in blood. Interestingly, in women, half of their production of testosterone occurs in the skin.

In older, genetically susceptible males, normal or high blood testosterone can lead to hair loss on the scalp, i.e., androgen dependent balding. Those hair follicles are slightly different. Testosterone still gets into the cell and is still converted to DHT. DHT still binds to the androgen receptor and still activates it. But scalp hair follicle cells have different accessory proteins than face or body hair follicle cells. The upshot is that the hair follicles don't make as much hair and the man becomes bald.

Steroid binding proteins, carriers, and "free"
I just want to spend a bit more time explaining free vs. bound. This is an important concept for steroid hormones, in fact, for all lipid hormones. Before I can do that, however, we need to make sure you understand some properties of hydrophilic and hydrophobic compounds.

Steroids are hydrophobic compounds[1], they do not like being in water and cats could be an analogy. Many hormones, such as insulin, are water soluble. So are most ions (sodium, chloride…) and sugars (sucrose, glucose, fructose..). Our analogy for water soluble compounds is ; they like fish, like being in water but need help to cross land.

A cat (fat) will have no problem getting across the piece of land (the cell membrane), but a cat does have a problem being in the water on either side of the land. There are proteins that are like rafts that help the fat-soluble compounds when they are in the water phases. (Figure 9)

---

[1] Remember, hydrophobic and lipophilic mean "water fearing" and "fat loving"..

Figure 9. Cats have no problem getting across land (the dark stripe in the middle) but they need a raft in order to be on the water (the light gray on either side). This is an analogy for fat-soluble compounds which don't need assistance in crossing membranes, but do need carrier proteins when they are in water compartments (inside or outside the cell).

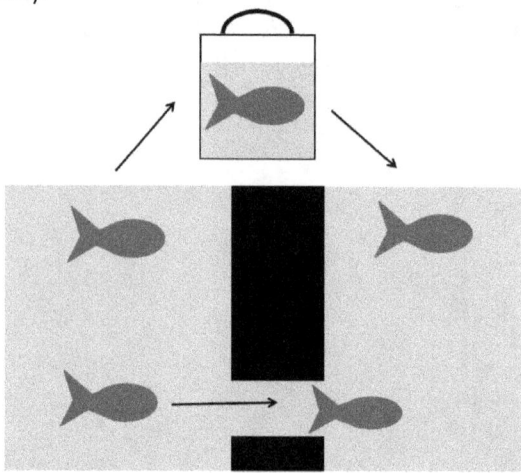

Figure 10. Transport of water-soluble compounds across a lipid (oil) membrane. A bucket or a tunnel can serve as a carrier or transporter that helps water-soluble compounds (fish) cross the fat, or lipid, cell membrane (land). These carrier or transport proteins are found in membranes, such as the cell surface membrane and in cell organelles.

Physiology

Now we can return to free vs. bound. Most steroid hormones are bound to proteins (rafts) in the blood. But it is the free concentration that is important. Because steroids are hydrophobic and lipophilic they need binding proteins when they are in blood. It is the free concentration that is able to cross cell membranes. Of course, the binding protein can release the hormone, but that is in response to a decrease in the free hormone concentration. This is a tough concept and I have two partial analogies. In the first analogy the free hormone is like unwrapped gum; this is the form you can actually chew. The ability to bind to a receptor and activate it is like only chewing unwrapped gum. If a steroid hormone is going to activate something, it has to be free (unwrapped). The binding protein is like the gum wrapper and the bound steroid is like wrapped gum. You can get access to it easily, but you do have to remove the wrapper. One real problem with this analogy is that you put the gum in your mouth but nothing has to carry the steroid across the membrane. Hence the need for my other analogy. In this analogy, the free hormone is like a dodge ball flying through the air. The ability to bind and activate a receptor is like only the balls flying through the air can hit an opponent and send them off the floor. The membrane is like the line between the teams; only the free dodge balls (hormone) can cross the line. This analogy fails a bit with the binding proteins, but it does have a little bonus: if you consider the opponents court as the cell cytoplasm, when the steroid hormone hits the opponent (transcription factor), the opponent (transcription factor) leaves the floor (leaves the cytoplasm and goes to the nucleus).

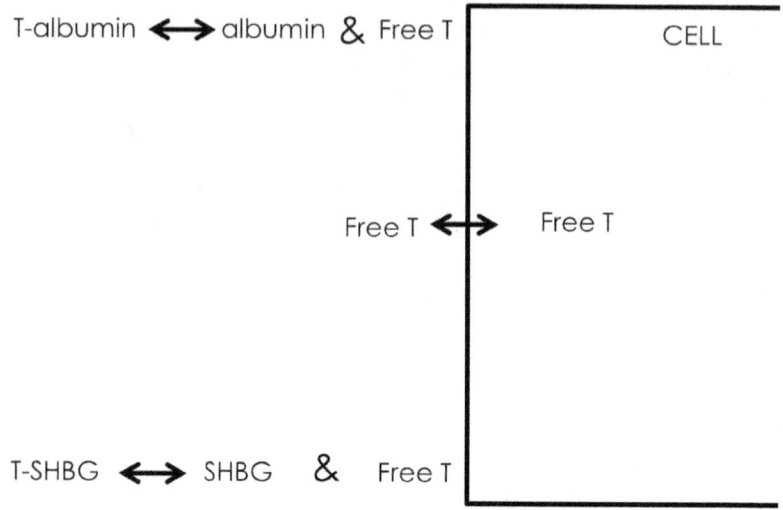

Figure 11. Only the free testosterone is able to cross the cell membrane. Most of the testosterone outside of a cell is bound to albumin or to SHBG (steroid hormone binding globulin). When testosterone is bound, the testosterone cannot cross the membrane. Our analogy is that the cell membrane is like the dividing line in a game of dodge ball. The dodge ball is testosterone and only the free ball can cross the membrane. If the ball is bound to someone, then it cannot cross the membrane. But clearly the ball can move from bound to free pretty easily.

To understand why the total testosterone can be misleading, consider this situation.
The total-testosterone equals the free-testosterone and SHGB-testosterone + albumin-testosterone.

total T = free-T + T-SHBG + T-albumin

If the total-testosterone is held constant, then an increase in SHBG leads to an increase in T-SHBG and so free T decreases.

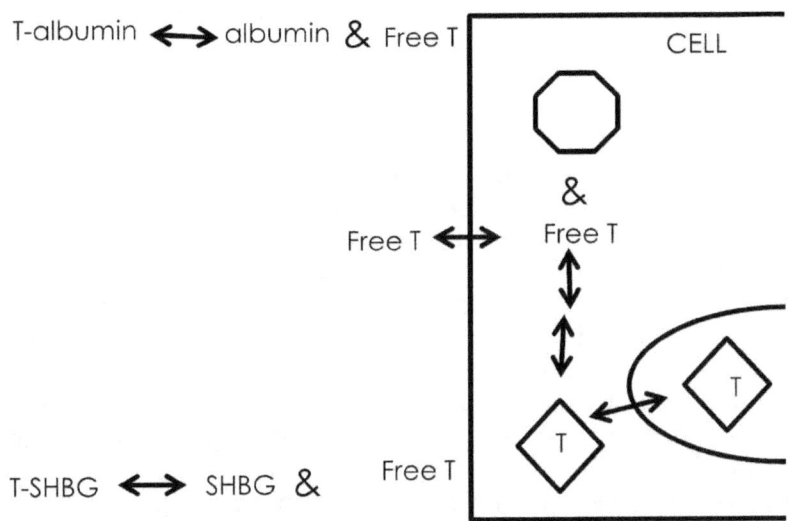

Figure 12. Here I've combined parts of Figures 4 and 11, that is the free hormone crossing the cell and the fact that it is the testosterone-receptor (transcription factor) complex the enters the nucleus to regulate the transcription of some genes.

Tori and testosterone creams

Tessa and Tori did not want facial hair, but many teen males do want facial hair. Thalassemia is a blood disorder and one of the side effects is poor beard growth. Many people, and especially teens, have a difficult time coping with a chronic, congenital disease, after all, we all want to fit in and we don't like to face our mortality. For the teen males with thalassemia, the poor beard growth might be the only observable difference between them and "normal" teens. Thus they often put testosterone cream on their face, which provides the testosterone to activate the receptors and get hair growth.

Why does a cream work? Because testosterone is lipid soluble it can diffuse out of the cream and through the skin, without a transporter. (There is a gradient from cream to blood, and the lipid cell membrane provides the pathway.)

There are several cases in the literature of women, and even children, getting inappropriate hair growth. In these types of cases, the clinician first inquires about whether someone else in the household is using testosterone cream. Sometimes children can get into the medicine cabinet and put the cream on themselves. But it seems unlikely that a female partner would put the cream on and not realize that it might lead to excess facial hair.

A few cases are a reminder about the importance of surface area and about the gender difference in circulating levels. Men have about 7 times more blood testosterone than women, but men make about 20 to 40 times more testosterone per day than women because of the increased breakdown rate in men. A man that needs testosterone (e.g., after testicular cancer) will spread a cream on his body to get the testosterone. He can get a higher dose by spreading the cream over a large surface area; the larger the surface area the more testosterone can be absorbed, just like the larger the screen the more air can come in. Surface area is increased by foldings in the lung and small intestine. However, only a fraction, often ~ 10%, is absorbed. In a few cases, it appears that the male's female partner got enough testosterone to have facial hair growth. In some cases, it was presumably skin-to-skin contact. In at least one case, this occurred even if the man showered before the contact. In Tori's case, the best guess is that the couple shared a towel and that transferred the testosterone. Because the female normally produces about 20 times less testosterone per day, it would take only 1/20th of the amount of cream on the man to have a significant effect on the woman.

Paul as well as phytoestrogens
There are sporadic cases of children less than 5 years old having breast development and/or pubic hair growth. Sometimes this reflects the steroids secreted by a tumor or other endogenous steroid problem. But sometimes the steroid comes from outside. A series of cases were reported from one institution where about 4 or 5 girls were all being given scalp massages by their mothers. The scalp lotions contained estrogen. In fact, in one case, the lotion was made from placentas and so contains both estrogens and progesterones. (see http://www.amazon.com/Hask-Placenta-No-Rinse-Treatment-Original/dp/B00091V6EO)

In another case, it appears that the child's diet was high in fennel, which is known to have high estrogen-like levels. Finally, there were 3 young girls who played together and had breast development. The clinician discovered that all 3 had eaten lot of strawberries. Don't worry-strawberries do NOT have phytoestrogens[1]. But the clinician felt that the farmer taking care of the field sprayed it with too many growth compounds and that one or more of them had estrogen like effects.

Caveat on case studies
One has to be careful when analyzing these types of case studies. Obviously the person (or the cat) is their own control. There is a good correlation between hormone or chemical levels or exposure and symptoms: Before the exposure or tumor, there was no problem. Then either a tumor, exposure to a cream, or a novel change in diet occurred. During this time, symptoms developed. Then surgery removed the tumor, or the person prevented cream exposure or stopped the novel dietary constituent and the problem went away. In these cases, it is reasonable to conclude that the middle piece contributed to the problem, but I would not say the deduction is air-tight.

Testosterone and muscle strength and aggression
It has been known for a long time that, in adult men with low testosterone, raising it to the normal level can fix many problems. Athletes have used testosterone-like compounds for a long time, but most of the early studies were poorly designed. A clear one was done in 1996[2] and it clearly showed that testosterone lowered percent body fat and increased muscle mass.

In a later study[3], the scientists wanted to study the dose dependent response. This was a very clever study. One problem with most

---

[1] Phytoestrogens are plant compounds that mimic estrogen.
[2] Bhasin S, Storer TW, Berman N, Callegari C, Clevenger B, Phillips J, Bunnell TJ, Tricker R, Shirazi A, Casaburi R. The effects of supraphysiologic doses of testosterone on muscle size and strength in normal men. N Engl J Med. 1996 Jul 4;335(1):1-7. PubMed PMID: 8637535.
[3] Bhasin S, Woodhouse L, Casaburi R, Singh AB, Bhasin D, Berman N, Chen X, Yarasheski KE, Magliano L, Dzekov C, Dzekov J, Bross R, Phillips J, Sinha-Hikim I, Shen R, Storer TW. Testosterone dose-response relationships in healthy young men. Am J Physiol Endocrinol Metab. 2001 Dec;281(6):E1172-

testosterone studies is that one is giving testosterone in addition to what the body is making. When this happens in some men, the extra testosterone feeds back and inhibits the endogenous testosterone production, so there is actually only a small increase in blood testosterone. In other men, the extra testosterone does not feed back and their total testosterone is equal to the exogenous testosterone taken and their normal testosterone, so they have much higher levels. Thus it is not fair to lump these two types of men together, but most studies that do this don't measure the blood testosterone levels and so their experimental group really has two different testosterone levels. And of course, the men may all start at different levels. To get around this problem in this later study, all the men were given gonadotropin releasing hormone (GnRH), a hormone which turns off testosterone synthesis. Then the divided the men into groups and gave each group a different amount of testosterone. The results look pretty convincing. The authors pointed out that some of the men at high testosterone developed acne and so the study could not be completely blinded. They also found no affect of these levels of testosterone on mood and behavior as judged by both the subjects and their live-in partners.

In a recent study[1] about whether testosterone increases aggressiveness in women, the scientists found no effect of the testosterone. But they did ask the women whether they thought they had received the testosterone or the placebo. The ones that thought they had received the testosterone did behave more aggressively. This study does have some problems, but it is interesting that the idea of testosterone had an apparently larger effect than the testosterone itself.

Three of the effects of testosterone in men are a decrease in fat mass, an increase in lean body mass and an increase in red cell production. Both a decrease in fat mass and an increase in red cell production are due to testosterone (or DHT) binding to the androgen receptor and activating it. The mechanism for the increase in muscle mass is harder to sort out. In fact, some sources are still dubious that

---

81. PubMed PMID: 11701431.

[1] Eisenegger C, Naef M, Snozzi R, Heinrichs M, Fehr E. Prejudice and truth about the effect of testosterone on human bargaining behaviour. Nature. 2010 Jan 21;463(7279):356-9. PubMed PMID: 19997098.

testosterone increases muscle mass; they feel the increase in muscle mass in athletes taking testosterone is due to increased exercise.

Some scientists think that the extra high levels of testosterone increase muscle mass but don't do it by activating androgen receptors. Rather, the theory is that the high testosterone inhibits the ability of cortisol and related glucocorticoids to activate their receptor. Glucocorticoid receptor activation tends to activate the catabolic or breakdown pathways in muscle, so inhibiting them would tip the balance to anabolic, or build-up pathways.

One reason a direct androgen receptor effect on muscle mass is not favored is that the androgen receptor is nearly saturated with testosterone at normal circulating levels. Thus a modest increase in testosterone would saturate the receptor but the doses some athletes use are extremely high. This is shown in Figure 13. As an analogy, consider the amount of testosterone as analogous to how damp or wet one's clothes are. Men with low testosterone levels are like men with damp clothes; giving extra testosterone or adding more water will have an effect. But for men with normal testosterone levels, their clothes are almost already soaked, so adding more water or testosterone won't have much of an effect on their clothes' wetness or the amount of receptors that are activated.

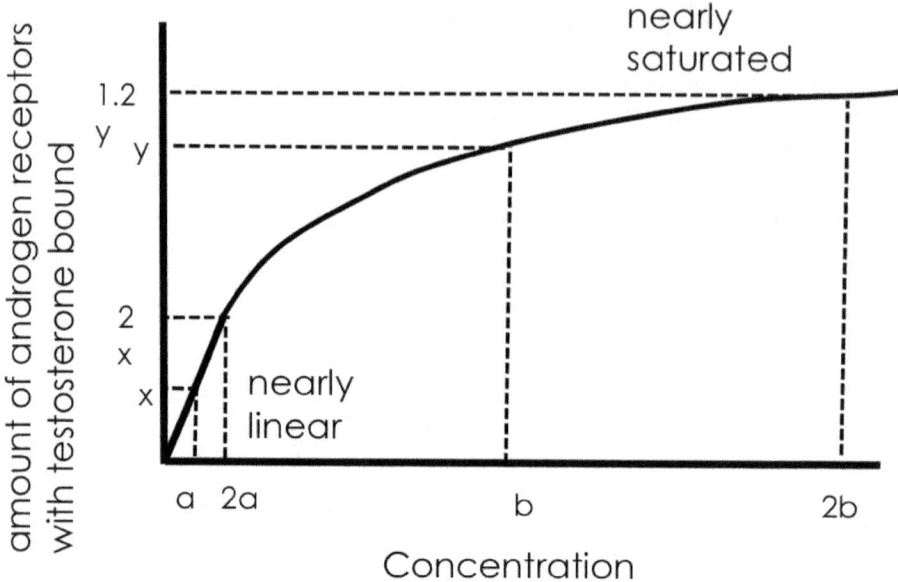

Figure 13. On the horizontal (x) axis is concentration; on the vertical (y) axis is the amount of androgen receptor with testosterone bound. Normal men are at about point b, so increasing their testosterone by 2 fold (to 2b) will have only a modest effect on the amount of receptors that are activated. In contrast, men with low testosterone (a) will have a nearly 2 fold increase in response when the testosterone concentration is doubled to 2a.

The testosterone levels in **Andy, the cat,** were high. The change in the shape of the face was due to an increase in the muscles of the face which altered how it looked. I presume that in humans whose faces change with high testosterone it is the increase in muscle mass that causes the shape change.

Physiology

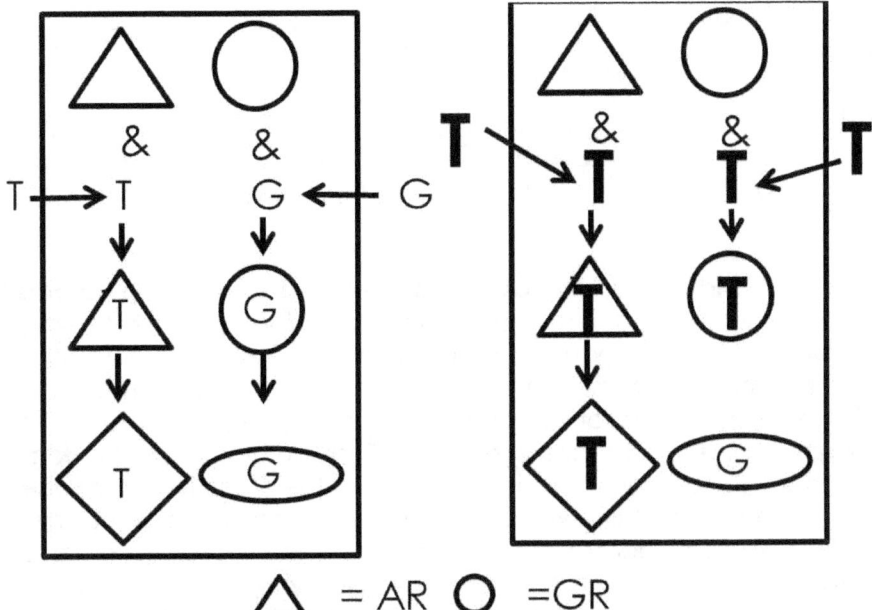

△ = AR   ○ = GR

Figure 14. On the left is a cell with exposed to modest testosterone, T. T binds to the androgen receptor (AR) and activates it (shape change) which can lead to changes in which genes are expressed. This cell also responses to circulating glucocorticoids (G) which bind to the glucocorticoid receptor (GR) and activate it. *Activated GR promotes catabolic effects.* On the right is a cell exposed to high testosterone. It can now also bind to the GR and prevents G from binding and activating it. This thus prevents the catabolic effects, leading to anabolic effects.

Do high concentrations of androgens cause cardiac side effects? I think the news media and sports groups make this claim, but the scientific evidence is not clear. It is certainly true that some young athletes who were taking steroids have died. And it is also true that their hearts were abnormal. But since we don't have any data on their heart structure before they started taking steroids, we cannot establish cause and effect.

Body builders do tend to have larger hearts but most of these studies were poorly designed; they did not have appropriate controls or were not randomized. It is unfortunate that the 2 studies I showed above did not include measurements of the subjects hearts as those studies seemed better designed.

I want to stress that I am not claiming that steroids are safe or that they don't affect the cardiovascular system. I'm saying that we don't have the quality of data that would be convincing, "beyond a shadow of a doubt".

What about androgens and psychological and behavior problems? This runs into similar problems. Someone on steroids acts violently; do we know if the steroids caused the behavior? In individual instances it is hard to establish cause and effect. I think it is hard to do high quality experiments because one is almost caught in a catch-22. The best experiments would take a set of similar volunteers and randomly assign them to take testosterone or not. In the 2 studies mentioned above, no affect on behavior and mood was noted. But, while these studies used doses there are higher than normal (supraphysiological) the reported incidents tend to be in people who used much higher doses than supraphysiological. Is it ethical to ask volunteers to take doses this high? If not, then the best one can do is poll users and try to find an appropriate set of non-users matched controls.

Bumps on the Penis

One of the other effects of the tumor in Andy, the cat, was that his penis developed bumps, often called spines. It turns out that penile spines are standard in many species, including cats, mice and chimps, but obviously not humans. And penile spines development is the result of testosterone activating the androgen receptor. Experiments like what happened to Andy support this theory-in animals that have penile spines, castration leads to a penis without bumps or spines. And giving them exogenous testosterone leads to the growth of spines. In addition, mice which have their androgen receptor knockout out don't get spines.

Recently, a molecular reason for why chimps have penile spines and humans don't, has been worked out. The scientists compared chimp and human DNA and looked for pieces of DNA that chimps had but humans did not. One of the pieces was just upstream of the androgen receptor. They suggest that this DNA provides the information to express the androgen receptor in the cells that are responsible for spine formation. In support of this, they put this DNA with other DNA that codes for a gene reporter gene. When expressed in mice, the reporter gene was expressed in the cells that made the penile spines (as well as the cells that make whiskers on the face). In addition, if they put this DNA and the reporter into cultured human foreskin cells, they also saw expression of the reporter, which suggests that all the other machinery for expressing the androgen receptor is present in these human cells. We will see if others can repeat these experiments and whether this is the only explanation for penile spines in chimps and mice. I want to look at DNA from other animals that don't have penile spines and see if this piece is also missing.

To summarize the penile spine issue: In animals that have penile spines, the cells responsible for it have a piece of DNA that tells that cell to express the androgen receptor. Thus when the animal secretes testosterone, the receptor is activated in these cells (as well as the "usual" cells) and spines develop. Humans are missing this piece of DNA, so these cells don't express the androgen receptor. Thus even when humans secrete testosterone (or take it) even though testosterone goes into these cells, it cannot activate the androgen receptor because there is no androgen receptor there. Obviously, human still make androgen receptors in many other cells and it remains responsive.   I think this story nicely illustrates the point that testosterone responsive cells are cells that express androgen receptors; which cells express the receptor depends upon the DNA of that animal.

Chapter summary
- Steroid refers to a chemical structure and is the "family" name; the "first name" is often **one** of their physiological effects, e.g., glucocorticoid steroids for raising blood glucose, mineralocorticoid steroids for regulation sodium and potassium, sex steroids for regulating egg and sperm production and secondary sexual characteristics
- steroids are lipophilic and can easily cross membranes; steroid creams are an effective way to deliver steroids locally
- steroids are bound to binding proteins; the small concentration that is not bound is the free concentration and it is the free concentration that is directly able to enter cells and bind to steroid receptors
- "Typical" steroid response
    - the steroid hormone(S) diffuses into the cell
    - The steroid hormone binds to its inactive receptor, e.g., androgen receptor, estrogen receptor, progestin receptor, aldosterone receptor, or cortisol receptor
    - The hormone-receptor complex changes shape and becomes active
    - The active hormone-receptor complex moves into the nucleus
    - After binding to DNA, the transcription from DNA to RNA increases for specific genes
    - The RNA moves to the cell cytoplasm
    - The RNA is translated to protein
    - Note that there are accessory factors specific to different cell types that can also alter which genes are expressed or turned off
- Steroids bind to "their" receptor specifically at low concentrations; at high concentrations, a steroid might bind to another type of steroid receptor

Physiology

Cycles in Synch?

Four new students arrive at their dorm suite, Andrea, Barb, Cathy, and Donna. They unpack and settle in for their first semester-what an experience! Their mothers come to pick them up for Christmas break. As they carrying stuff down to their cars, Andrea tells her mom, "Remember when I had my first period, you told me how all the women in your sorority ended up with their menstrual cycles in synch-well it happened to us, too!" Barb's mom overhears this conversation and says, "That idea was all the rage when I was in college, but it's been discounted now. The only way you all would have synchronized periods is if you all are the pill and take pill #1 on the same day."

In this section, we are going to discuss the events of the menstrual cycle, how the pill (oral contraceptives) work, and some thoughts about whether women living together end up with synchronized periods.

What are some differences between ovulation in animals and humans?

First I want to make sure we understand several different words and features of the female reproductive physiology that are common to all mammals. We will then consider features that are rare, or even unique, to humans. Human women are well aware that they periodically bleed during their child-bearing years, hence "period". The blood comes from the shedding of the lining of the uterus, the endometrium, and this shedding is called menstruation. Not all mammals menstruate, that is, in some mammals the endometrium is reabsorbed and not shed. However, all mammals do ovulate—they release mature eggs that are capable of being fertilized by sperm. In many mammals there are visible and/or physiologically detectable changes in the female when she is ovulating. These changes cannot only be signals, but attractants, since this is the time when mating is most likely to result in offspring, allowing the genes of the parents to be passed on.

Ovulation occurs between two consecutive times of menstruation. It makes sense that the ovulation and menstrual cycles would be offset, since the development of the uterine lining allows for growth of a fertilized egg. If ovulation and menstruation occurred at the same time, it would be difficult for the fertilized egg to develop; hence these genes would be unlikely to be passed along (that is, would be selected against).

What occurs during the ovulation cycle in humans?

There are a number of hormones whose levels change during the menstrual cycle and their levels are regulated by intertwining negative and positive feedback loops. While we can clearly measure the rise and fall of hormones and have some good ideas of some of the regulatory pathways, other parts are still not fully understood on a molecular level.

Historically, the 4 primary hormones that vary are two steroids, estrogen and progestrone, and two peptide hormones, lutenizing hormone (LH) and follicle stimulating hormone (FSH). All four of these hormones are also found in men, though estrogen and progesterone are in low concentration. In men, LH and FSH regulate, among other things, sperm maturation.

I am going to start when ovulation occurs, not when menstruation occurs, in contrast to most books. A surge in LH precedes ovulation. Dipsticks, similar to those that test for pregnancy, are available for monitoring LH surges for those who want to know when they are about to ovulate. Many textbooks have ovulation occurring midway between 2 consecutive menstruations, but there is great variation in what time in the cycle ovulation occurs as well as in the length of the cycle.

The LH surge causes one mature follicle to rupture; this rupture releases the egg cell and also other cells that develop into the corpus luteum. The corpus luteum (which literally means yellow body[1]) is present only temporarily. During the many days that the corpus luteum exists, the corpus luteal cells secrete progesterone, estrogen, and inhibin so this is an endocrine gland. Inhibin is a relatively recently discovered peptide hormone. All three of these hormones turn off the release of FSH and LH. FSH and LH are two hormones of a group call gonadotropins. [2]

[1] The yellow color is apparently due to the accumulation of pigments like carotene.

Physiology

FSH and LH are released by the anterior pituitary. Estrogen, progesterone, and inhibin act on both the pituitary and on the hypothalamus. The hypothalamus secretes gonadotropin releasing hormone[1] (GnRH) which is one of the signals to release FSH and LH (gonadotropins).

Progesterone also acts on the endometrium, causing it to grow, make new blood vessels, and store lipids and glycogen, potential energy sources. If the egg from ovulation had been fertilized, then it would implant in the endometrium and these new blood vessels and energy sources would sustain the fertilized egg for a bit.

The corpus luteum is a bit like a self-destructing CD; without intervention, it starts to undergo regulated cell death (call apoptosis) after about a dozen days. A fertilized egg is the intervention that stops the self-destruction; fertilization causes the release of signals, such as chorionic gonadotropin[2]. Chorionic gonadotropin is a peptide that overrides the self-destruct signal for the corpus luteum. Conveniently, chorionic gonadotropin is small enough that it is filtered when the blood goes through the kidney. The kidney does not reabsorb it, so it is lost in the pee. This is convenient because an increase in chorionic gonadotropin is the basis for most dipstick tests for pregnancy and

---

[2] Gonadotropins. Gonad: testes or ovaries. Tropin-to grow. So these hormones regulate the growth of the ovary or testes.

[1] Gonadotropin releasing hormone: a name that makes pretty good sense, but it would have been even easier if it were called LH-and-FSH-releasing hormone; "easier" in the sense of remembering its function; harder to say.

[2] chorionic gonadotropin. You already know what a gonadotropin does. Chorionic means this hormone is made by the chorionic villi which are cells that will become the placenta.

checking pee levels tends to be more convenient than checking blood levels.

As the cells of the corpus luteum die, they make less progesterone. So now the signal that tells the endometrium to grow is going away, and the endometrial cells die and are sloughed off. This material then flows from the uterus to the cervix and exits the vagina. This ends the luteal phase (which started at the end of ovulation); luteal phase because this is the phase when the luteum grows.

Now that the corpus luteum is gone, we've lost it as an important source for estrogen and progesterone and those levels decline. Estrogen and progesterone decline enough that the pituitary starts to secrete FSH; to turn it around, the high levels of estrogen and progesterone in the previous phase turned off FSH secretion. With low levels, the signal to turn off is absent, hence the signal is turned on and FSH is secreted. Shortly afterward, some of the endometrium (including blood) is discharged.

The first day that blood is discharged is the traditional first day of the cycle-it is a very visible marker. While the uterus is sloughing off the endometrium, FSH is signaling the ovary to have some of its follicles start to mature. In addition, FSH and LH cause other cells in the ovary to start making estrogen and progesterone.

This is a part where it gets a bit more complicated and confusing. The ovary produced estrogen does two things: first it negatively feeds back to the hypothalamus and pituitary, and in turn reduces FSH and LH secretion. You might be thinking, aha, negative feedback loop, less LH, less estrogen. That would certainly be the expectation from other endocrine feedback loops, but estrogen has a second effect: it positively feeds back on the ovary cells to increase estrogen production even more. So estrogen levels stay high, but FSH and LH fall. Low FSH means that there is no signal for more follicles to mature, which seems like a good thing, since we already have a few follicles maturing, remember, a mature follicle releases an egg, which has the possibility of being fertilized. Therefore, if two follicles mature at the same time two eggs would result. If fertilized then this would result in fraternal twins. Estrogen is also signaling the endometrium to grow.

Now we are almost ready for the ovulation step. Estrogen is high, one follicle is nearly mature and ready to release its egg, FSH and LH are

Physiology

low. At this point, estrogen switches from a negative feedback on GnRH secretion to positive feedback; part of the different response might be due to progesterone. This positive feedback loop leads to a surge in LH. And then ovulation occurs and the cycle continues.

To summarize the main steps:
- At the right state, with high estrogen, there is an LH surge
- A mature oocyte is released from the follicle
- Other released cells form the corpus luteum which secretes estrogen, progesterone, and inhibin
- Estrogen, progesterone, and inhibin negatively feedback to lower LH and FSH
- Progesterone also acts on the endometrium, causing it to grow, get new blood vessels, and store lipid and glycogen, potential energy sources.
- Without intervention, the corpus luteum self-destructs
- Progesterone (and estrogen) decrease
- Less estrogen and progesterone releases in a negative feedback and FSH increases
- FSH signals some follicles to start to mature
- The endometrium is sloughed off and is lost as menses as the follicles continue to mature
- FSH and LH cause other cells in the ovary to start making estrogens and progesterone.
- estrogen
    - negatively feedbacks to the hypothalamus and pituitary to reduce FSH secretion
    - positively feedbacks to the ovary to increase estrogen secretion
- High estrogen, low FSH, LH, and nearly mature oocyte
- Estrogen becomes a positive feedback regulator of LH release
- LH surge
- Ovulation

Hypothalamus & Pituitary	Ovary	Follicle	Corpus Luteum	Endo-metrium
	1. producing E			
2. **+ve fdbk** LH surge				
		3. release EGG & cells for CL		
			4. CL grows	
			5. Increase E&P	
6b. -ve fdbk decrease LH, FSH				6a. endometrium grows
			7. CL dies so	
			8. decrease E&P	
9b. -ve fdbk increase LH, FSH				9a. endometrium sloughs off-menses
	10 a. increase E&P		10 b. few follicles start to mature	
11. -ve fdbk so decrease FSH				
	12. E continues to increase			

Physiology

Summary of some of the key steps in the ovulation cycle
1) Ovaries are producing estrogen and the blood levels of estrogen are high.
2) At some point the blood levels of estrogen go even higher and estrogen suddenly exerts positive feedback, leading to a surge in release of LH.
3) The surge in LH causes the breaking of the follicle, the release of the egg, and the release of cells that because the corpus luteum.
4) The corpus luteum grows.
5) The cells of the corpus luteum secrete progesterone and estrogens.
6) a) Progesterone stimulates the growth of the endometrium. b) At the same time, the high progesterone and estrogen negatively feedback on the hypothalamus and pituitary to decrease the release of LH and FSH.
7) The corpus luteum dies.
8) Since the corpus luteum cells die, they no longer make estrogen and progesterone, thus those concentrations decrease.
9) a) Without progesterone, the endometrium dies and is sloughed off and its excretion is noted as menstrual discharge. b) At the same time, With less estrogen and progesterone, the negative feedback onto the hypothalamus and pituitary is lessened, therefore there is an increase in FSH and LH secretion.
10) a) The increase in FSH, follicle stimulating hormone, causes a few follicles to start to mature. b) At the same time the increase in FSH and LH also causes other cells in the ovary to increase secretion of progesterone and estrogen.
11) The increase in progesterone and estrogen decrease FSH release which means that no more follicles start to mature this month.
12) The ovary continues to increase the production of estrogen.
13) Eventually estrogen gets high enough and something else changes so that the feedback on the hypothalamus and pituitary is that there is a surge in LH release and away we go for another ride.

How do oral contraceptives work?

Oral contraceptives can be a misleading term, as these drugs often do several things, including reduce acne, reduce premenstrual systems as well as other positive effects. The name, oral contraceptives, refers to the fact that they are oral drugs. One of their effects is to prevent conception, that is, preventing a sperm from fertilizing an egg. There are, in development, drugs to interfere with sperm function and thus someday there may be male contraceptives.

Current contraceptives contain chemicals that are similar to estrogen or progesterone or a combination of the two. At the right levels, estrogen with or without progesterone from the pills, provide negative feedback, reducing pituitary FSH and LH secretion. Thus there is no ovulation. No egg, no chance of a fertilized egg. Progesterone leads to mucous secretion by the cervix that is thicker, and this provides a barrier reducing the chance that sperm will get to a possible egg and fertilize it.

Some birth control packets have 21 pills and some have 28 pills and they often have a similar cost. In fact, the 28 pill-packet has 21 pills of the hormones and 7 pills without hormone in the form of sugar pills. With the 21 pill packet, a person takes a pill a day for 21 days, and then waits 7 days and starts a new packet. With the 28 day packet, a person takes a pill a day for 28 days, and immediately starts the next packet. Personally, I would find the 28 packet easier to deal with. But is it too bad that the packets don't have the same number of pills as there are days in the month or that all months are simply 28 days. [1] Sometimes women chose to skip the sugar pills and continue with the hormonal pills. In this case, they avoid the monthly bleeding.

---

[1] The organization part of my brain wishes we would go to a calendar of 13 months of 28 days and an ExtraDay (not a day of the week, for example Sunday June 28, ExtraDay, Monday July 1....).

Physiology

Synchronization?
There is a controversy in the scientific literature about whether menstrual cycles can be synchronized. The debate is on several levels.
1) Consider the "classic" 28 day cycle with 5 days of menstruation. Woman A starts her period on Jan. 1. Woman B starts her period on Jan. 14. Are they 14 days out of synch? Typically the research only asks if someone is menstruating, not which day of menstruation it is. So Woman A is menstruating from Jan 1 to 5 and Women B from Jan 14-18. So they are only off by 13-14 days. And that is the most they can be off. Some claim that the synchronization is due purely to random chance, and that the studies that claim otherwise contain too much variability.
2) Women's cycle vary considerably. Some women have regular cycles, but the length is not 28 days. If Woman C has a 30 day cycle and Woman D a 20 day, they could overlap on Jan 1, then again 60 days latter (2 cycles for C, 3 for D).... Is that synched? And what about cycles that are closer, and odd, like 29 and 31. They can only be in "synch" every 29*31 days. (Or more often if we allow for the 5 day overlap above.)
3) Other women have different numbers of days between their periods each month. If they really become in synch, then presumably living in close proximity to other women means that their cycles regularized and became in synch.
4) What is a possible mechanism for becoming in synch? A popular idea is that Woman E secretes an odor compound at some day in the cycle, say the day of ovulation. Woman F then smells that compound. Now what happens? The smell signal goes to the brain. Someplace in the brain a signal then goes to the hypothalamus and overrides the estrogen feedback (or lack of feedback). And then the cycle restarts in a way that is makes Woman F's cycle closer to Woman E.
5) Most studies are only of a few women for a few cycles, and the opponents to synchronization think it all too likely that only data that show no effect by the proponents are discarded and/or that they stop the study when the women are in synch for 1 month, and that that "in synch" is just a coincidence of 25 and 30 day cycles overlapping.

I am a bystander in this debate, which is a two edged sword. It means I may be more impartial and it definitely means I am less knowledgeable than the experts. But there are experts on both sides of this issue. I think the mechanism outlined in #4 is plausible, but no one has any suggestions, let alone data, for any of the connections from the menstrual cycle chemical at the nose to the brain to the hypothalamus and any mechanism for the resetting. There is a wee bit of data on a possible chemical odor triggers.

The data I've seen do seem to be weak and I think could be explained by random chance. But the objections in #1-3 seem to me to cut both ways. That is, it would take awfully good data to rule out random chance for the effect and it may be hard to get enough women who are interested enough to get careful data for long enough to actually see an effect. In other words, the effect could be there, but it is so hard to get good enough data to rule out a random effect. In other areas of physiology, when one is trying to argue for no effect, the argument is more convincing if one has a positive control in addition to the test drug. A positive control is one where one can measure the effect.

To summarize this section,
- in humans, there are cyclic changes in the endometrium and the ovary
- the ovary and corpus luteum secrete the steroids progesterone and estrogen
- the hypothalamus secretes GnRH
- high GnRH leads the anterior pituitary to release the peptide hormones FSH and LH
- high FSH leads to follicle maturation and endometrial growth
- high LH leads to ovulation
- at most times in the cycle, estrogen and progesterone, have negative feedback effects on GnRH and LH, and estrogen has positive feedback on ovary estrogen production
- Just before ovulation, estrogen has positive feedback on LH release
- The bleeding at the start of a period is due to the excretion of the dying corpus luteum
- Pills containing estrogen or progesterone like compounds can suppress LH release, thus preventing ovulation and preventing pregnancy; they also have other positive and negative effects

Physiology

Skin Games

Columbia University has an advice column, Ask Alice. One student asked for advice about whether she should be worried because she heard that another girl on her hall had chlamydia. She was particularly worried about whether she could catch it from the communal toilet seats.

Several health organizations are advocating for male circumcision in Africa to reduce the spread of the HIV virus. On the other hand, there are activists in the U.S. working to decrease the number of circumcisions performed on babies in the U.S.

Blood products are screened for HIV, but not for Chlamydiae. Why the difference?

What are the properties that make some microbes only transmitted by intimate contact and cause Sexually Transmitted Infections? Other microbes are primarily transmitted by kissing, others by poor hand washing and by fecal contamination; why the difference?

In this chapter, we will discuss the physiology of ways in which sexually transmitted infections are transmitted. This will involve both an understanding of some microbiology as well as skin and genital physiology. We will NOT discuss how the STIs cause symptoms or death. However, many STIs will show no overt symptoms.

The sections in this chapter are:
- Skin microbes
- Why does sex spread HIV?
- What about circumcision and the spread of HIV?
    - What is circumcision and foreskin?
    - Epidemiology and correlation between circumcision and STD transmission.
    - Possible mechanisms by which having foreskin might make one more susceptible to STD infection.
- What are the odds of getting HIV?
- Summary of HIV
- Herpes Simplex I and II
- Human papillomavirus

- Mononucleosis
- Analogy for virus entry and selectivity
- Bacteria: Chlamydia and Gonorrhea.
- Crabs
- What can be caught from a toilet seat?
- Intimate behavior

Why don't our skin microbes hurt us? There are 3 main categories of microbes that can affect us: viruses, bacteria and fungi. In this chapter, we will only study the first two. A **virus** consists of a few proteins, mostly making a coat around the virus, and nucleic acid (either DNA or RNA) which codes for its key proteins. The virus cannot reproduce on its own. It can only harm us if it can get into our cells. A virus on our skin is not in direct contact with our cells, it is sitting on a layer of keratin[1] and other key proteins and sugars. In order for a virus to invade a cell, it needs to touch the cell. It usually attaches to some membrane proteins. Often this activates the cell machinery for endocytosis[2].

A layer of dead cells, keratin and other extracellular proteins protect us. So long as we have this layer of protection between our living skin cells and the virus, it can't get it. AIDS is caused by *Human Immunodeficiency Virus, HIV*. Even if HIV were to touch our skin cells, it could not get in. That is because HIV attaches to a particular membrane protein that is not produced by skin epithelial cells. However, this entry protein is produced by the Langerhans cells and dendritic cells in the skin, see below. When HIV is in blood that is

---

[1] Keratin is a protein that provides a tough coat on the outside of the skin.

[2] Endocytosis means to take something from the outside of the cell and bring it in, endo- being a prefix for in. Exocytosis is to take internal vesicles and fuse them with the outside membrane, releasing their contents. The release of insulin, parathyroid hormone, antidiuretic hormone, and neurotransmitters all involve exocytosis. A similar process occurs when insulin increases skeletal muscle glucose transport and the antidiuretic hormone increases renal water transport by inserting aquaporins The process is the same, but the physiological function is different. In the first examples, the cell is releasing molecules it has stored and the released molecules are to signal other cells to change their behavior. In the second example, the cell is putting more transporters into the membrane, which affects what that cell can import (or export).

transfused to another person, then the HIV has access to that person's blood cells. A subset of white cells does express the HIV entry protein and this allows the HIV to infect the host. As you might expect, scientists are trying to find ways to cover up this protein so that HIV cannot attach to the entry protein.

Why does sex spread HIV?
There are two parts to the answer. One is, how does HIV leave the infected person's blood? Let's call this person A. Two is, how does HIV enter the other person's blood? We'll call them person B. I think it should be obvious that if the sex is so vigorous that A bleeds and B has an open wound, some of the HIV in A's blood can enter B, get into their blood, and infect them. But most HIV is spread during sexual acts in which no one is obviously bleeding. However there are often microabrasions that are not noticeable to the eye but which are large enough to allow a bit of blood to leak out. HIV is primarily found in genital secretions and blood. You probably know that semen can contain HIV. Vaginal secretion and pre-ejaculate can also contain HIV. Thus any of these fluids can potentially transmit HIV from person A.

There are several ways HIV can gain entry to the body. First, is when the skin barrier is broken, for example, by ulceration or microabrasions. Second, it can be captured by endocytosis from the external part of the cell and can be carried in vesicles and released by exocytosis on the body side of the skin cell, a process termed transcytosis. This second process is only thought to occur in the rectum[1] and endocervix, because they are the only epithelial monolayers. Third, it can bind to HIV entry proteins on the surface of specialized immune cells, the Langerhans[2] cells and dendritic cells. While HIV can live

---

[1] Ask Alice has 2 columns that clearly and bluntly address the issue of anal sex and homosexuality at these links. The bottom line is that some heterosexual and homosexual couples enjoy anal sex and some heterosexual and homosexual couples do not.
http://www.goaskalice.columbia.edu/1490.html
http://www.goaskalice.columbia.edu/0900.html

[2] The Islets of Langerhans are the beta cells in the pancreas. Langerhans also described specialized cells in the skin and it is the latter that are involved with HIV. Paul Langerhans was a medical student when he discovered them.

inside Langerhans cells and dendritic cells it primarily reproduces inside white blood cells.

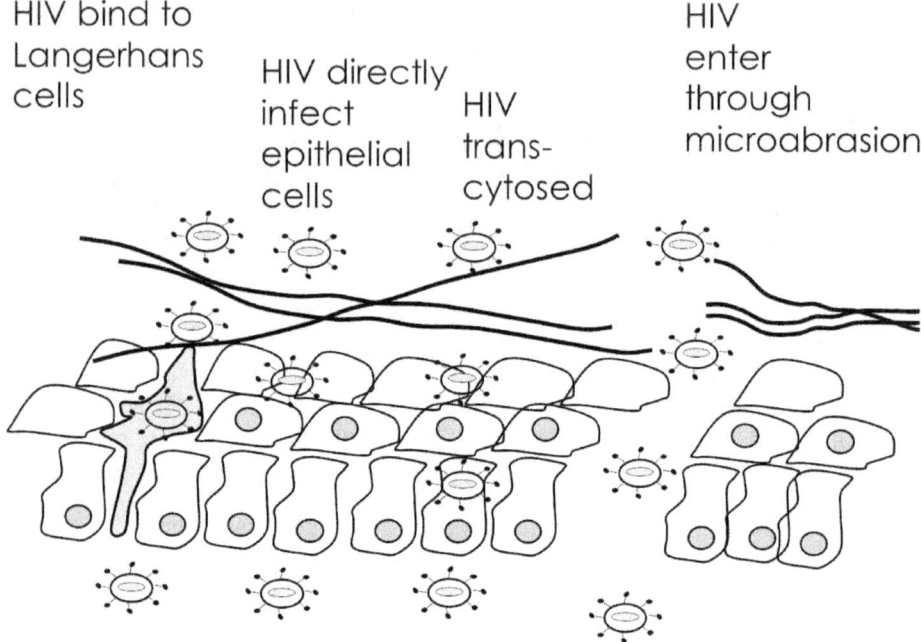

Figure 1. The four ways HIV can enter the body. From left to right:
HIV binds to a Langerhans cell (gray) and gains entry to the lymph.
HIV binds to an epithelial cells and infects it.
HIV is trancytosed across the epithelial cell layer (endocytosed at the skin surface and exocytosed at the interior surface, thus bypassing the epithelial security system.
HIV enters through a microabrasion, a place where the epithelial layer has been mechanically disrupted.

Langerhans cells often serve as a protection against HIV; they bind the virus, bring it inside the cell and destroy it. But if there are too many HIV, then they overwhelm that system, and HIV survives inside the Langerhans cells. Langerhans cells are known to migrate from the surface to the lymph, so if they contain HIV they are delivering HIV to the lymph, which contains some white blood cells. The lymph fluid eventually gets access to the blood and so HIV has now moved from the surface to the blood.

Physiology

A condom will cover the area on the penis that has the Langerhans cells, so in addition to preventing the pre-ejaculate and semen from touching the other partner, it also protects the condom wearer. But clearly in order to do this it must be worn whenever there is the potential for the penis to be in contact with the other partner's genital and fecal excretions and epithelia.

The body has several defenses again HIV. The vaginal wall, ectocervix, glans penis, and foreskin have several layers of epithelial cells, so these are relatively thick layers and that helps reduce the chance that HIV can enter by crossing these epithelial cells. In contrast, the endocervix and the rectum have a single layer of epithelial cells, so if HIV is transported to this region, they have less of a barrier to cross. The vagina also has other protection; the pH is normally slightly acidic and HIV do not survive well in acid.[1] However, semen is alkaline so the combination of semen and vaginal fluid is neutral and won't harm HIV.

There are also good bacteria normally present in the vagina; one of these is lactobacilli and they produce hydrogen peroxide which destroys the virus.[2] Taking antibiotics can kill these and other

---

[1] Recently, one scientist has suggested that applying lemon juice to the vagina or penis shortly before or after intercourse would lower HIV transmission because of the added acidity of the lemon juice. After one of his lectures, he asked the audience to volunteer and test whether adding lemon juice to the penis or vagina was painful and he claims they found it was not painful. However, the effect of long term application of lemon juice to these tissues has not been scientifically studied. He did point out that sex workers in Jos, Nigeria, have used it for years in the belief that it reduced the chances of pregnancy and STIs. However he and other scientists are cautious about trying new techniques too fast. Some detergents were found that destroyed the virus' membrane coat. These were used in human trials and it was found that the use of the detergent increased the risk of infection, because the detergent also destroyed the epithelial cell membrane and actually allowed surviving HIV to have easy entry. Whether lemon juice also reduces the barrier function of genital epithelia is not known.
[2] Applying topical hydrogen peroxide to the vagina is NOT a good idea; the lactobacilli only secrete small amounts of hydrogen peroxide, but HIV virus is

beneficial microbes in the vagina (and colon) and therefore has the potential to increase the chance of getting HIV from sex. The vagina also produces mucus. The mucus is a physical barrier; it is hard for the tiny virus to make its way through the thick, viscous mucus.

What about circumcision and the spread of HIV?
There have been a number of studies that claim that circumcised males have a lower risk for female to male transmission of some STDs, including HIV. I will try to summarize both sides of the argument. One major concept that I hope this discussion illustrates is that different scientists require different amounts/kinds of evidence for "proof", or stated another way, different scientists can look at the same data and interpret the effect differently.

No one keeps track of what fraction of babies, or teens, get circumcised so it is difficult to get reliable estimates on the fraction of men who are circumcised; it varies depending upon culture and religious background.

I've broken this discussion into 3 parts:
1. What is circumcision and foreskin?
2. Epidemiology and correlation between circumcision and STD transmission.
3. Possible mechanisms by which having foreskin might make one more susceptible to STD infection.

1. What is circumcision and foreskin?
All male mammals are borne with a flap of skin that covers the end (glans) of the penis. Females have an analogous structure, often called the clitoral hood. In case you want more background, there is explicit information about foreskin including graphic animations and pictures at these web sites:
http://en.wikipedia.org/wiki/Foreskin
complete anatomical description and explicit pictures
http://www.circumstitions.com/Works.html
an animation of retraction of foreskin
http://www.noharmm.org/anatomy.htm
more explicit photos

---

small. A person cannot apply so little hydrogen peroxide.

In circumcision, this foreskin is removed. Many religions and cultures consider this an important thing to do and in some cultures there is a religious ceremony that accompanies the circumcision, for example, in the Jewish faith, often the rabbi performs the circumcision. In some groups it is done shortly after birth and in some groups it is done at puberty. Adult men can also choose to be circumcised. While getting a piece of skin cut never sounds pleasant, particularly in the very sensitive genital area (physically and psychologically), body piercing in America has become more popular and so apparently many people find the temporary pain worth it for the "decoration".

One theory about the origins of the practice of circumcision is that it occurred in desert cultures where it would be easy to get sand caught in the foreskin and this would lead to infections; in the days before antibiotics, infections were often fatal. On the other hand, some consider circumcision of babies as inappropriate because of the claim that lack of foreskin reduces sexual pleasure.

One of the debates about circumcision is that it should be a personal choice; obviously a baby can't make that choice. But many think teen or adult males will resist circumcision because of the fear of its pain. However, there are some men that voluntarily have their genitalia pierced merely for cosmetic reasons as illustrated in this Ask Alice column: http://www.goaskalice.columbia.edu/0289.html?utm_source=Get+Alice%21+In+Your+Box&utm_campaign=f29ba7ee8b-090310_Non_CU&utm_medium=email

2. Epidemiology and correlation between circumcision and STD transmission.
There have been a number of studies looking at potential risk factors for STD transmission. Many found that circumcised males had a lower rate of HIV infection than uncircumcised males. (Remember that the "natural" state is uncircumcised, which is a bit like saying the "natural" state is to have a beard and uncut hair for men, and not removing armpit and leg hair for women.)

All these early studies have been criticized as merely correlational. And many were retrospective: the researchers would examine a population of HIV positive men and then identify a matching population of HIV negative men, and then determine the fraction of each that were circumcised. A major problem is what constitutes the matching population. This type of evidence is not as strong as a prospective study where one starts with a random population sample that is identical in all the key factors, and randomly assigns the volunteers to the control or treatment group. The prospective study avoids, or at least reduces, any bias in the selection. For example, in the retrospective study, it could be argued that the men who were circumcised also had a different sexual behavior than the uncircumcised and it was this behavior difference, not the circumcision, which accounted for the different rates of transmission.

If the circumcision occurred when the males were born, then there could well be culture differences in the upbringing of the child that could account for the different transmission rates as adults. If the circumcision occurred when the men were adults, you can easily imagine that the mindset and behavior of men volunteering to be circumcised would be different from those who decide to continue to be uncircumcised.

Those that are against circumcision as a way to decrease the rate of STD transmission point out the flaws I mentioned; those that support circumcision often agree that these are potential flaws, but argue that they matched the samples to avoid this bias or that there is no evidence that circumcision itself actually does alter behavior.

In the last 5 years, there have been several studies where adult men were randomly placed in a control group and an intervention group. The intervention group had their penises circumcised at the beginning of the study and the control group agreed to be circumcised at 24 months, at the end of the study. These studies also showed that the circumcised men had a lower rate of getting an STD. You can view this as the prosecution's case; they are claiming that this evidence is adequate to support the call for male circumcision to reduce STD transmission (from female to male). The other side, like the defense case, points out weaknesses in these studies. I'll detail those below; you are the jury and have to decide whether you think the weaknesses are significant enough to say the defendant is not guilty, that is, that these weaknesses, you think, are important enough that you don't think it is warranted, with the current evidence, to support the call for male circumcision to reduce STD transmission.

a. Control group not matched. The volunteers in the intervention group had a circumcision and were given the instructions to not have sex for 6 weeks in order to allow for healing of the penis. They also received additional medical attention. The volunteers in the control group were given no special instructions. One scientist's letter argued that a better control group would have been one in which a cut was made elsewhere on the penis, also sutured, and then the group given the exact same instructions and medical attention as the intervention group.

b. The rate of HIV transmission was reduced by 60% in one group. While this is a good reduction, it is only for 24 months. If the men continue to have the same type of sex at the same frequency, then over their lifetime, they will probably get an HIV infection. An analogy: In a basketball game, you can foul a good foul shot shooter (say 90%) or a bad shooter (say 30%). While it is true that fouling the bad shooter reduces the chance that the other team will score points, if you continue to foul the bad shooter, sooner or later, they are likely to score some points.

3. Possible mechanisms by which having foreskin might make one more susceptible to STD infection.
Prosecution side (that is, foreskin is a risk factor for STD transmission): When the foreskin is retracted during intercourse, that exposes both the underside of the foreskin as well as the glans penis to the fluids and potentially infectious microbes in the vagina. (As an analogy, imagine wearing a long sleeve shirt. On a sunny day, you rolled the shirt back on itself and now the underside of the shirt is exposed to sun as well as your arm skin. The underside fabric of the shirt might fade in the sun and the sun might sunburn your arm.) In the circumcised male, there is no foreskin to be exposed (the shirt is short sleeved.) In addition, in the circumcised male, the part of the penis now always exposed changes a bit and becomes tougher and also has less Langerhans cells. (If you are always wearing short sleeves, you arm becomes gradually tan and you are less likely to sunburn.) Langerhans cells are known to bind to microbes and take them into the body.

Uncircumcised males have more surface area exposed to allow for the transmission of HIV.

Defense side: The studies on the surface properties of circumcised and uncircumcised penises have been mostly done on cadavers (so these may not reflect the properties of living tissue) or in only a limited number of samples, about 10 each. (Of course, how many living, sexually active men are likely to volunteer to have skin samples removed from their penises? Well, maybe the guys that are willing to get pierced? Would that be a "representative" sample?)
Secondly, there is recent evidence that Langerhans cells also make a protein that binds to the surface of HIV and allows the cell to destroy the virus.

What are the odds of getting HIV? This is difficult to estimate. From our basic science perspective, the rate of transmission depends upon how many viruses leave person A. It also depends on a lot of different factors in person B: Do they have open wounds/ulcers? Do they have microabrasions? What is the pH of the fluid the HIV comes in contact with? Which epithelial layer(s) must HIV get through? How thick is the keratin layer, if present? Is HIV in the vicinity of Langerhans or dendritic cells? Are there so many HIV that they overwhelm the Langerhans or dendritic cell defense system?

The rate for transmission of HIV from an infected person via a needle stick is about 5 out of 1,000. A commonly given, but misleading, figure is that there is about a 1 in 1,000 chance of contracting HIV from an infected person during heterosexual sex. This figure apparently primarily applies to heterosexual couples in a monogamous relationship (with one HIV infected). A recent study looking at a series of previous studies looked at different populations to see how other factors influenced the rate. In 4 of the 5 studies examining penile-vaginal contact the rate varied from about 0.7 to 3 chances in 1000 contacts of transmitting HIV. In the 5th case, the rate was between 10 and 30, which was about the same as the one study looking at penile-anal contact, about 25 to 60 per 1000 contacts. There were two studies examining circumcision status. In the first study, the rates for circumcised men were between 3 and 200 per 1,000 and for uncircumcised between 90 and 500, both pretty large ranges. In the second study, the range for circumcised men was between 5 and 8 and for uncircumcised men between 8 and 20.

Physiology

Of course, a key factor in the risk for HIV is whether the partner is infected. Current data suggest that most people have a limited number of lifetime partners but a few people have many partners. If a person with many partners has HIV, then it can be widely spread. Perhaps an analogy would be useful. Consider the spreading of cold germs from the hand of an infected person to the hand of the "partner" and then from the hand to the nose. Imagine a church where most people only shake hands with a few people, but the usher shakes hands with very many people. If the usher has a cold, then it is likely to spread widely. But it also depends upon whether every one else washes their hands before touching their noses.

Summary of HIV. HIV is fragile and does not survive long outside the human body. It is primarily found in blood and genital secretions. Epithelial cells form a tight barrier to keep out microbes. The skin has layers of dead skin and keratin to help prevent microbes from touching live cells. HIV and other microbes can cross this barrier if it is torn; tearing does not need to be large enough to see blood cells, microabrasions are large enough. HIV can also be transcytosed across single epithelial layers like those found in the anal/rectal area and the endocervix. Langerhans and dendritic cells can prevent HIV infection at low doses, but at high does of HIV, these cells probably assist in HIVs movement to the blood.

HIV can be spread from any behavior where infected genital secretions or infected blood, via microabrasions, are deposited in a "sensitive" region. A "sensitive" region is one that has cells that contain the entry receptor for HIV or which has only a single layer of epithelial cell or otherwise allows HIV access into the lymph or blood. <u>HIV can occur during intimate contact; behaviors that some might not call "sex".</u>

Herpes Simplex I and II can both cause genital ulcers, though HSV I more often causes oral infections. Since the ulcer has disrupted the epithelial layer, people with active ulcers are at higher risk for catching other STIs, including HIV. However, even without active ulcers people with HSV can transmit the disease to their partners because the virus can be shed even in the absence of ulcers. In many studies, less than ¼ of the people were aware they had HSV; the clinicians deduced they had HSV because the patient had antibodies against HSV.

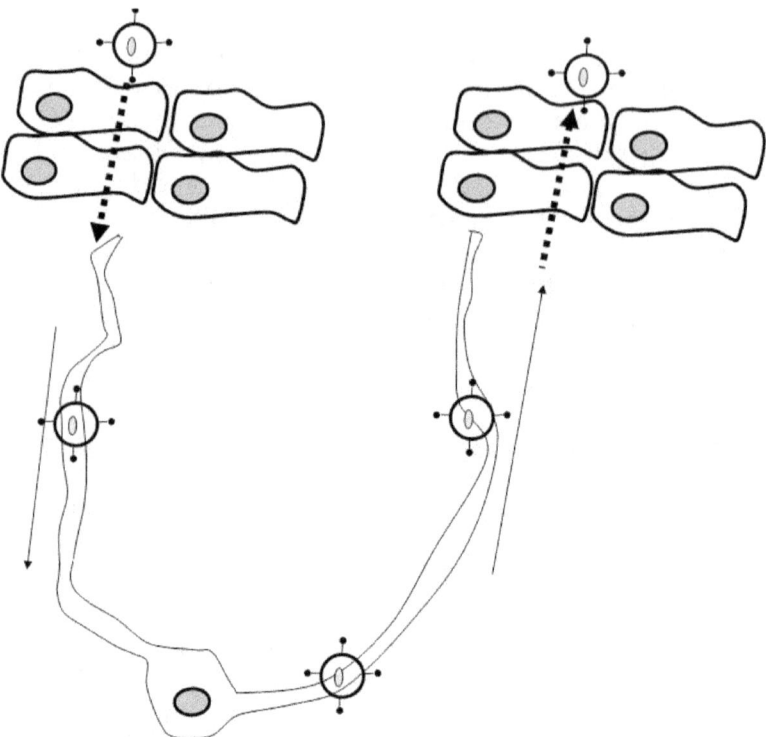

Figure 2. HSV infection and release. Starting at the top left. HSV has been shed from the skin of person A and is now in contact with the skin of person B. HSV can cause ulcers. Even in their absence, HSV can be shed from the skin and transferred to a partner. The initial infection can cause ulcers. The virus can also move into a nerve and stay latent for some time. Then it can get reactivated, move up the nerve (perhaps on a different axon) and infect the skin and then be shed.

HSV needs skin-to-skin contact. If the infected person is shedding virus (and most of the time this is not observable) then the virus can get on the partner's skin. The virus still needs to get across the epithelial layer. Once it does that, it multiples and causes inflammation.

Something unusual about this virus is that when it comes in contact with a nerve, it can get inside. The axon transport system will then carry the virus to the cell body. While the virus can reproduce in the nerve, it does not kill the nerve. The virus can hang out in the nerve for some time, so the infected person, even if they had ulcers, may see the ulcer disappear and appear fine. Then sometime later the virus will travel out of the cell body and attack some epithelial cells again. When it is going back out of the nerve, it might go along a different axon than when it came in, so it can cause ulcers in a new location.

We don't know all the causes of why the virus in the nerve suddenly becomes active again, but some of the causes include fever, loss of immune function, surgery, and ultraviolet light exposure. The new reproduction in the epithelial cells may result in virus shedding whether or not there is ulceration.

Another sexually transmitted virus is human papillomavirus (HPV). Papillomavirus is a common cause of intimate contact infection (ICI) and it can cause cervical cancer as well as neck and throat cancers. It enters basal cells in the genitourinary track and anorectal surface through microabrasions. Transmission requires skin-to-skin contact. It is not clear if condoms reduce the transmission; most HPV studies have not focused on condom use and have relied on self-reports which may not be too accurate. In addition, HPV can be released from skin areas not covered by condoms. There is a correlation between alcohol drinking and HPV transmission but alcohol drinking and unsafe sex practices are also correlated and it seems likely that unsafe sex practices do increase the risk of HPV transmission. HPV can definitely be transmitted by penetrative sex. Non-penetrative sex transmission is harder to verify because of doubts about volunteers honestly reporting their behavior, but virgins who have had non-penetrative sex do get infected with HPV. HPV not only causes cervical cancer but can also cause throat and neck cancers.

Mononucleosis is also caused by a virus, but it is not transmitted by sexual activities. It is primarily transmitted by kissing. Why is this? The virus tends to invade cells in the oral cavity. This means that the protein receptor that the virus binds to, in order to gain entry into the cell, is present primarily in cells in the oral cavity and not in cells around the genital area. The mononucleosis virus is also secreted in saliva, not genital fluids.

Here's an analogy for virus entry and selectivity.

- In this analogy, the body is a dormitory. The cells are the rooms. The virus is an invader. The door colors are the different types of entry proteins.
- If the invader just leans against the dorm wall or doors, they can't get in. If a virus or bacteria is just lying on top of the skin, it can't get in.
- If the outside door is secure the invader can't get it. If the skin barrier is intact, the virus can't get it.
- If the virus gets inside the dorm, then its ability to invade depends upon the door color. HIV needs red doors. HSV needs green doors. HPV needs blue doors. Mononucleosis needs yellow doors.
- In the dorm, maybe most of the yellow doors are near one outside entrance (for example, the oral cavity). But there are also some green and blue doors. There are green and blue and red doors near one of the exits (the genital area).

We now turn to 2 bacteria, those that cause Chlamydia and Gonorrhea.

Chlamydiae are bacteria that need to be inside a cell to reproduce. Chlamydiae exist in two stages. The infectious stage consists of elementary particles and these can migrate from cell to cell. If the elementary particles contact the right kind of cell, they can get inside. Once inside and reproducing, they are in the reticulate body stage. Eventually, elementary bodies are produced and released from the cell. When Chlamydiae are inside a cell, the bacteria are protected from attack by antibodies as antibodies cannot cross the cell membrane.

Presumably, transmission occurs when an infected person rubs their skin on another person. If that skin has cells containing Chlamydiae or has Chlamydiae in the elementary particle stage have recently been released onto its surface, then some of the Chlamydiae can be transferred to the partner. If enough Chlamydia can enter the partner's cells, then they may get an infection.

Chlamydiae can enter a cell by binding to the cell surface.

Physiology

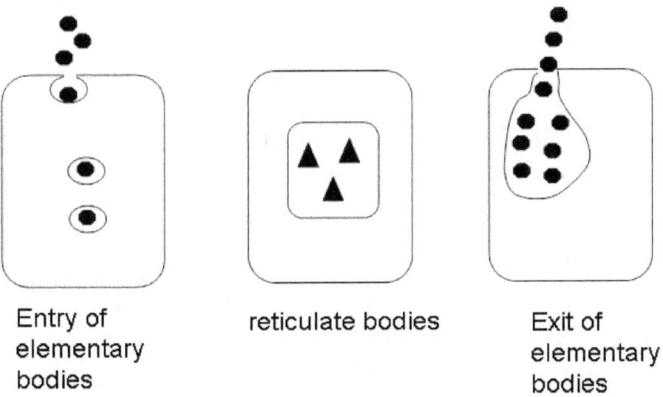

Figure 3. The stages of Chlamydia. Left: the elementary particles infect a cell. Middle: The bacteria change from the elementary body stage to the reticulate body stage. Right: The bacteria change back to the elementary body stage and are released.

Chlamydia are pretty fragile. They can only survive outside the body for a short time. Thus, it is highly unlikely that one can get Chlamydia from a toilet seat.

Gonorrhea is caused by the bacteria Gonococci. Gonococci have proteins on their surface that allow them to bind to columnar epithelial cells that lack cilia. Somehow this binding triggers an endocytosis event at the body surface membrane. The Gonococci are then transported to the basal membrane where they are exocytosed. Once at the inside surface, they grow on the basement membrane. Which surfaces of the body are covered with columnar epithelial cells? The reproductive and excretory tracts, primarily cervix, urethra, and rectum. Columnar epithelial cells are also found in the throat and eye and so Gonococci can also infect there. Prepubescent girls can get gonorrhea in their vagina, as the vagina does not become keratinized until the pubertal rise in estrogen. It is thought that the cervix is a more likely location for transmission in females after puberty.

Both Gonococci and Chlamydiae are able to infect the cervix. The cervix is often divided into the ectocervix (closer to the outside) and the endocervix (closer to the inside). The ectocervix has a thicker epithelial layer then the endocervix. Between these two is a transition zone. The endocervix and transition zone are most likely to be infected by Gonococci and Chlamydiae. Since oral contraceptives can increase this zone they do increase the risk of getting gonorrhea and Chlamydia.

Rectal infections can happen in those that have anal intercourse. Rectal infections also happen in some women when infected cervical discharge inadvertently gains access to the rectal area.

Like most organisms, Gonococci need iron to grow. Gonococci get their iron by binding transferrin and lactoferrin. Interestingly, they can only bind the human forms of these proteins. Presumably this accounts for why Gonococci only infect humans.

Finally, we will discuss crabs, which are NOT microbes. **Crabs are pubic lice.** These are very small insects that like to live in the pubic hair area; however they are not limited to their namesake as they can also live in armpit hair, eyebrows and head hair. Usually lice in head hair are a different species. Pubic lice are primarily spread by intimate sexual activity. If hair from two different people comes in contact, the lice move from one person to another.

Can you get pubic lice from sheets, towels or bathing suits? Some report this to be highly unlikely. However, lice can survive for about 24 hours without being on a human. An NIH web site http://www.nlm.nih.gov/medlineplus/ency/article/000841.htm
suggests that pubic lice can be transmitted by sheets, towels or clothing. I think a sheet, towel or piece of clothing may resemble hair enough that pubic lice may be dislodged from hair and move to the fabric and then onto a person's hair. Presumably it is illegal to try on underwear in a store because that could transmit the lice; in fact, according to the NIH website, some women have reported getting pubic lice from trying on bathing suits at the store. However, pubic lice tend to be couch potatoes, they don't often get off someone's hair, and they are not designed to walk on smooth surfaces. So getting lice from toilet seats is very unlikely.

If you get a pubic lice infection, what can you do about it? Will antibiotics help? No, antibiotics[1] fight bacteria and pubic lice are an

---

[1] The specialized use of the term "antibiotics" annoys me. Just looking at the

insect; one uses treatments that are essentially insecticides. Apparently some college students with pubic lice try to self-medicate with antibiotics and that only prolongs their infection.

Lice feed on blood, they inject their saliva into the skin and something in the saliva causes a hypersensitivity reaction, which can cause itching.

What can be caught from a toilet seat? This information is taken from http://shc.osu.edu/blog/can-you-contract-an-sti-from-a-toilet-seat.
Most sexually transmitted microbes are very fragile. They cannot live for long on their own. They generally require skin-to-skin contact or fluid-to-skin contact. It is possible for someone to leave some non-urine fluid on a toilet seat. This fluid could contain intimate contact infection (ICI) microbes. If there were enough of these microbes, they have the potential to cause an infection, if they can get into the body. Presumably the only skin surface touching the toilet seat is the butt. The intact butt skin is an excellent barrier, but if you have an open sore, then there is a theoretical possibility of transfer this way. While pubic lice can live outside the body for about 24 hours, they like a warm and hairy environment. They are highly unlikely to be on a smooth surface such as a toilet seat.

- **In summary, sexually transmitted infections may be a misleading term, because of the ambiguity of the word sex.** Maybe a more accurate term is intimate contact infection (ICI).
- For infections that are secreted in bodily fluids, the infected fluid needs to be in contact with the partner.
- For infections that are shed from the surface of the genitals, the genital tract, or the anal rectal area, then these surfaces need merely contact the partner in an area that has the potential for invasion.
- While the skin and most genital, anal, rectal surfaces provide a barrier against infection, any small tear can allow bacteria or virus to enter. These microbes are less than 1 um, well below

---

word, anti-against, and bio-life, I think the term should apply to any form of life: virus, bacteria, and lice. We could use the term, "anti-bacterial" for those drugs that target bacteria. I think this would avoid a lot of confusion.

the resolution of the human eye, so microtears or microabrasions need not be visible to allow infections.
- Often intimate contact involves rubbing and some friction and that seems likely to create microdamage to the barrier. Microdamage is large enough for microbes to enter.
- Many microbes attach to specific protein "receptors" to gain entry into cells and the body; the types of contacts that transmit the microbe depends in part on which parts of the body have cells with these "receptors".

Black & White (Banning Tanning??)

Imagine that you and your parents are going on a cruise with several of their college classmates and their children. As a few complain that they have already had a skin cancer removed, you look closely at them. Their hair color and their skin color have a great deal of variation. Hair color is often grouped into 3 categories, blonde, brunette and black. What about red hair? It is much more difficult to delineate distinct groups of skin colors, light, medium and dark don't even begin to describe the range of colors. Several have distinct tan lines. One parent bickers with their child about sunscreen. A few children suggest that a parent is addicted to tanning.

Maybe you wonder about what causes all the differences in skin color, hair color and eye color. Like many characteristics, skin and hair color reflect both genes and the environment. Genes primarily determine eye color. In the last 10 years there has been much progress made determining some of the genes that are involved in different skin and eye colors in humans, but there remains a lot that is unknown. Studies in mice have been going on for much longer and there are probably over 100 genes that have an effect on hair and skin color in mice.

The sections in this chapter are:
- Chemicals for color
- Genes for Color?
- Melanocytes
- Ultraviolet light and DNA damage
- Why do UVA wavelengths penetrate more deeply than UVB wavelengths?
- How Do Sunscreens Work?
- Sunburn
- Skin Cancers
- Sunless Tanning
- Addiction

Chemicals for color
Melanin is a term that refers to the major pigments that our skin produces. These are complex molecules and fall into 2 general classes. Pheomelanins tend to give yellow, red, and reddish brown colors. Eumelanins tend to give black and brown colors. The final

color that we see is depends upon the ratio of these two classes of pigments. It also depends upon exactly which chemicals are produced for each pigment as well as how they are distributed in the skin.

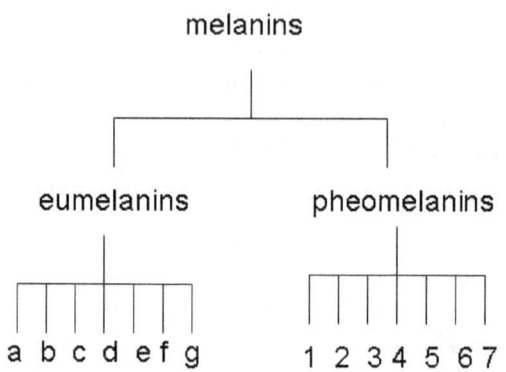

Figure 1. Melanins refers to a whole class of compounds. Two subgroups are Pheomelanins which tend to be yellow, red, or rust and eumelanins, which tend to be black or brown. Each of these subgroups is composed of many different chemicals (a-g or 1-7).

Genes for Color?
The current theory is that the first humans had dark skin. As you know, now there are humans with a variety of different color of skin. Evolutionary biologists have wondered whether the mutation that gave rise to light skin color arose once or more than once. Recent analysis suggests that light skin color is due to the influence of different genes in peoples whose ancestors are Europeans and those whose ancestors are East Asians. Because different genes are apparently involved, it appears that light skin of both groups is a result of convergent evolution; light skin color apparently being a selective advantage when one lives far from the equator.

One popular theory about why lighter skin is an advantage far from the equator relates to vitamin D. Light has to get to the vitamin D precursor in keratinocytes in order to convert it to previtamin D3. All people have to balance the need of light for Vitamin D synthesis with the damage light can cause to the DNA in skin. Near the equator, there is so much strong sunlight that dark skin provides the protection from the damages of the sun and just enough light gets through to allow enough vitamin D to be made. But far from the equator, especially in the winter, there is less chance of sun damage and for making vitamin D; so having light skin might be an advantage. Thirty years ago, a few scientists suggested that hunters/fishers in ancient Europe would have acquired enough vitamin D in their diet from eating animals and fish that they wouldn't need sunlight to generate vitamin D in the skin[1].

Even within ancient Europe, people had different skin color. But recent analysis of DNA from hundreds (but not even thousands) of individuals suggests that many of the people whose ancestors were from Europe have a one amino acid difference in their SLC24A5 gene compared to almost all people whose ancestors were from Asia and Africa. This difference, at least in mice, results in light colored skin. The researchers clearly state that this one difference does not account for all the variation in skin color, but it does appear that this mutation is "necessary but not sufficient" to get lighter skin in Europeans, though other genes clearly contribute.

Interestingly, recent analysis of the variability in European SLC24A5 gene and nearby genes suggests that this mutation arose about 6,000 to 12,000 years ago, long after humans are thought to have migrated to Europe (~40,000 years ago). You might recall that 6,000 years ago is about when Europeans began to have access to domesticated cows (and when the lactase persistence trait first became selected for). So it might be that light skin color is also a recent development in European history and the changing eating habits of Europeans provided the selection.

---

[1] I wanted to check whether a high meat diet had enough vitamin D. According to the Office of Dietary Supplements at NIH, http://ods.od.nih.gov/factsheets/vitamind the daily RDA for vitamin D is 600 International Units. Four ounces of salmon has this much vitamin D. But four ounces of venison is only 14 IU. Beef liver is apparently the highest part of the cow for vitamin D and 4 ounces is only 50 IU. So if ancient Europeans ate fish, they easily got enough Vitamin D. But it would take a lot of meat to meet their needs as forty ounces of beef (2 ½ pounds) is a lot, about 4000 calories. On the other hand, they may well have needed that many calories as Lewis and Clark's group ate around 9000 calories a day per person.

There are a number of candidate gene mutations to account for the light skin color in people whose ancestors are from East Asia. So far as I can tell, no one has identified one gene that is common to East Asians, like SCL24A5 seems to be so common in Europeans.

No one has a detailed theory about how a mutation in SLC24A leads to lighter skin, but an understanding of the process of how melanin is made will help us understand why it is plausible. Understanding this process will also allow us to better understand how red hair and light skin occurs as well as what happens in tanning.

Melanocytes

Melanins are produced in cells called melanocytes.[1] Within the cell, the melanin is stored in the melanosomes, which are vesicle-like structures. Vesicles are basically just a lipid membrane surrounding cellular fluid and are found in many cell types. Melanosomes are exported from the melanocytes to the keratinocytes. One method of export is exocytosis, in which the vesicle membrane fuses with the surface membrane and releases its contents. Over 90% of the cells in the skin are keratinocytes and one of their functions is to produce keratin; the synthesis of keratin is partly under the control of vitamin A[2]. Melanocytes are the next most common cell type in the skin and make up about 1% of skin cells.

SLC24A5 is localized to the membranes of structures inside the cell that are on the pathway to forming melanosomes, so it is plausible that an alteration in SLC24A5 function could alter the pigments in the melanosomes.

Some of the mutations that results in red hair and light skin color occur in a receptor call the MC1 receptor, for **melanocortin** receptor 1. When this receptor is inactive, the cell's rate of conversion of tyrosine to dopaquinone is slow. Most of the dopaquinone is able to react with the amino acid cysteine (or the related peptide glutathione). The

---

[1] Cyte is the suffix for cell, for example, erythrocyte is a red (blood) cell.
[2] We talk about vitamin A as a hormone regulating epithelial function in the chapter, Orange you glad?

Physiology

resultant compounds then form a variety of polymers which we classify as pheomelanin, that is, the pigments that look yellow or red.

When the hormone, melanocyte stimulating hormone, (MSH) binds to the melanocortin receptor, the activated receptor increases the production of cAMP in the cell. This has several effects, but one is to increase the amount of the enzyme that converts tyrosine to dopaquinone. If the cell has low cysteine, then there is more dopaquinone than cysteine. Dopaquinone then forms polymers that are called eumelanin and give a dark brown or black color. Some people with red hair and light skin have a mutation in the melanocortin receptor and they have much more pheomelanin than eumelanin. The different mutations have additive effects so if someone has several of these mutations they have even redder hair and lighter skin.

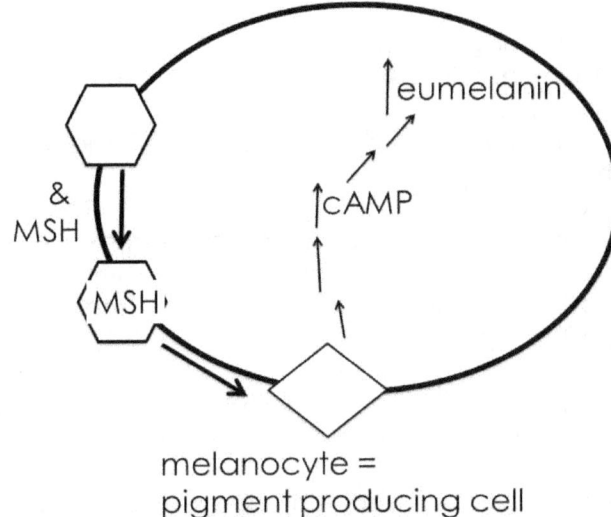

melanocyte =
pigment producing cell

Figure 2. The MSH receptor is inactive and has a stop sign shape in the absence of MSH. When MSH binds, then the receptor changes shape (to a diamond). This shape change eventually leads to an increase in cAMP and this eventually leads to an increase in the amount of enzymes that convert dopaquinone to eumelanins. You might find it interesting to compare this figure with Figure 6 for calcium altered parathyroid hormone release in the chapter, Bubbling Beverages Bad for Bone?

An increase in melanocyte stimulating hormone would increase the amount of dark pigment made. Interestingly, ACTH can also bind to melanocortin and activate it. ACTH is a pituitary hormone that regulates cortisol production from the adrenal glands. In Addison's disease, you may recall that the adrenal gland can't make much cortisol. The pituitary senses the lack of cortisol which alters the feedback signal from the pituitary to the adrenal. This signal is the hormone ACTH. So Addison's patients often have very high blood ACTH levels which activate the melanocortin receptor in the melanocytes and can lead to darker pigmented skin, at least in some places on the body.

Ultraviolet light and DNA damage
Melanocyte stimulating hormone is also released by keratinocytes. One of the signals for the release is an increase in DNA damage. So this explains tanning:
- ultraviolet light damages some DNA
- damaged DNA leads to keratinocyte release of melanocyte stimulating hormone
- melanocyte stimulating hormone activates melanocortin receptors on melanocytes
- cAMP goes up
- more eumelanin is produced and stored in melanosomes
- more melanosomes go from melanocytes to keratinocytes and spread out over the nucleus, thus protecting the remaining good DNA and giving a the person a tan.

Since a red haired/light skinned person doesn't respond very well to the melanocyte stimulating hormone, they can't tan very well.

Because melanins have color, you know that they absorb visible light. They also absorb ultraviolet light. Ultraviolet light causes damage to skin, and particularly vulnerable is DNA. Most of the melanin in the keratinocytes is placed between the nucleus and the outside (i.e., the sun side). In this way, the melanin can absorb ultraviolet light and reduce the amount that hits the DNA.

Ultraviolet light causes the damage. There is a debate about whether some types of UV light are ok. The UV light is classified as UVA and UVB, a bit like dividing the class into tall and short as where you draw the line is a bit arbitrary. UV B has higher energy than UVA; with more energy it can cause more damage. But UVA penetrates into your skin a bit deeper, so it can damage more cells.

Why does UV light increase the chance for skin cancer? The short answer is that UV light has more energy than visible light and UV light can mutate the bases in DNA. Some DNA mutations cause the cell to have uncontrolled growth, that is, are cancerous.

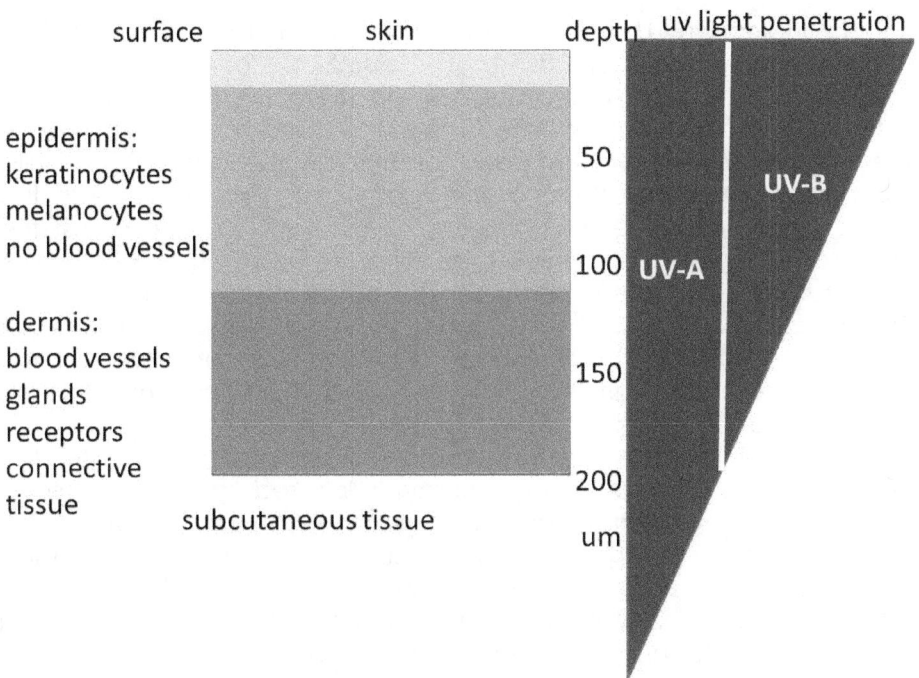

Figure 3. On the left are the 3 major layers of skin and some of the cell and tissues types found there. On the right is the approximate depth that different wavelengths of ultraviolet light penetrate in the skin. Note that the division between UVA and UVB is arbitrary in the sense that wavelengths vary continuously in length as well as depth of penetration.

Why do UVA wavelengths penetrate more deeply than UVB wavelengths?
The wavelength of UVA is longer than UVB. Remember the wavelength is the distance between 2 peaks on a wave. The longer the wavelength that more likely it is to miss hitting something. Did you

ever try to walk on a sidewalk and not walk on the cracks? Walking on a crack would be like UV light hitting a molecule and being absorbed. The distance between two steps is like the wavelength. The longer your steps, the less likely you are to step on a crack. To take an extreme case, if your steps are very short, you are almost always going to step on every crack (of course, many steps are one the crack-free part of the sidewalk).

Understanding why UVB has more energy is a bit more complicated. First let's clarify one more thing. The amplitude of the wave is the amount of energy it has. Imagine you have two light bulbs, one emitting UVA and another UVB. Furthermore, imagine that you could see the light. You can adjust how much electricity goes to each bulb; this increases the amplitude of the light wave. The waves with the higher amplitude have more energy. But now, say you adjust the bulbs so the amplitude of the waves is the same. Then UVB has more energy than UVA. This is because UVB has more peaks in 1 second than does UVA. That is, the frequency of UVB is greater than UVA (or the wavelength of UVB is shorter.) Imagine an energy detector counting the number of peaks. In 1 second, it will have more peaks from UVB than UVA and each peak transfers energy.

This is true of all light, the longer the wavelength, the greater the penetration but the lower the energy. Infrared light has an even longer wavelength that UV light. So which light do you think penetrates the skin to greater depth? Suppose you wanted to image some particular cells in your body, say some cancer cells. You have some antibodies that will only bind to the cancer cells. You attach a color molecule to the antibodies, put the complex into the blood and let the antibodies bind to the cancer cells. Now you want to detect the light from the color molecule. Would you want to use infrared light or UV light? You'd like the light to penetrate as far as possible so you can see as much as possible.[1]

---

[1] You may wonder about X-rays-they have even smaller wavelengths and yet penetrate further-this does not seem consistent with what I just said about UV light. The wavelength of UV light is about 400 nm or 0.4 um or about 1/10th the diameter of a cell nucleus or the thickness of a red blood cell. X rays have a wavelength that is about 100 times smaller or about the size of an atom. Thus it is probably easier to think about X rays behaving more like a particle than a wave in terms of whether they are absorbed. And since most of an atom is

Physiology

A mechanistic answer to how UV light causes damage requires a bit more background. UV light is absorbed by aromatic compounds. Aromatic chemicals initially got their name because they had an aroma, that is, they smelled, such as the chemicals in moth balls or modeling glue. The structure of these compounds was determined and most of them have a ring structure with double bonds, that is, a series of atoms that are all connected in a particular fashion. For a chemist, the term aromatic now refers to structures with the rings and doesn't necessarily mean the structure smells. Only 3 of our 20 amino acids are aromatic. But all 4 bases of DNA are aromatic.

When we say that an aromatic compound absorbs light that means that the energy has transferred from the light to the molecule. This extra energy can change the molecule, just like when a car hits a fence. The fence absorbs the energy from the moving car and in the process the fence breaks. So the more UV light a cell absorbs the more likely that light is to be absorbed by a DNA base and the more likely the DNA will mutate.

This photochange can also be illustrated with Vitamin D.

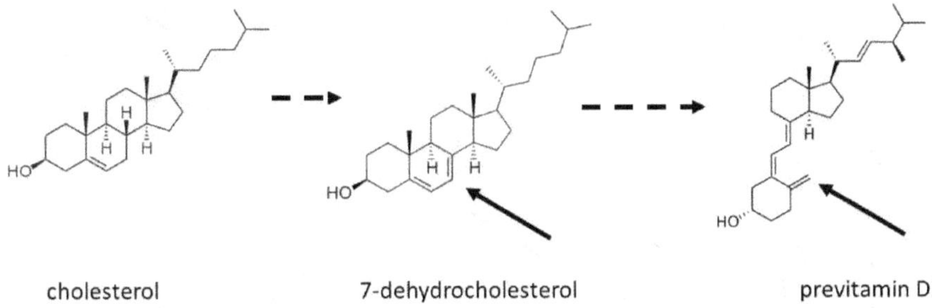

cholesterol   7-dehydrocholesterol   previtamin D3

Figure 4. Cholesterol is converted to 7-dehydrocholesterol; note the double bond at the arrow which now makes that ring essentially aromatic. When light hits 7-dehydrocholesterol that the "aromatic" ring absorbs the light and it breaks the ring.

---

empty space, they can penetrate right through many atoms.

How Do Sunscreens Work?

Before we go on to DNA mutations and cancer, let's talk about how sunscreens work. The idea of sunscreens is to put chemicals onto your skin that absorb the UV light before it can hit and damage your skin cells. As you might guess, many of the sunscreens contain aromatic chemicals as they absorb UV light so well. Many of these compounds work better if they are applied about a half hour before one gets into the sun. This allows time for the chemicals to react with some of the molecules in your skin; this allows them to work better. How much sunscreen is required? An average sized body needs about 1 ounce of sunscreen. As the sunscreen molecules absorb the sunlight, they are destroyed so one needs to reapply about every two hours.

Some inorganic compounds also work well in protecting keratinocytes from UV light, titanium oxide is an example. Unfortunately, it most of its formulations, it appears white on the skin and is not considered attractive. Newer formulations can avoid this.

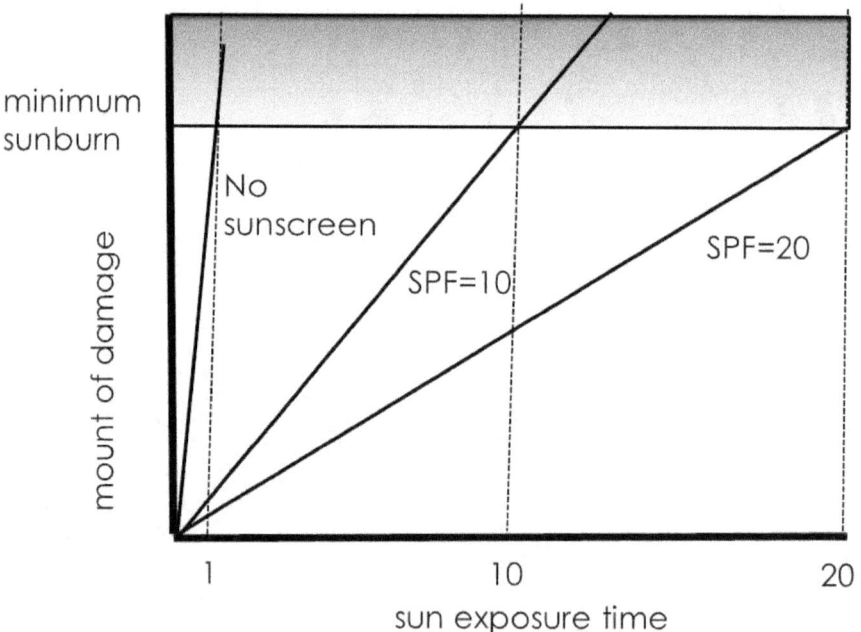

Figure 5. Sunscreens are often rated by their SPF numbers. How to interpret the numbers: Let's say you spend 1 hour in the sun today. The next day, you have a bit of a sunburn. If you had used a sunscreen on the first day with a SPF of 10, you could have stayed in

the sun for 10 hours and had the same, little bit of sunburn. This assumes that you reapplied the sunscreen every 2 hours (and whenever swimming or sweat washed it off). If you had used a sunscreen on the first day with a SPF of 20, you could have stayed in the sun for 20 hours and had the same, little bit of sunburn One important caveat about SPF numbers is that it only refers to how well they block absorbs UVB, not UVA.

Sunburn
If enough sun hits the cells that some keratinocytes are damaged enough that they essentially commit suicide, also known as programmed cell death or apoptosis. When this happens they lose their melanin and so don't absorb light. People of any color can get sun burned, just as any skin color can get burned by fire. For people that have a light complexion, the area looks red because one can now see a bit of the hemoglobin in the blood close to the skin surface. Before the burn, the blood was there, but there was enough melanin one couldn't see it or didn't notice it.

Skin Cancers
There are 2 kinds of dangerous DNA mutations. The regulation of growth in the cell is a balance of pro-growth and anti-growth factors. Obviously, if UV light damages the DNA of an anti-growth factor and inactivates it, then the balance is shifted to pro-growth. It is also possible for UV light, either directly or indirectly, to activate pro-growth DNA, again shifting the balance to pro-growth. Both of these types can happen to the skin.

Basal cell carcinomas and squamous cell carcinomas are the result of keratinocytes growing uncontrollably. Melanomas are unregulated growth of melanocytes. Unfortunately, at the moment about 10-20% of melanoma cases result in death.

Sunless Tanning
Dihydroxyacetone is a small sugar molecule that is approved as an externally applied tanning agent. Within a few hours of being applied to the skin, dihydroxyacetone reacts with proteins on the skin surface and turns brown, thus looking like a tan. However, this "tan" does not provide much protection from the sun, so sunscreens are still needed. It does appear to be a reasonably safe way to cosmetically achieve a tan, if one is careful to also apply sunscreen when in the sun.

Canthaxanthin is a coloring agent added to foods in small amounts. It is similar to carotene. It is NOT approved as a tanning pill. Some people take large amounts of it. It can make them turn an orange or orange brown color. There have been reports of eye damage and liver damage from using such high concentrations, so it does not seem as safe as carotene.

Addiction

Could tanning be addicting? There is anecdotal evidence that some people behave as if they are addicted to going to tanning salons. Some social scientists have done surveys using questions similar to questions used to determine if someone is addicted to alcohol, the so-called CAGE questions:
- C, Have you ever felt you should Cut down on drinking?
- A, Have people Annoyed you by criticizing your drinking?
- G, Have you ever felt Guilty about your drinking?
- E, Have you ever had a Eye opener drink to help you feel better?

Answering two of these questions as yes is consistent with an addiction.

For opiates, at least 2 things contribute to the addiction. First, our body has receptors for opiates. It should seem strange that the human brain and other parts of the body have receptors for plant compounds. (Opiates were first isolated from poppies.) In the 1970s scientists also thought this strange and began looking for compounds our bodies made that would also activate these receptors. Endorphins and enkephalins are two classes of endogenous compounds that activate opiate receptors. One location of these receptors is in the brain's so-called reward center. Activation of this center leads to a pleasurable feeling. Incidentally, opiate receptors are also found in the smooth muscle cells in the intestine. This accounts for the side effect of opiate medication-constipation. Since it is much easier to measure isolated muscle contraction than trying to measure pleasure, scientists used isolated intestinal muscle to purify enkephalins and endorphins.

A second contribution to addiction is that the brain circuitry remodels itself. Part of this remodeling is in response to environmental cues or associations with the pleasurable experience, much like Pavlov's dogs circuits were modified so that they salivated when they heard a bell.

Physiology

So it seems possible that some people have developed a Pavlovian response to tanning beds. Their first few times using the beds resulted in pleasurable experiences and for some this might have become an expectation. Some surveys suggest that some teenagers start going to tanning salons with their mothers. It seems possible that the pleasure from the improved relationship with their mother gets associated with the tanning salon trip. Also, if some trips to the tanning salon are done with good friends and get associated with that pleasure as well as the anticipatory discussion of either a dance or a fun trip, which could add to the association.

It is possible that there is direct activation of the pleasure pathway by tanning, independent of the environmental cues. When the keratinocytes signal the melanocytes, the keratinocyte releases melanocyte stimulating hormone. It turns out that the melanocyte stimulating hormone is just a piece of a longer protein called POMC; sunlight increases the production of POMC. One of the other pieces of POMC is endorphin. Some have speculated that this might account for the apparent addiction to tanning in some people. I think there is a potential flaw in this suggestion. The current theory is that endorphins cannot cross the blood brain barrier. If keratinocyte release of endorphin were the cause of the addiction, then the endorphin would need to have access to the brain reward center and would need to cross the blood brain barrier.

Some scientists have been trying to develop a way to safely tan, since most scientists now think that any tanning by sunlight or UV light results in DNA damage. One approach has been to make the melanocytes think that the keratinocytes have been exposed to too much UV light. What's the signal? Melanocyte Stimulating Hormone. Injection of Melanocyte Stimulating Hormone into the blood does stimulate tanning. But most people don't like to take injections. Melanocyte Stimulating Hormone is a protein, so if one took it orally, it would be digested by the proteases in the digestive track before it was absorbed. So scientists have worked on making an active melanocyte stimulating hormone protein that is shorter and that resists proteases. A group at Arizona was able to make a protein, Melanotan I (MT I) that was the size of Melanocyte Stimulating Hormone and mimicked Melanocyte Stimulating Hormone effects in cell culture. Here is what one of the chemists wrote about his next discovery.

This MC (MT I) is well on its way toward development for use as a potential "therapeutic tan," a "tan from inside-out," with minimal need for prolonged sun exposure.
During the development of MTI, I served as a proverbial "human pincushion" (a.k.a., guinea pig), that is, I tested the efficacy of the peptide to produce a tan on myself.
...
One mistake in my deliberations was made, however. MTI had previously been administered at a dose as high as 10 mg without physiological consequences (other than tanning). I forgot, however, that MTII was only about half the molecular weight of MTI. Therefore, when I took an equivalent (10 mg) dose of MTII, I inadvertently received about twice the number of molecules of the peptide. Unlike MTI, however, MTII caused a rather immediate, unexpected response: nausea and, to my great surprise, an erection (no figure provided). While I lay in bed with an emesis pan close by, I had an unrelenting erection (about 8 h duration) which could not be subdued even with a cold pack. When my wife came upon the scene, she proclaimed that I "must be crazy." In response, I raised my arm feebly into the air and answered, "I think we may become rich."
Realizing the importance of the observation, I was determined to find out what a lower dose of MTII might do. ... MTII may prove equally effective in women wherein it might be useful in the treatment of female sexual dysfunction (FSD). In both men and women, MTII apparently works at the level of the CNS. from Hadley ME. Discovery that a melanocortin regulates sexual functions in male and female humans. Peptides. 2005 Oct;26(10):1687-9.

I wonder if part of the appeal of tanning relates to the possible sexual arousal. This might be difficult to sort out since beaches and swimwear probably have this connection independent of any keratinocyte responses.

One of my students asked, why can Melanocyte Stimulating Hormone can cross the blood brain barrier and beta-endorphin cannot? Beta-endorphin is a much larger peptide than melanocyte stimulating hormone so it seems reasonable to me that it can't get across. I rechecked the literature and there is only 1 clear report stating that beta-endorphin can't cross the blood brain barrier; there is also one report that it can be secreted into the cerebral spinal fluid and thus eventually, if it isn't broken down, get access to brain neurons. I also found that there are other peptides around the same size as beta-endorphin that might cross the blood brain barrier. In conclusion, I think the fact that tanning releases beta-endorphin and melanocyte stimulating hormone is enough evidence to suspect that tanning could be addiction in some people. Personally, for me to indict tanning as addictive, I would want clearer evidence that beta-endorphin released by skin cells was able to get to brain neurons or that inhibitors of melanocyte stimulating hormone interfered with the addiction.

I

In summary, we have learned
- that many genes regulate the color of the skin and hair.
- That UV light damages DNA
- that damaged DNA is one of the signals to release Melanocyte Stimulating Hormone.
- Melanocyte Stimulating Hormone acts on receptors in melanocytes to increase eumelanin production.
- The eumelanins are packaged in melanosomes and exported to the keratinocytes, where they are placed on the sun side of the nucleus, to absorb UV light.
- UVB light certainly damages DNA and UVA may as well.
- The more UV light that gets to the DNA in the keratinocytes, the greater the chance of non-melanoma skin cancer.
- Melanomas are rare, but much more aggressive skin cancers.
- It is much safer to get adequate vitamin D from dietary sources than from sun exposure.
- Sunburn occurs when there is so much UV exposure that many keratinocytes commit suicide; the eumelanin is lost and the red color in people with light complexion reflects one's ability to see the hemoglobin in the blood supply just below the skin.

Physiology

Is there Proof?

In this chapter, we are going to travel with ethanol (alcohol) as it travels from a glass into your body and finally out. What does ethanol do along the way? We'll cover the following topics:
- Absorption
- Liver, #1
- Lungs
- Brain
- Alcohol breakdown/liver #2
- Hangovers
- Alcohol and mice and flies
- Fetal Alcohol Syndrome and data scrutiny

First, let's start with the alcoholic drink. Most likely, a microbe, usually yeast, produced the alcohol. The microbe breaks down the carbohydrates and sugar in hops, grapes, potatoes, and corn, to make beer, wine, vodka, and bourbon, respectively. In the absence of oxygen, one of the end products is ethanol. In the presence of oxygen, acetic acid is produced; acetic acid (vinegar) is essentially oxidized ethanol. Both ethanol and acetic acid have 2 carbon atoms[1].

Some "hard" liquors refer to their "proof". For example, Captain Morgan has a rum that is 100 proof which means it is 50% alcohol. The term "proof" is shorted from 100 % proof positive that the rum was 50% alcohol. It refers to the fact that sailors were concerned that their captains were watering down their daily ration of rum. So just before the ration was passed out, the captain had to pour some of the rum onto black gun powder. If it could be lit, then the water content had to be less than 50%, so the ethanol content had to be more than 50% and that meant the captain had 100% proofed positive the rum.

If the ethanol is in a glass of wine, then the drinker might swirl the glass first and sniff the drink. You probably know that the vapor pressure of alcohol is less than water which means that alcohol will boil at a lower temperature than water or that, at the same temperature below boiling, more alcohol will evaporate than water. The vapors from the

---

[1] Ethanol can be symbolized as CH_3CH_2O and acetic acid CH_3COOH.

wine also include other molecules that bind to receptors in the nose and activate the epithelial cells that activate the neurons that send the signal to the brain. This is similar to the steps between food in the mouth and the sensation of taste (The Good, the Bad and the Spicy) or between sound waves in the cochlea and the sensation of noise (Do you hear what I hear?). One key difference between taste and smell is that we currently think that there are probably less than 10 different receptors for taste, but hundreds of receptors for smell. We can detect more than 10 tastes, but that is because each taste receptor cell probably has a unique set of the 10 taste receptors, so different foods activate different sets of taste receptor cells to different degrees. This is similar to the eye, there are only 3 different cones, but we can detect much more than 3 colors (wavelengths) of light. In fact, I would say that taste and sight have a lot in common. The closest system to smell is the immune system at least in terms of number of "receptors".

When the alcohol is sipped, it might bind to the capsaicin receptor or to related channels in the TRP family. When alcohol binds to the receptor protein or channel protein, the protein changes shape which leads to activation of the cell. This signal is then transmitted to the brain and the brain interprets this as warmth. Certainly, when the drink is swallowed, this is a common perception, the "warmth" of the drink as it goes down to the stomach. The drink can also make one feel warm all over; but again, this is a perception not a reality. The perception occurs because alcohol causes vasodilation and skin vasodilation creates the sensation of warmth.

Absorption
Alcohol can easily cross membranes and so a fair amount of alcohol is absorbed in the stomach. But the stomach is not designed as an organ of absorption, the small intestine is. One difference is that the small intestine has the membrane transport proteins that transport glucose, amino acids, and other nutrients in the cell membrane adjacent to the food. But since alcohol does not need these transporters, that is not important in this chapter. The other specialization of the intestine for absorption is that the small intestine has infoldings of the layer of epithelial cells called villi. And each cell, on the surface facing the "food", has small infoldings of its membrane called microvilli. This greatly increases the small intestine surface area, which coincidently has about the same surface area as the lung.

Physiology

Whether you absorb more of the alcohol from the stomach or the small intestine depends upon how much time the alcohol spends in each organ. For the same amount of time, the small intestine would absorb more because of its larger surface area. If the alcohol spent more time in the stomach, a larger fraction would be absorbed there, but it would take a longer time because of the small surface area. An example of this is a story I heard during the Cold War. The Soviets and the Americans would be working out the details of some treaty. This would include an excellent dinner at the end of the day and lots of drinking of vodka. And then further negotiations during the night. The Soviets would swallow a stick of butter before going in to dinner. It takes the stomach a while to churn up the butter. Small droplets of fat squirt into the small intestine, where they activate specialized cells that release the hormone cholecystokinin[1]. Two actions of cholecystokinin is to slow down the contractions in the stomach and to keep the pylorus, the valve between the stomach and intestine, closed. This makes some sense as there is no need to overwhelm the intestine with too much food as it can be stored nicely in the stomach. So the Soviet stomachs would be releasing their content to the intestine much more slowly than the American stomachs and this includes the alcohol. So for the same amount of alcohol ingested, the Americans would have more absorbed sooner than the Soviets and this has the potential to influence their decision-making processes.

Suppose someone had a few quick alcoholic drinks and 5 minutes later felt fine. Would it be ok to drink some more? A recent study suggested that novice drinkers don't understand the delay between drinking alcohol and blood levels. It takes time for alcohol to be absorbed by the stomach and by the intestine. So even when someone stops drinking, there is alcohol in their stomach and intestine. Thus it is quite likely that their blood alcohol levels will continue to rise after they stop drinking, unless they are only taking occasional sips so that they are drinking at about the same rate that the alcohol is absorbed. In this later case, there is only a little alcohol in their GI track when they stop drinking. The article suggested that it would be

---

[1] Cholecystokinin got its name from the fact that it made the gall bladder contract. Choli-cyst refers to gall-bladder, and kinin has the same root as kinetic-movement.

worthwhile explaining to novice drinkers about this delay. They proposed a funnel analogy. If you have a wide mouthed funnel with a very narrow opening at the end, after you fill up the wide mouth, it takes time for the funnel to empty into the jar. If you want the jar full, but not overflowing, you have to be careful. (If one wants to drink to a slight level of intoxication, one has to be careful.) If you keep filling up the funnel with water until the jar is full, then the water will overflow as the funnel continues to empty, just like the blood level of alcohol will continue to increase after one finishes drinking while the stomach and intestine continue to empty the alcohol into the blood.

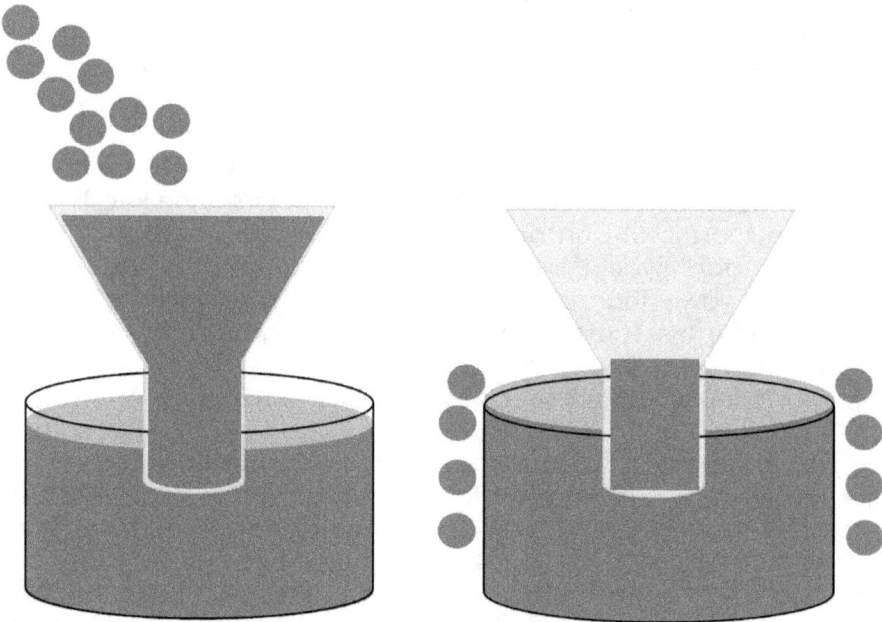

Figure 1. There is a delay between drinking alcohol and the final highest alcohol concentration in the blood. As an analogy, if you fill a funnel to its top, there is a delay before the can is filled and water will overflow the can.

In cases of extreme drinking, not understanding this delay can mean that bystanders don't take appropriate actions when someone has drunk too much. When someone drinks too much too fast and passes out, sometimes their friends <u>mistakenly</u> think the best thing to do is to

Physiology

let them sleep it off. But if the drinker is not arousable, then they are at risk for a coma, or death, from alcohol poisoning.

Let's explore what happens when the person drinks until they pass out. They pass out because the blood alcohol levels have gotten so high, the conscious centers in the brain shut down. But they still have alcohol in their stomach and intestine, so after they pass out, even though they are not drinking, the blood alcohol levels will continue to increase. This means more brain areas will shut down. The most dangerous one is the respiratory center because if that shuts down, it's over. But there can also be other issues. This much alcohol often stimulates the vomit center, since the body, rightfully, registers this much alcohol as a toxin. Vomiting will hopefully get rid of all the alcohol in the stomach (but not the intestine). But the danger is that the drinker will end up choking and suffocating on the vomit because their motor responses are impaired.

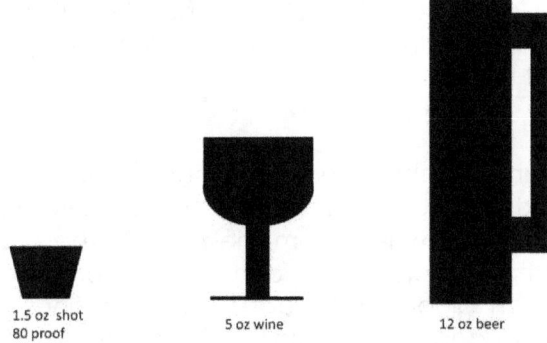

Figure 2 A "standard" drink in the US is defined as 0.6 ounces of pure alcohol (17 mls), which is the amount found in the above volumes of drinks, since 80 proof is 40% alcohol, wines are typically about 11% alcohol and beer typically 5% alcohol.

Typically after a drink or two, the alcohol concentration in the blood is about 5 to 10 mM. To put this in context, this is higher than the concentration of potassium (5 mM) or calcium (2 mM) in the blood. Most drugs are present at one thousand to one billion times **lower** concentrations! So alcohol is not a drug in the traditional receptor sense; because it is also chemically a solvent, this combination has made it difficult to determine exactly how it has its effects, particularly to determine which protein(s) it affects, either directly by binding to them, or for membrane proteins, indirectly by altering the membrane lipid layer.

Liver
The blood that takes away the nutrients is in the portal circulation and the blood travels to the liver first. The blood that came to the intestine was distributed in capillaries so there is a lot of blood vessel surface area to absorb all the nutrients. This blood is then collected and "portaged" to the liver, where it goes through a second set of capillaries before returning to the heart via the portal circulation[1]. The second set of liver capillaries provides a great surface area so that the liver can remove a lot of the nutrients, including some alcohol and glucose, before going on to the heart.

The liver absorbs some of the alcohol. As you know, in the long term, alcohol can damage the liver. The liver is also important for breaking down the alcohol. These are fascinating stories and we will come back and cover them a bit later in the chapter, but I think we ought to move on to other organs at this point.

Lungs
From the liver, the blood and alcohol flow to the heart. I know you want to go to the brain next, but actually the blood returning from the liver (and other organs) first goes where from the heart? The blood next goes to the lungs, so that the blood can dump its carbon dioxide and get fresh oxygen. Even blood that went to the intestines has burned up oxygen and produced carbon dioxide. The intestines are doing a lot of work absorbing all the nutrients. In order to get all the nutrients into the cell, the cell has to move the nutrients up a concentration gradient, or uphill. This requires energy. As you know, the cell uses ATP as its energy source. So it might be natural to expect that each of the uphill nutrient transporters would be an ATPase to harness the energy of ATP to move the glucose, or amino acid, or other nutrient uphill. Just as a water pump converts the energy from electricity (or a gasoline motor) into moving the water uphill. But that is not the design adopted by nature.

---

[1] I guess portal has the same root as portage. When you portage a canoe you carry it over land from one creek to another.

Physiology

To transport molecules uphill, the cell first moves Na uphill using an ATPase, the Na pump (or the Na,K pump because it actually moves Na uphill, out of the cell and K uphill into the cell). The other nutrient transporters are designed to move glucose (or other molecules) uphill by using the energy in the Na gradient as the Na runs back into the cell. So it is a bit like using electricity (ATP) to move water uphill (Na pumping by a Na pump) and then using the water gradient, as it runs downhill, to run a water wheel (the Na glucose cotransporter) to drive the flour grinding wheel, to move glucose uphill.

So anyway, the intestine has used ATP to absorb its water soluble nutrients. To replenish the ATP, it has burned oxygen to convert the energy in glucose to ATP with carbon dioxide as the waste product. This intestinal blood, after passing through the liver, has gone to the right side of the heart and is pumped into arteries heading to the lung.

Let's get back to alcohol. When the alcohol gets in the capillaries in the lung, some of it can escape and diffuse across to the air in the lung and then be exhaled. The concentration of alcohol in breathe is about 2000 times less than the concentration in the blood, but still measurable! Hence the use of breath alcohol analyzers.
The alcohol remaining in the capillaries returns to the left side of the heart and is then pumped out to all the other (non-lung) organs.

Brain
You've been patient long enough; let's take the path that allows alcohol to travel from the heart to the brain. Alcohol gets to the brain and affects many systems.

One of the effects of low doses of alcohol in many people it that it makes them more loquacious, more gregarious, and more talkative. People will often do some activities that they wouldn't do in the absence of alcohol. So the results of taking alcohol seem to be a stimulation of some activities. But in the brain, the main effect of alcohol is inhibition. Ah, but the inhibition of what? A lot of the neurons in your brain are actually turning off other neurons in your brain. It would be like your older sibling always telling you to be quiet. If alcohol is like your parent distracting your older sibling then you become more active. It is not that your parent stimulated you, but that your parent inhibited your older sibling allowing you to now be active.

Some of the centers that are most sensitive to alcohol are those involving decision making. Higher doses can cause several types of issues: irritability, abusiveness, and aggression and modification of some of the centers for motor coordination. This is the basis for some of the tests for sobriety, walking a straight line or closing one's eyes and touching your finger to your nose. In addition to alcohol intoxication, there are a number of neurological disorders that also affect these processes. I went to the doctor a few years ago because I had found myself a bit clumsier than usual for a few days and then was sore in places I didn't expect to be, without changing my exercise routine. Two of the tests were to walk in a straight line and the touch my nose test. I couldn't do either! It was quite embarrassing. The doctor at first thought I had a virus, but then I mentioned to him that I had fallen on the ice a month previously, so he sent me for another CAT scan. Sure enough, I had some blood leaking out of my vessels inside the skull and it was putting pressure on the part of my brain involved in coordination. Apparently, this is fairly common; to have a blow to the head, to initially be ok, but then about one month later, to have a slow blood leak develop. Fortunately, I had surgery to drain the blood and as far as I can tell, I'm "normal" now.

Some sources say a blood alcohol content of 0.04% leads to measurable impairment in driving. Legally, the limit in Missouri is 0.08%. If one compares male drivers age 16-20 that had blood alcohol levels of 0.02%, they had double the risk of a fatal accident to sober age matched controls. Personally, I think this statistic is ambiguous. If the sober controls never drank alcohol (they are under 21, after all), then they might not have the same risk-taking/cautious personality ratio as those that did drink. In that case, as a scientist, I would have to say there are 2 variables: personality and alcohol. The next sentence in my source gives additional numbers, but the sentence does not indicate an age range, so I don't know if this applies to all drivers or just young ones.

alcohol range	fold increase in accidents
0.08-0.1%	50
0.1-0.15%	240
>0.15 %	15,000

Physiology

Four Loko, a drink that combines alcohol and caffeine, was in the news in 2010 because of several students drinking too much and having to go to the emergency department. It has been hard to get definitive numbers on the caffeine content, but my memory is that 1 can of Four Loko was equivalent to 2 to 4 shots of rum and 5-7 cans of coke. Rum and coke has been a popular drink for some people for quite some time, so what happened with Four Loko? I think Four Loko had more caffeine than some drinkers realized. Alcohol not only impairs one's judgment and coordination, it impairs one's ability to realize one is impaired. For some inexperienced drinkers it apparently is the lethargy and tiredness that they use as a signal to stop. But caffeine prevents the lethargy and tiredness, but not the other problems of the alcohol. So they ended up drinking more than they expected. For many drugs, the higher the concentration, the faster it is metabolized, but not for alcohol, in part because we are talking about concentrations 1000 times or more higher than for other drugs. This means the enzyme that breaks down the alcohol is saturated even after a very small drink.

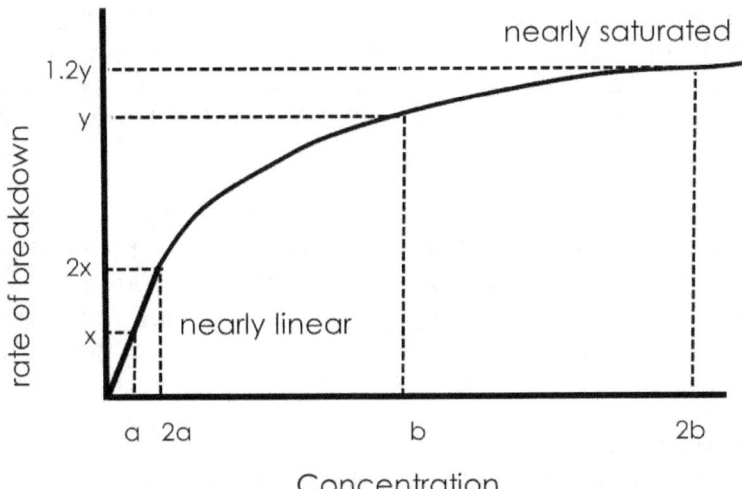

Concentration

Figure 3. Saturation of the rate of ethanol breakdown. For many drugs, the concentration range is on the linear part of the curve. If one doubles the concentration from a to 2a, the rate increases about 2 fold, from x to 2x. In contrast, at many alcohol concentrations, one is on the non-linear part of the concentration range, so doubling the alcohol from b to 2b only has a modest effect on the rate, from y to 1.2 y.

As an example, the small brackets could be analogous to you reading one easy book at 10 pages per minute in class A. If you are assigned 2 books in class B, you can easily increase your reading to 20 pages per minute and finish the 2 books in about the time you were initially reading the first book for class A. The large brackets are you reading one very difficult book for class C about as fast as you can, 10 sentences per minute. Suddenly you are assigned to read two of these books for class D and the best you can do is read about 12 pages per minute. It will take you about twice as long to read the books for class D as for class C.

Hangovers often occur when someone has too much alcohol. In order to understand the theories about what causes hangovers, it will help to understand how we get rid of alcohol. So let's temporarily leave the brain and find out how we get rid of alcohol. From the brain, of course, the blood goes back to the heart and then off to the lungs to get more oxygen. The brain accounts for about 20% of our resting metabolism (oxygen consumption). This is primarily to fuel the Na pump (and perhaps the Ca pump). Every time a nerve fires, Na moves downhill, so the Na pump is needed to put the Na gradient back in place. In fact, it is estimated that only about 2% of our neurons can be working at one time due to the limitation of oxygen delivery to the brain. Exercise appears to improve brain function. One reason is that exercise probably also increases blood flow to the brain and of course it makes oxygen loading in the lung more efficient. Exercise also increases the production of a peptide, Brain Derived Neurotrophic Factor, and this peptide improves neuronal functioning.

Alcohol breakdown-liver #2
After getting re-oxygenated in the lungs, the blood goes back to the heart and then out to the other organs. Which organ is most responsible for getting rid of alcohol? In general, for any substance, the two major excretions are urine and feces; for some compounds expired air and sweat may also be important. For alcohol the most important organ is the liver; urine and breath excrete about 10%. For drugs in general, the liver is a key organ. For fat-soluble compounds, the liver attaches a water-soluble compound to the fat-soluble

compound so that the fat-soluble compound becomes water-soluble. Another major way of dealing with drugs is to convert them. Oxidation is a key reaction and it is the important one for alcohol. The enzyme alcohol dehydrogenase converts alcohol to acetaldehyde[1].

Acetaldehyde is a much more dangerous compound than alcohol. Fortunately, the reaction from alcohol to acetaldehyde is slower than the reaction from acetaldehyde to acetic acid. So even though alcohol can be present at 5 to 10 mM, acetaldehyde is normally present at concentrations at least 1000 times less.

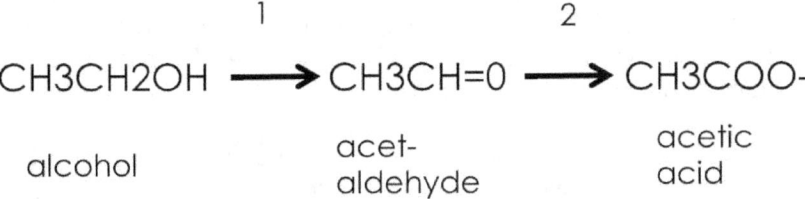

Figure 4. Alcohol breakdown proceeds in two steps. Enzyme 1 is usually alcohol dehydrogenase and enzyme 2 is acetaldehyde dehydrogenase. Modified from http://pubs.niaaa.nih.gov/publications/AA72/AA72.htm

Some people drink antifreeze; so do some pets. Antifreeze is highly toxic, but apparently it tastes sweet. The major ingredient in antifreeze is ethylene glycol. Perhaps you recognize the prefix ethyl, like ethanol, it has 2 carbons. This compound itself is not very toxic, but it is

---

[1] Acetaldehyde is a contraction of acetyl and aldehyde. We have two parallel terms going on in chemistry. One set is systematic, so for example, 1 carbon compounds are meth---, 2 carbon compounds are eth---, 3 carbon are prop---, 4 carbon are butyl---, 5 carbon are penta---, and so on. In this terminology, ethanol is converted to ethanal which is converted to ethanoic acid. In the other set of names are based on the original words before we knew the chemical structure. So the final produce, ethanoic acid is usually called acetic acid, and it is the major component in vinegar. Going back one step we have acetyl-aldehyde. So acetyl means 2 carbons and an aldehyde which has a double bonded oxygen to the terminal carbon, in symbols C=O. An alcohol is C-OH, so alcohol dehydrogenase removes a hydrogen (H) from C-OH to form C=O. Aldehyde dehydrogenase converts C=O to COOH, which is an acid. Oxidation and dehydrogenation mean the same thing in this situation.

converted by alcohol dehydrogenase to glycolate and oxalate, which are toxic compounds. The major treatment for ethylene glycol ingestion is to inhibit the enzyme alcohol dehydrogenase. This means ethylene glycol concentrations stay high, but it also means the metabolites of ethylene glycol, that is, glycolate and oxalate stay low. Since glycolate and oxalate are much more toxic, the patient has a better chance of surviving. For a long time, the major method of inhibiting the enzyme was to give the patient alcohol because then the alcohol would keep the enzyme alcohol dehydrogenase occupied and it wouldn't be able to break down ethylene glycol to quickly. One of the problems with this therapy is that one of the major symptoms of ethylene glycol ingestion is behaving like one is intoxicated. Obviously, if someone is intoxicated from alcohol, giving more alcohol is not a good idea.

You are probably aware that different ethnic groups tend to have a higher or lower probability of metabolizing alcohol. The aldehyde dehydrogenase has several different forms or polymorphisms due to some mutations. Consider 3 people, each of who has 2 genes for aldehyde dehydrogenase, one from their mother and one from their father. Suppose the standard genes works at 10 units per hour and the mutated gene at 2 units per hour. For person X, both genes are the standard one, so their rate is 2 x 10 = 20 units per hour. Person Y has one "standard" gene and one modified gene. Person Y's rate is 10+2 = 12 per hour. Person Z has 2 copies of this modified gene, so their rate is only 2 + 2 = 4 units per hour. If person Z drinks too much alcohol, their acetaldehyde concentration will increase and this will cause an unpleasant facial flushing, nausea and other unpleasant symptoms. Disulfiram is a drug sometimes used to help alcoholics kick their habit. It inhibits aldehyde dehydrogenase and so causes the same unpleasant effects.

Not only does the liver metabolize alcohol, but it can also be damaged by alcohol. One of the first effects is increased fat in the liver. The sympathetic system is apparently activated, which stimulates the fat cells to release fatty acids. These are transported to the liver. Normally they would be oxidized, but oxidation is slowed by the presence of alcohol, and so fatty acids begin to accumulate. Even one large dose of alcohol can increase the fat in the liver. If this continues long enough, one develops a fatty liver. This can then progress to inflammation of the liver. Eventually, the liver cells begin to die and the tissue becomes more fibrous, like scar tissue. This last step is irreversible, fortunately, the liver has excess capacity and some damage is ok. But obviously, if you damage the liver enough, you are going to get very ill.

Hangovers
Hangovers are complicated and our scientific understanding is poor. One problem is how to define hangover as all the measures so far are subjective. What symptoms are required to be considered a hangover?

Many earlier studies were done on only a small group of people. I don't think a study on a small group of people necessarily is a poor study; it depends upon how large and how variable the effect is. Hypothetically, I would consider a study in which all 5 people in the test group died whereas all 5 people in the control group lived as pretty convincing as long as the control group and the test group were very well matched. But there is great individual variation in the response to alcohol, so in a small study about hangovers, it can be difficult to determine whether the difference is due to chance or to the test substance, alcohol.

Part of the hangover is presumably due to an effect of alcohol but it is not directly related to blood alcohol levels at the time of the symptoms. When drinking alcohol in the evening, as blood alcohol levels increase, one does not notice any hangover symptoms. Rather, the symptoms occur when one wakes up in the morning. At that time, one's blood alcohol content is often close to zero. Since some of the symptoms of hangovers overlap with some of the symptoms of alcohol withdrawal, some have proposed that hangovers reflect acute alcohol withdrawal, but most researchers think hangovers and acute alcohol withdrawal reflect different processes.

Four of the main factors that people initially consider as contributing to the hangover are: dehydration, sleep alterations, alcohol, and congeners[1].

---

[1] Congeners are other compounds found in some alcoholic beverages; there is some evidence, for example, that hangovers are worse when one drinks A rather than B and some of the hangover affects can be mimicked by giving a person

We'll examine each of these 4 factors in turn.

Dehydration
Dehydration and hangovers both involve not feeling well and the list of symptoms is similar. But is the headache from dehydration the same as the headache from a hangover? If dehydration is the cause of hangovers, then the headaches (and other symptoms) need to be the same. In addition, alcohol consumption needs to cause dehydration. This initially seems counterintuitive: one is drinking, why would one become dehydrated? It does appear that people drinking alcoholic beverages urinate a larger volume of fluid than they drink; this would mean that they are out of balance and would have less water in their body at the end of the evening than before they started drinking. Some studies show that blood levels of antidiuretic hormone[1] decrease after drinking alcohol (compared to an equal volume of non-alcoholic fluid). If antidiuretic hormone is low, then the signal to the kidneys to reabsorb salt and water has decreased. That is a double negative, so the result is that there is more salt and water excretion.

Let's summarize the results so far. At least superficially, dehydration results in the same symptoms as a hangover. And drinking alcoholic beverages seems to result in extra fluid loss, that is, it seems to cause dehydration. So is a hangover simply due to dehydration? Suppose it is, what is a key prediction? How could one prevent a hangover? In this theory, one merely needs to remain properly hydrated to avoid a hangover. I don't think this has been well studied by scientists; it seems to me one easy preliminary experiment would merely be to have volunteers take a picture of the toilet after their first urination in the morning and compare the amount of yellow color with the degree of symptoms[2]. Since it is commonly thought that hangover is due to dehydration (see Wikipedia for example) and since people try all sorts of remedies to avoid hangovers, I would have guessed that if maintaining hydration effectively prevented hangovers, it would be

---

some of the other compounds in A.
[1] ADH, see the chapter Soy and Licorice as Medicine.
[2] See the chapter Soy and Licorice as Medicine, Figure 5.

widely used, at least by experienced drinkers who need to go to work the next day.

Sleep. Alcohol itself can alter sleep patterns and sleeping physiology. In addition, I suspect many people stay up later when they are drinking alcohol than they do on other nights. And certainly in the movies, after a lot of drinking, people end up sleeping in uncomfortable or unfamiliar locations, which can alter the quality of the sleep. But people that stay up all night, or have a poor night sleep, don't seem to have quite the same quality of ill feeling that hangovers produce. So the current theory is that the quality of sleep can impact the severity of the hangover, but poor sleep, in the absence of alcohol, does not cause a hangover.

Alcohol and its metabolites
Since blood alcohol levels are low or zero when one wakes up with a hangover, two general theories have been suggested. One is that alcohol leads to the release of a variety of signaling molecules (cytokines and other immune system modulators are currently the most popular ideas). These take some time to have their action on cells, including neurons, so that there is a lag between when the alcohol levels were high and the eventual not feeling good symptoms. In support of this, in small studies, giving people other drugs that release similar cytokines and immune system modulators does led to symptoms similar to hangovers in the absence of alcohol. I don't know if anyone has given alcohol drinkers drugs that either inhibit the cytokine release, or inhibit their downstream effects to see if the symptoms are gone.

Another theory is that the hangover is caused by the elevated levels of acetaldehyde. This has two appealing aspects. First, even when blood alcohol levels are close to zero, acetaldehyde levels can remain elevated, since acetaldehyde breakdown is slow. Second, acetaldehyde is a reactive species and can modify proteins and other molecules and remains irreversibly bound. So even when blood levels of acetaldehyde go to zero, many molecules in cells may remain modified with acetaldehyde.

Congeners
Maybe it is not actually the alcohol, but other parts of the drink that cause the hangover. Can non-alcoholic beer lead to a hangover? If

so, then congeners are a possibility. If not, is the only difference between alcoholic and non-alcoholic beer the alcohol or are other substances also missing?

Hangovers have not been well studied scientifically. Those that do study it are caught in a dilemma. If you want to know what causes the hangover, the usual scientific approach is to study people under controlled conditions and just vary 1 parameter between groups. However, in real life, many different factors contribute to the intensity of the hangover. Which is better to study if one wants to find a remedy for hangovers?

Alcohol and mice and flies
Scientists have tested ethanol effects in mice. To find out which protein is involved, scientists have knocked out genes. Over 100 knocked out genes change the mouse's response to alcohol. However, this does not mean the protein is the primary target. For a primary target, ethanol would bind to the protein and changes its function. But the knock out results could also occur if ethanol didn't bind to the knocked out protein. For example, if protein A binds ethanol and that complex alters protein B that alters protein C that causes the mouse to stumble, knocking out protein C might stop the mouse from stumbling.

Work is also being done in fruit flies. They have smell receptors that can sense alcohol. It is important to the fly to avoid high concentrations of alcohol because it is potentially toxic. One of the receptors has the cute name, LUSH[1]. This type of work may help us determine what protein binding sites for ethanol look like. This might allow us to identify the alcohol binding sites on human proteins and perhaps lead to some therapeutic drugs.

Fetal Alcohol Syndrome and data scrutiny
Alcohol in the pregnant mother can cause fetal alcohol syndrome in the baby. Research suggests that one of the effects is that alcohol inhibits a protein involved in neural adhesion. Babies that have a mutation in this protein have many symptoms similar to babies with

---

[1] The fly protein LUSH got its name because when it was mutated, the flies were attracted to higher concentrations of alcohol than wild type flies.

fetal alcohol syndrome. The protein is important for how nerves grow and connect to each other and it does have an alcohol binding site.

I think some of the research on the levels of alcohol that cause fetal alcohol syndrome is politically motivated, from both the side that wants to control women and the side that wants to empower women. One side says that there is no conclusive[1] evidence of a known safe dose of alcohol during pregnancy. The other side says that there is no conclusive evidence that an occasional drink causes fetal alcohol syndrome. Both statements could be true. But both are also probably a bit misleading.

Alcohol and fetal health are both emotional issues; let's analyze analogous statements for whether drug X affects adults. Side A: There is no conclusive evidence of a known safe dose of drug X. Side B: There is no conclusive evidence that small doses of drug X cause damage. Both sides could be looking at the graphs in the Figures 5, 6 and 7,

---

[1] When I use conclusive, I mean that 12 random people would find the evidence, beyond a reasonable doubt, supports the claim.

In Figure 5 top left are some possible data. One can extrapolate these data assuming a linear curve below the last data point as in the top right. In this case, even a small amount of drug does have an effect. On the other hand, side B could say that, looking at the other extreme of the error bars, that there is a threshold and below a certain dose there is no effect, as in the bottom left.

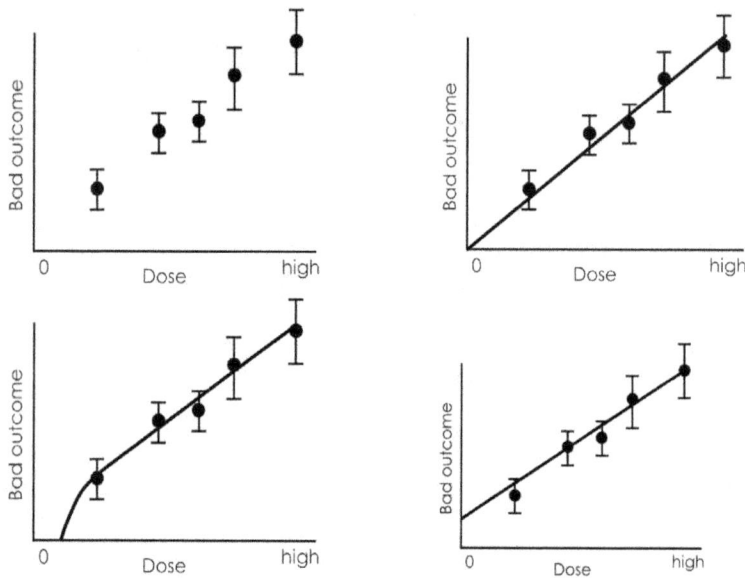

Figure 5. Top left. Data from a hypothetical study.
Top right. A reasonable line is drawn in which the effect goes to zero at the lowest dose with the assumption that the response remains linear below the lowest dose. Linearity below the lowest dose is an assumption. (I would say the data are consistent with the response being linear for the doses shown, but without any data at lower concentrations, there is no experimental evidence that it is linear below the lowest dose tested.)
Bottom left. A non-linear curve with a threshold dose. In general, the only way to rule out a non-linear response between the lowest dose and zero is to do doses in that range, but that is exactly the range where the signal is usually hardest to detect.
Bottom right. A reasonable line is drawn in which there is an effect even at the lowest dose and this makes it even harder to detect an effect at low doses because of the "background" signal.

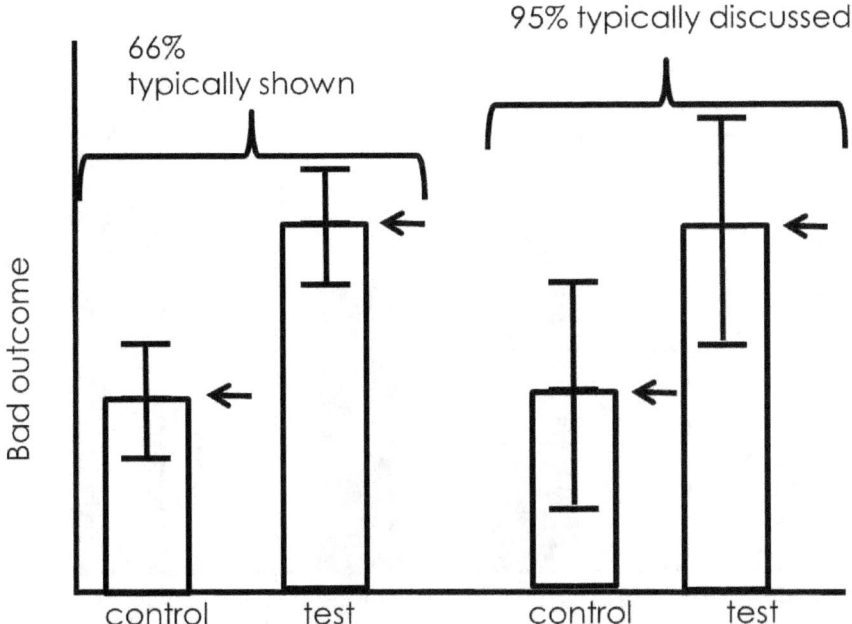

Figure 6. How to read error bars, part 1. On the left is a typical plot showing the control and test responses. The arrow points to the horizontal bar for average value. The other two shorter horizontal bars show one standard deviation from the average; if the response is statistically "normal" [1] then 66% of the values are within the limits between the two shorter bars. However, scientists typically consider only things outside a 95% window as unlikely to happen by chance, so most articles discuss the 95% range, which is 2 standard deviations from the mean; that range is shown on the right.

Within the error, the smallest dose of radiation or drug X had no significant effect, that is, results like this could be obtained if one studied a drug that really had no effect and had normal, random variation.

---

[1] See the chapter From Anemia to Vampires, for a discussion of "normal"

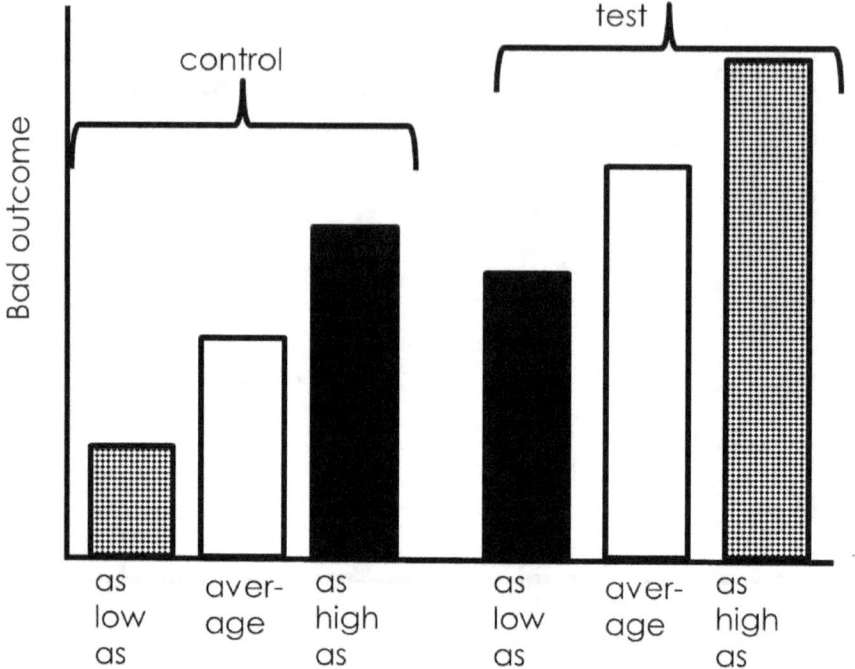

Figure 7. How to read error bars, part 2. Here I have separated the columns from Figure 6 into 3 columns:
the lowest possible value that could occur with a 5% chance,
the average value, and
the highest value that could occur with a 5 % change.
Since the solid bars are reasonable possibilities for the data, one could interpret the data as consistent with the idea that the drug, at this dose, was at least as safe as the control.
On the other hand, the checkered bars are also possible given these data. In this case, one could interpret the data as consistent with the idea that the drug had several fold more bad outcomes than the control.
Which interpretation one picks depends upon the initial bias before the experiment was done: Do you need conclusive evidence that the test is safe or do you need conclusive evidence that the test is dangerous?

Physiology

My personal evaluation of the risk is that if I thought I might be pregnant, I would forego the alcohol; for me, the short-term pleasure does not outweigh the long-term risk of damage to the fetus. But if I had been drinking and then found out I was pregnant, I wouldn't stress out about my previous behavior. I would immediately stop, of course. And I would talk with a trusted health care provider. But it is my understanding that intense stress also increases the risk of harm to the fetus.

There is some research that suggests that improved nutrition can decrease the risk of fetal alcohol syndrome in alcoholics and others who do not stop drinking during pregnancy. Trying to implement a policy based on this work is tough as one wants to decrease any potential harm, but one does not want the message to be misinterpreted that drinking a lot during pregnancy is ok.

In summary
- We have reviewed why blood alcohol levels peak well after drinking has stopped. It is because of the delay in absorption by the stomach and intestine. And that this delay depends upon the other contents of the stomach and intestine.
- The liver is the major organ for metabolizing alcohol and the first metabolite, acetaldehyde is also harmful.
- As the level of blood alcohol increases, different brain functions are affected.
- We also discussed that fact that the interpretation of the data requires a starting assumption. Suppose one wanted to determine whether one alcoholic drink had an adverse effect on the fetus. Does one assume that one alcoholic drink has an effect and require evidence to disprove that statement? Or does one assume that one alcoholic drink has no effect and require evidence to disprove that statement? In addition, each person has a different level of requirement for what "disprove" means.

Appendix
References for cases

Orange you glad?

Vincent
Collins CE, Koay P. Xerophthalmia because of dietary-induced vitamin a deficiency in a young Scottish man. Cornea. 2010 Jul;29(7):828-9. PubMed PMID: 20489600.

Valerie
Braunstein A, Trief D, Wang NK, Chang S, Tsang SH. Vitamin A deficiency in New York City. Lancet. 2010 Jul 24;376(9737):267. PubMed PMID: 20663549; PubMed Central PMCID: PMC2929009.

Frank
Roueche B. The Medical Detectives. Truman Talley Books. New York. 1980.

Gerrit de Veer
Lips P. Hypervitaminosis A and fractures. N Engl J Med. 2003 Jan 23;348(4):347-9. PubMed PMID: 12540650.

Can Viagra Kill You?

George
Arora RR, Timoney M, Melilli L. Acute myocardial infarction after the use of sildenafil. N Engl J Med. 1999 Aug 26;341(9):700. PubMed PMID: 10475830.

Marshall
McLeod AL, McKenna CJ, Northridge DB. Myocardial infarction following the combined recreational use of Viagra and cannabis. Clin Cardiol. 2002 Mar;25(3):133-4. PubMed PMID: 11890373.

Soy and Licorice as Medicine
Solomon
Chong L, Baikunje S, Poller DN, Roberts IS, Venkat-Raman G. An unusual cause of acute renal failure with volume depletion due to renal losses. Am J Kidney Dis. 2008 Aug;52(2):366-9. Epub 2008 Jun 30. PubMed PMID: 18585834.

Victor
Ettinger R, Goodson JE, Clancy J. How a broken tooth saved a life. Spec Care Dentist. 2010 Sep-Oct;30(5):183-4. doi: 10.1111/j.1754-4505.2010.00157.x. Epub 2010 Aug 17. Erratum in: Spec Care Dentist. 2010 Nov-Dec;30(6):276. PubMed PMID: 20831735.

Alex
Moon SS, Kim HJ, Choi YK, Seo HA, Jeon JH, Lee JE, Lee JY, Kwon TH, Kim JG, Kim BW, Lee IK. Novel mutation of aquaporin-2 gene in a patient with congenital nephrogenic diabetes insipidus. Endocr J. 2009;56(7):905-10. Epub 2009 May 20.PubMed PMID: 19461158.

Sue
Cooper H, Bhattacharya B, Verma V, McCulloch AJ, Smellie WS, Heald AH. Liquorice and soy sauce, a life-saving concoction in a patient with Addison's disease. Ann Clin Biochem. 2007 Jul;44(Pt 4):397-9. PubMed PMID: 17594790.

Lily
van Uum SH. Liquorice and hypertension. The Netherlands Journal of Medicine 63:119-120, 2005.

Anemia to Vampires
Iris B. Doce
Chen SH, Hung CS, Yang CP, Lo FS, Hsu HH. Coexistence of megaloblastic anemia
and iron deficiency anemia in a young woman with chronic lymphocytic thyroiditis.
Int J Hematol. 2006 Oct;84(3):238-41. PubMed PMID: 17050198.

Physiology

Steroids Make the World go Round

Andy the cat
Millard RP, Pickens EH, Wells KL. Excessive production of sex hormones in a cat with an adrenocortical tumor. J Am Vet Med Assoc. 2009 Feb 15;234(4):505-8. PubMed PMID: 19222361.

Tessa
Eliakim A, Cale-Benzoor M, Klinger-Cantor B, Freud E, Nemet D, Feigin E, Weintrob N. A case study of virilizing adrenal tumor in an adolescent female elite tennis player--insight into the use of anabolic steroids in young athletes. J Strength Cond Res. 2011 Jan;25(1):46-50. PubMed PMID: 21116197.

Tori
de Ronde W. Hyperandrogenism after transfer of topical testosterone gel: case report and review of published and unpublished studies. Hum Reprod. 2009 Feb;24(2):425-8. Epub 2008 Oct 23. Review. PubMed PMID: 18948313.

Paul
Edidin DV, Levitsky LL. Prepubertal gynecomastia associated with estrogen-containing hair cream. Am J Dis Child. 1982 Jul;136(7):587-8. PubMed PMID: 7091084.

Physiology

Chapter Summaries

Orange you glad?
- Vitamin A is like many nutrients. We need enough, but not too much.
- Vitamin A deficiency is a cause of night blindness, because a vitamin A derivative is part of the photoreceptor.
- Too much vitamin A is a problem, because vitamin A is also a transcription factor and high levels at the wrong time and place alter cell growth and death; this can account for birth defects as well as bone defects in adults.
- Vitamin A has a narrow therapeutic index. We are unable to regulate its absorption because no proteins are directly involved.
- While we also cannot control carotene absorption, carotene is a much better way to supply the body with vitamin A. We can control the rate of conversion from carotene to vitamin A because an enzyme is required. Too much carotene only results in orange skin, so carotene's therapeutic index is large
- Proper fat digestion is important for proper vitamin A absorption and carotene absorption (and the absorption of other fat soluble vitamins).

Blood Sweat and Tears
- Nerves communicate to adjacent nerves or muscles by releasing a chemical, a neurotransmitter.
- Acetylcholine and norepinephrine (noradrenaline is a synonym) are two neurotransmitters.
- The signaling is halted by either breaking down the neurotransmitter (e.g., acetylcholine) or by transporting it out of the synaptic space (e.g., norepinephrine)
- Acetylcholinesterase breaks down acetylcholine. Inhibitors of acetylcholinesterase will lead to an increase of acetylcholine at active nerves.
- Specific proteins bind specific chemicals; these are often called receptors.
- There are two classes of acetylcholine receptors: one opens ion channels, one activates G proteins.
- In the body, neurotransmitters are specific because their concentration is high only in a very local region. Drugs that

mimic neurotransmitters have side effects because the drug concentration is high throughout the body and can activate all their receptors, not just the local ones that need "to be fixed".

Impressive Horses and Fainting Goats
- We have learned that mutations in sodium and chloride channel can alter the signaling from the nerve to skeletal muscle calcium release.
- These mutations can occur in humans and domestic animals.
- Skeletal muscle contraction involves the sequence:
- Acetylcholine activating the nicotinic acetylcholine receptor which opens its gate and allows positive ions to enter
- The positive ions shift the balance of charge between the inside and outside of the cell
- When a threshold in charge has changed, sodium channels open and an even larger change in charge occurs
- The change in the balance of charge between the inside and outside of the cell causes intracellular stores (sarcoplasmic reticulum) to release their calcium
- In Impressive, the sodium channels do not close tightly after opening and this slow leak causes more calcium entry and muscle contraction
- In the fainting goats, the chloride channels do not work well as rest, making the muscle cells more excitable

Do you hear what I hear?
- Sequence for hearing:
    - sound waves bend cilia
    - bent cilia open K channels
    - balance of charge inside and outside changes
    - cell calcium increases
    - cell releases glutamate
    - glutamate activates adjacent neuron
    - signal goes to brain via a series of neurons
- When sounds are too loud, extra energy is required to regenerate the K and Ca gradients; the extra energy leads to high levels of oxygen radicals which damage proteins and DNA
- Hearing cells do not regenerate so if damage is severe and too many cells die, hearing is lost

Physiology

- Decibels are a measure of how loud a sound it; decibels are based on a log scale so that going from 60 to 80 decibels means the pressure has increased ten fold
- wavelength and frequency are reciprocals, so high frequency sounds have small wavelengths which explains why bats use ultrasonic frequencies to find small insects
- **The general plan of the sensory system** is that you have a cell that is particularly sensitive to a specific input:
    - sound pressure at a distinct frequency for hearing cells,
    - light at certain frequencies for cones,
    - different chemicals for taste and smell, and
    - pressure or temperature for some touch systems.
- Sequence to brain:
    - specialized cells change charge balance
    - often cell calcium changes
    - cell releases a neurotransmitter
    - adjacent neuron is activated
    - a series of neurons take signal to appropriate part of brain
    - brain interprets signal as from the stimulus

The Good, the Bad, and the Spicy
- The same protein can respond to different stimuli.
- Some proteins can respond to both heat and the pepper chemical capsaicin.
- Some proteins can respond to both cold and the mint chemical menthol.
- Once a protein is activated, the downstream cellular changes are the same independent of the stimulus, for example whether the stimulus was heat or capsaicin
- The brain "knows" whether a cell codes for heat or sweetness or other taste.
- A receptor changes shape when it binds a chemical; the change in shape leads to changes of the cell interior. A cell can communicate with a neighboring nerve cell by releasing a neurotransmitter.
- Often, at high concentrations, a chemical will bind to more than one type of receptor.
- There is not necessarily a one-to-one correspondence between molecular, perception, and behavioral responses to a chemical.

- Opening a sodium channel allows the positively charged sodium ions to enter and alters the relative balance of positive charges between the inside and outside of the cell.
- Some proteins change shape in response to a change in the relative balance of positive charges between the inside and outside of the cell.
- **A sweet compound** binds to the sugar G protein receptor, which undergoes a shape change, which leads to changes inside the cell, which causes the release of ATP, which signals the adjacent nerve and ultimately the signal is identified in the brain as arising from a "sweet-sensing -cell".
- **A bitter compound** binds to one of the bitter G protein receptor types, which undergoes a shape change, which leads to changes inside the cell, which causes the release of ATP, which signals the adjacent nerve and ultimately the signal is identified in the brain as arising from a "bitter-sensing -cell".
- **An umami compound** binds to one of the umami G protein receptor types, which undergoes a shape change, which leads to changes inside the cell, which causes the release of ATP, which signals the adjacent nerve and ultimately the signal is identified in the brain as arising from a "umami-sensing -cell".
- **A salty food** raises the local sodium concentration outside "salty-sensing cells", which means more sodium enters the cell, which changes the relative amount of positive charge inside vs. outside the cell, other proteins sense this change of relative charge, a cascade occurs and the "salty-sensing cell" releases neurotransmitters from stored vesicles, exciting the adjacent nerve and ultimately the signal is identified in the brain as arising from a "salty-sensing -cell".
- **A sour food** raises the proton concentration, which effects "sour-sensing cells" in ways that are not yet fully defined. This leads to a cascade and the "sour-sensing cell" releases neurotransmitters from stored vesicles, exciting the adjacent nerve and ultimately the signal is identified in the brain as arising from a "sour-sensing -cell".
- **A spicy hot food** has chemicals that bind to TRPV1, the heat sensitive channel, causing it to open and allow more sodium into the cell, which changes the relative amount of positive charge inside vs. outside the cell, other proteins sense this change of relative charge, a cascade occurs and the "heat sensitive neuron" releases neurotransmitters from stored

vesicles, exciting the adjacent nerve and ultimately the signal is identified in the brain as arising from a "heat sensitive neuron".

Ouch That Hurts
- There are specific molecules that respond to injury- mechanical, heat, chemicals, inflammatory mediators.
- These activate appropriate receptors which lead to nerve activation and a signal to the brain, where pain is received.
- The hot pepper channel, TRPV1, is one key pain sensor.
- Gain-of-function mutations in specific sodium channels cause the pain sensing neurons to fire inappropriately and cause several diseases of inappropriate pain sensation.
- Another disease of inappropriate pain sensation is caused by a mutation in the nerve growth factor tyrosine kinase receptor.
- Both of these finds suggest possible new targets for drug development (sodium channel inhibitors and nerve growth factor tyrosine kinase receptor inhibitors).
- The absence of sodium channels in pain sensing neurons leads to a state where pain is not felt and reminds us that pain often serves a useful purpose.
- The main drugs for chronic pain relief are opiates, which activate pain inhibitory pathways in the brain, but these drugs are also addicting.

Sleep tight, don't let the bed bugs bite
- description of current theories on why we sleep
- description of the stages of sleep
- how sleep pills work which involved a review of neurotransmitters and cell charge balance
- theories about how much sleep we need
- observations of what happens if we are sleep-deprived
- description of different types of sleepwalking
- Some people who have unusual sleep patterns

Can Viagra Kill You?
- the effect of blood volume on blood pressure
- the effect of blood pressure and dilation on blood flow
- how actin and myosin filaments slide
- that enzyme activity can be regulated by adding a phosphate; kinases are enzymes that transfer phosphates

- that calcium activates calmodulin and the calcium/calmodulin complex activates MyosinLightChain-Kinase
- that MyosinLightChain-Kinase transfers a phosphate to MyosinLightChain which activates it
- that nitric oxide causes an increase in cGMP
- increases in cGMP promote smooth muscle relaxation
- Viagra inhibits cGMP breakdown
- ATP is important as both the source for phosphate which is a regulatory switch and as an energy source
- one mechanism of drug-drug interactions occurs when two drugs affect the same process, as both nitroglycerin and Viagra affect smooth muscle relaxation
- another mechanism of drug-drug interactions occurs when two drugs are broken down by the same enzyme and this kind of multitasking slows down each process
- the discovery on nitric oxide
- nitric oxide is made on demand, not stored

In A Heart Beat

- The amount of blood the heart pumps per minute, the cardiac output, is the product of the amount pumped per beat (stroke volume) times the heart rate.
- An increase in cardiac output, all else held constant, leads to an increase in blood pressure
- Heart rate is a balance of the influence of acetylcholine and norepinephrine.
- Acetylcholine decreases cAMP in the heart and slows the heart rate
- Norepinephrine increases cAMP in the heart and increases the heart rate as well as the force of contraction because of the increased calcium levels.
- The flow of blood is governed by the pressure gradient, always flowing from high pressure to low pressure.
- In order for flow to occur, there must be a pathway (open valve).
- The pressure volume curve for the heart summarizes the changes during the cardiac cycle.
- Stroke volume, the amount of blood pumped per heart beat, is the difference between the volume of blood in the heart at the end of the ventricular relaxation (end diastolic volume)

- and the volume of blood in the heart at the end of ventricular contraction (end systolic volume).
- The amount of blood in the heart at the end of the ventricular relaxation (end diastolic volume) depends upon the amount of blood returning from the venous system, the amount of time to fill, and the ability of the ventricle to stretch.
- The amount of blood in the heart at the end of the ventricular contraction (end systolic volume) depends upon the aortic blood pressure and the relation between cardiac muscle fiber length and force.
- Ventricular volume is a surrogate measure for the length of cardiac muscle fibers.
- Ventricular pressure is a surrogate measure for the amount of force cardiac fibers can generate.
- When we exercise, and pigeons fly, cardiac output increases primarily because of an increase in heart rate. Stroke volume also increases because of an increase in the force vs. length relationship, allowing the volume of blood remaining in the heart at the end of the contraction phase, the end systolic volume, to be less.
- When trout swim and elite endurance athletes run, there is an additional increase in stroke volume because the volume of blood remaining in the heart at the end of the relaxation phase, the end diastolic volume, is greater, because their hearts are able to stretch to greater lengths.

Simon Says "Move"
In summary,
- ATP is the fuel used by your muscles to contract. There are several different sources to replenish the ATP and some are available quickly and used up quickly, while others last longer, but also take longer to be used.
- Extra energy is stored in muscles as creatine phosphate
- Another source of local energy is glycogen
- An increase in muscle cell size occurs when the cells make more proteins than they break down. Insulin and anabolic steroids increase protein production; cortisol decreases protein production in muscles.
- Studies are ongoing to determine what types of exercise and activities promote good health; it may be that too much time

- sitting can lead to poor health outcomes even in those that regularly exercise a few times per week.
- Studies that have interesting or novel results are more likely to be pursued, and published, than experiments that show no effect, which can bias the literature towards effects.
- Several studies done by independent groups that reach the same conclusion are more convincing then many studies done by related groups.

Blowing in the Wind
- Movement of air in the respiratory system requires a pathway and a gradient
- Inhalation occurs when the diaphragm contracts, increasing lung volume, decreasing lung pressure. If there is a pathway open, then air moves in
- We have discussed some of the physiology of coughs, sneezed and hiccups, and related concepts from the chapter The Good, the Bad and the Spicy as well as the chapter Can Viagra Kill you?
- We have used what happens at high altitude and low oxygen to explore some of the physiology of the respiratory system
- Changes in hemoglobin-oxyen affinity can be modulated by anions such as diphosphoglycerate, inositol phosphate and bicarbonate and alter how much oxygen binds to hemoglobin in the lungs and how much is released in the tissues
- Carbon dioxide functions as a ph buffer
- Changes in carbon dioxide levels alter ph
- Oxygen binding to hemoglobin is not a linear response, but shows saturation
- More hemoglobin and more red cells increases the oxygen carrying capacity of blood but also increases blood viscosity

Soy Sauce and Licorice as Medicine
- If there is no intake of salt and water, then the kidney cannot correct the problem, one does need to drink or take in salt. But the change in kidney function can conserve salt and water while the person goes to find some.
- Low water simulates the thirst center
- Low water increases antidiuretic hormone secretion
- Antidiuretic hormone is a water soluble hormone
- ADH secretion increases the pathways for water movement in

the last part of the kidney tubular; because of the osmotic gradient, water is reabsorbed when ADH is high and the pathways are present
- Low salt stimulates the salt appetite
- Low salt increases aldosterone secretion
- Aldosterone is a steroid hormone and like all steroids is lipid soluble
- Aldosterone secretion increases the pathways for sodium reabsorption in the kidney
- These system also responds to too much water or salt by turning off the pathways.
- Overconsumption of real licorice can mimic an overactive aldosterone receptor because licorice inhibits the enzyme that breaks down cortisol in the kidney. Cortisol can then bind to the aldosterone receptor and activate it.

From Anemia to Vampires
- Movement of iron and B12 across cell membranes required both a concentration gradient and a pathway. The pathways for the two compounds are different.
- Iron absorption across the intestine requires transporters in the apical as well as the serosal membranes.
- Hepcidin is a hormone secreted by the liver that regulates the amount of serosal membrane iron transport to the blood.
- When body iron stores are high, hepcidin increases.
- An increase in hepcidin leads to a decrease in serosal membrane iron transport by ferroportin.
- When iron stays inside the intestinal cells, it does not get to the blood but is lost in the feces.
- B12 absorption requires the stomach to secrete Intrinsic Factor; Intrinsic Factor is needed to cloak B12 in the intestine and allow it to be recognized in the terminal part of the small intestine where it is absorbed.
- Heme is broken down inside our intestinal cells therefore we are unable to absorb the heme from our diet into our blood stream.
- Scientists often use the word "normal" is a statistical sense. If a particular parameter has a bell shaped curve, then often the 5% largest and smallest values are "abnormal". This is different than the lay use of the word abnormal which often implies a "defect"

Got Cows?
- Complex molecules are broken down into parts before being absorbed, for example starches to glucose, lactose to galactose and glucose, proteins to amino acids
- Absorption of water soluble compounds normally occurs in the small intestine
- If substantial amounts of nutrients arrive in the large intestine, bacteria digest them, often generating hydrogen gas, protons and short chained fatty acids. The latter create an osmotic gradient.
- In the GI tract, aquaporins provide a pathway that allows water to move in response to osmotic gradients.
- Epithelial cells use sodium to drive uphill transport of glucose and amino acids.
- Epithelial cells move Na and Cl to create osmotic gradients to drive water movement; in the small intestine sometimes cells move Na and Cl and water from blood to lumen and sometimes the other direction.
- Lactase is an enzyme in the intestine that breaks down lactose; it is present in all mammalian children. Some human adults have a mutation that allows lactase production to persist into adulthood and these adults avoid the problems associated with eating and digestion milk.
- Cholera activates the pathways that move Na and Cl from blood to lumen, creating an osmotic gradient to move water; the activation is so strong that there can be tremendous loss of water and minerals.
- Salt/starch/water solutions can be used to help maintain hydration in cholera patients and restore minerals
- Cystic fibrosis is a disease caused by a defective chloride channel, CFTR. This channel is important in lung, GI, and sweat epithelia for the movement of chloride, which drives sodium movement and this osmotic gradient drives water movement. When defective, water movement is inappropriate and allows for bacterial colonization of the lungs and fibrosis if the pancreas and cystic duct.

How Sweet it is
- several hormones regulate blood sugar levels including insulin, glucagon, cortisol, epinephrine and norepinephrine

Physiology

- an increase in blood sugar leads to an increase in insulin
- an increase in insulin causes a drop in blood sugar by increasing sugar uptake into skeletal muscle and liver, for storage as glycogen, and also by regulating fat synthesis in liver and fat cells
- glucagon, cortisol, and epinephrine all tend to increase blood sugar by promoting glycogen breakdown and/or conversion of amino acids to glucose
- fats are not really converted back to glucose but are metabolized to fatty acids, which can be used by skeletal muscle and the heart
- under low insulin and low glucose conditions, fats are also converted to ketones which can fuel the brain
- high blood glucose levels lead to cardiovascular problems, often related to the resulting high blood fat levels
- high blood glucose levels also lead to capillary, nerve, eye, and kidney damage.
- type I diabetes is caused by an autoimmune response that damages pancreatic beta cells; eventually pancreatic alpha cells are also destroyed.
- type II diabetes is caused by insulin resistance and can eventually also involve beta cell damage
- control of blood glucose levels appears to be a key to maintaining good health
- regular exercise and diets that minimize simple sugars are thought to reduce the risk for type II diabetes as well as improve the outcome for patients.

Bubbling Beverages Bad for Bones?
- once we stop growing, calcium intake needs to match calcium excretion
- if there is more excretion than intake, we are losing calcium, probably from bone
- the blood levels of calcium are sensed by the parathyroid gland, which releases parathyroid hormone
- the effects of parathyroid hormone that we discussed are to increase kidney calcium reabsorption and to produce more of the active form of vitamin D.
- the active form of vitamin D leads to an increase in the proteins required to absorb more calcium from our intestines

the net effect is that when blood calcium is low, the intestine absorbs more dietary calcium and the kidney excretes less calcium, thus blood calcium hopefully increases.

Steroids Make the World go Round

- Steroid refers to a chemical structure and is the "family" name; the "first name" is often **one** of their physiological effects, e.g., glucocorticoid steroids for raising blood glucose, mineralocorticoid steroids for regulation sodium and potassium, sex steroids for regulating egg and sperm production and secondary sexual characteristics
- steroids are lipophilic and can easily cross membranes; steroid creams are an effective way to deliver steroids locally
- steroids are bound to binding proteins; the small concentration that is not bound is the free concentration and it is the free concentration that is directly able to enter cells and bind to steroid receptors
- "Typical" steroid response
    - the steroid hormone(S) diffuses into the cell
    - The steroid hormone binds to its inactive receptor, e.g., androgen receptor, estrogen receptor, progestin receptor, aldosterone receptor, or cortisol receptor
    - The hormone-receptor complex changes shape and becomes active
    - The active hormone-receptor complex moves into the nucleus
    - After binding to DNA, the transcription from DNA to RNA increases for specific genes
    - The RNA moves to the cell cytoplasm
    - The RNA is translated to protein
    - Note that there are accessory factors specific to different cell types that can also alter which genes are expressed or turned off
- Steroids bind to "their" receptor specifically at low concentrations; at high concentrations, a steroid might bind to another type of steroid receptor

Cycles in Synch?

- in humans, there are cyclic changes in the endometrium and the ovary

- the ovary and corpus luteum secrete the steroids progesterone and estrogen
- the hypothalamus secretes GnRH
- high GnRH leads the anterior pituitary to release the peptide hormones FSH and LH
- high FSH leads to follicle maturation and endometrial growth
- high LH leads to ovulation
- at most times in the cycle, estrogen and progesterone, have negative feedback effects on GnRH and LH, and estrogen has positive feedback on ovary estrogen production
- Just before ovulation, estrogen has positive feedback on LH release
- The bleeding at the start of a period is due to the excretion of the dying corpus luteum
- Pills containing estrogen or progesterone like compounds can suppress LH release, thus preventing ovulation and preventing pregnancy; they also have other positive and negative effects

Skin Games
- **In summary, sexually transmitted infections may be a misleading term, because of the ambiguity of the word sex.** Maybe a more accurate term is intimate relation transmitted infections.
- For infections that are secreted in bodily fluids, the infected fluid needs to be in contact with the partner.
- For infections that are shed from the surface of the genitals, the genital tract, or the anal rectal area, then these surfaces need merely contact the partner in an area that has the potential for invasion.
- While the skin and most genital, anal, rectal surfaces provide a barrier against infection, any small tear can allow bacteria or virus to enter. These microbes are less than 1 um, well below the resolution of the human eye, so microtears or microabrasions need not be visible to allow infections.
- Often intimate contact involves rubbing and some friction and that seems likely to create microdamage to the barrier. Microdamage is large enough for microbes to enter.
- Many microbes attach to specific protein "receptors" to gain entry into cells and the body; the types of contacts that transmit the microbe depends in part on which parts of the body have cells with these "receptors".

Black and White (and Banning Tanning ?)
- that many genes regulate the color of the skin and hair.
- That UV light damages DNA
- that damaged DNA is one of the signals to release Melanocyte Stimulating Hormone.
- Melanocyte Stimulating Hormone acts on receptors in melanocytes to increase eumelanin production.
- The eumelanins are packaged in melanosomes and exported to the keratinocytes, where they are placed on the sun side of the nucleus, to absorb UV light.
- UVB light certainly damages DNA and UVA may as well.
- The more UV light that gets to the DNA in the keratinocytes, the greater the chance of non-melanoma skin cancer.
- Melanomas are rare, but much more aggressive skin cancers.
- It is much safer to get adequate vitamin D from dietary sources than from sun exposure.
- Sunburn occurs when there is so much UV exposure that many keratinocytes commit suicide; the eumelanin is lost and the red color in people with light complexion reflects one's ability to see the hemoglobin in the blood supply just below the skin.

Is there Proof?
- We have reviewed why blood alcohol levels peak well after drinking has stopped. It is because of the delay in absorption by the stomach and intestine. And that this delay depends upon the other contents of the stomach and intestine.
- The liver is the major organ for metabolizing alcohol and the first metabolite, acetaldehyde is also harmful.
- As the level of blood alcohol increases, different brain functions are affected.
- We also discussed that fact that the interpretation of the data requires a starting assumption. Suppose one wanted to determine whether one alcoholic drink had an adverse effect on the fetus. Does one assume that one alcoholic drink has an effect and require evidence to disprove that statement? Or does one assume that one alcoholic drink has no effect and require evidence to disprove that statement? In addition, each person has a different level of requirement for what "disprove" mean.

www.ingramcontent.com/pod-product-compliance
Lightning Source LLC
Chambersburg PA
CBHW051622170526
45167CB00001B/25